AF364155

Fundamentals of
Mechanical Engineering

Fundamentals of
Mechanical Engineering

Dr. Anil Kumar

&

K. Sudhakar

Assistant Professor, Department of Energy,

Maulana Azad National Institute of Technology,

Bhopal - 462051, MP.

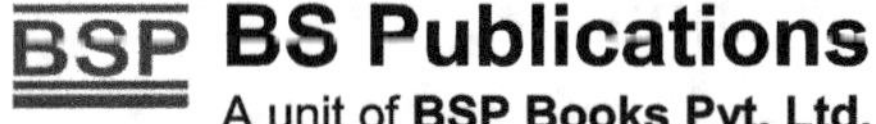

BSP **BS Publications**
A unit of **BSP Books Pvt. Ltd.**

4-4-309, Giriraj Lane, Sultan Bazar,
Hyderabad - 500 095
Phone : 040 - 23445605, 23445688

Published by :

 **BS Publications**

An unit of **BSP Books Pvt., Ltd.**

4-4-309, Giriraj Lane, Sultan Bazar,
Hyderabad - 500 095
Phone : 040 - 23445605, 23445688
e-mail : info@bspbooks.net

ISBN : 978-93-85433-19-1 (HB)

Dedicated to

**Teachers, Parents
&
Family**

Preface

"Fundamentals of Mechanical Engineering" has been included as 1st Year common paper for all branches of Engineering and Technology by Indian Technical Universities and Institutes.

This book is written to provide a clear and authoritative introduction to the subject of "Fundamentals of Mechanical Engineering". In order to know all aspects of the subject, more reference and text books have to be referred. It is the daunting task for the students and faculties of all institutions. To avoid the complexities, this book entitled "Fundamentals of Mechanical Engineering" has been brought out. It follows the sequence of topic, prescribed in the course syllabus.

This book has been divided in to five chapters. Chapter 1 deals with materials, Chapter 2 deals with measurements, Chapter 3 deals with fluids, Chapter 4 deals with thermodynamics and Chapter 5 deals with reciprocating machines.

Apart from that, numerical problems, formulas and review questions are given at the end of each chapter to help the students to prepare for university examination.

We have consulted various numbers of books, journals, and websites during the preparing the text and also a large number of works were discussed. Hence we can not declare very much distinctiveness. It is the compiled work with novel presentation.

It is not possible to claim perfection in such a work. Many sort of errors must have crept in. We shall be thankful to all those who bring any of these into our notice.

We hope this book shall meet the scope of the engineering students.

- Authors

Acknowledgement

It is a pleasure for us to acknowledge the help, advice and encouragement which we have received from Our Chairman Shri. A. N. Singh, Director. Dr. R. P. Singh and Dr. Saroj Rangnekar, Head, Department of Energy.

We are highly indebted to our colleagues and students of Department of Energy, MANIT, Bhopal.

We are highly grateful to Prof. G. N. Tiwari, CES, IIT Delhi, Prof. S. N. Varma, Ex. Vice Chancellor, RGPV, Bhopal, Dr. O. P. Chaursiya, Associate Professor, Dept. of Mech. Engg., National Institute of Technology, Patna and Dr. Subhash Solanki, Associate Professor, Dept. of Mech. Engg., Ujjain Engg. College, Ujjain for their moral support.

We extend our acknowledgement to Mr. Rajesh Mandapati and Mr. Himanshu Kalra for their assistance.

I (Dr. Anil Kumar) deeply acknowledge the patience, co-operation and support of my wife Ms. Abhilasha and my son Master Tijil Kumar during the preparation of the book.

Last but not the least, we express our sincere thanks to BS Publications, Hyderabad for their encouragement.

As authors we bear responsibility for all interpretation, opinions and errors in this work however, many have helped us and we express our gratitude to all of them.

We would thankfully welcome further suggestion and comments for the improvement of the book.

- Authors

Contents

Chapter 1

Materials

Chapter 2

Measurement

Chapter 3

Fluids

Chapter 4

Themodynamics

Chapter 5

Reciprocating Machines

CHAPTER 1

Materials

> **Materials:** Classification of engineering material, composition of cast iron and carbon steels on iron-carbon diagram and their mechanical properties; Alloy steel and their applications; stress-strain diagram, Hooks law and modulus of elasticity. Tensile, shear, hardness and fatigue testing of materials.

1.1 Engineering Materials

All engineers need to know about materials. Innovation in engineering often means the clever use of a new material for a specific application. For example: plastic containers in place of age-old metallic containers. It is well learnt lesion that engineering disasters are frequently caused by the misuse of materials. So it is vital that the professional engineer should know how to select materials which best fit the demands of the design – economic and aesthetic demands, as well as demands of strength and durability. Beforehand the designer must understand the properties of materials, and their limitations. Thus it is very important that every engineer must study and understand the concepts of Materials Science and Engineering. This enables the engineer

- To select a material for a given use based on considerations of cost and performance.

- To understand the limits of materials and the change of their properties with use.

- To be able to create a new material that will have some desirable properties.

- To be able to use the material for different application.

1.2   Classification of Engineering Materials

1.2.1 Metals

Valence electrons are detached from atoms, and spread in an 'electron sea' that "glues" the ions together. Strong, ductile, conduct electricity and heat well, are shiny if polished.

1.2.2 Semiconductors

The bonding is covalent (electrons are shared between atoms). Their electrical properties depend strongly on minute proportions of contaminants. Examples: Si, Ge, GaAs.

1.2.3 Ceramics

Atoms behave like either positive or negative ions, and are bound by Coulomb forces. They are usually combinations of metals or semiconductors with oxygen, nitrogen or carbon (oxides, nitrides, and carbides). They are hard, brittle, and insulators. Examples: glass, porcelain. Examples of ceramic materials are ranging from household and lab products to high performance combustion engines which utilize both metals and ceramics.

Crystalline ceramics (a) and non-crystalline glasses (b) are shown in Fig. 1.1. Open circles represent nonmetallic atoms, solids represent metal atoms.

1.2.4 Polymers

Polymers are bound by covalent forces and also by weak Van der Waals forces, and usually based on C and H. They decompose at moderate temperatures (100–400 °C), and are lightweight. Examples: plastics rubber. Polymers or commercially called "Plastics" need no introduction.

Polymer composite materials and reinforcing glass fibers are in a polymer matrix.

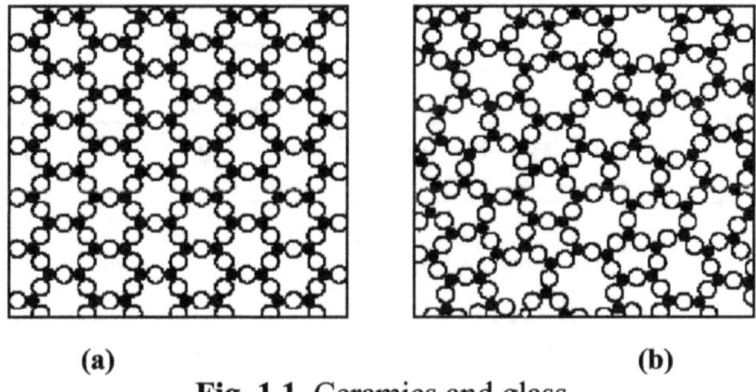

(a)   (b)

Fig. 1.1 Ceramics and glass

1.2.5 Glasses

Depending on the material structure, the glass can be opaque, transparent, or translucent. Glasses can also be processed to yield high thermal shock resistance.

The Table 1.1 concludes the materials with their applications and properties.

Table 1.1 Properties and Applications of some Engineering Materials.

	Application	**Properties**
Metals		
Copper	Electrical conductor wire	High electrical conductivity, good formability
Gray cast iron	Automobile engine blocks	Castable, machinable, vibration-damping
Alloy steels	Wrenches	Significantly strengthened by heat treatment
Ceramics		
SiO_2-Na_2O-CaO	Window glass	Optically useful, thermal insulating
Al_2O_3, MgO, SiO_2	Refractories for containing molten metal	Thermal insulating, melt at high temperature, relatively inert to molten metal
Barium titanate	Transducers for audio equipment	Converts sound to electricity (Piezoelectric behaviour)

Table 1.1 *Contd...*

	Application	**Properties**
Polymers		
Polyethylene	Food packaging	Easily formed into thin, flexible airtight film
Epoxy	Encapsulation of integrated circuits	Electrically insulating and moisture-resistant
Phenolics	Adhesives for joining piles in ply wood	Strong, mosisture resistant
Semiconductors		
Silicon	Transistors and integrated circuits	Unique electrical behaviour
GaAs	Fiber-optic systems	Converts electrical signals to light
Composites		
Graphite-epoxy	Aircraft components	High strengths-to-weight ratio
Tungsten carbide-cobalt	Carbide cutting tools for machining	High hardness, yet good shock resistance
Titanium-clad steel	Reactor vessels	Has the low cost and high strength of steel, with the corrosion resistance of titanium

1.3 Material Properties

Some of the most important properties of materials are grouped as follows:

1.3.1 Density

It is defined as mass per unit volume. It is expressed as kg/m^3.

1.3.2 Specific gravity

It is the ratio of density of a material to density of water.

1.3.3 Porosity

The term porosity is used to indicate the degree by which the volume of a material is occupied by pores. It is expressed as a ratio of volume of pores to that of the specimen.

1.3.4 Strength

Strength of a material has been defined as its ability to resist the action of an external force without breaking.

1.3.5 Elasticity

It is the property of a material which enables it to regain its original shape and size after the removal of external load.

1.3.6 Plasticity

It is the property of the material which enables the formation of permanent deformation.

1.3.7 Hardness

It is the property of the material which enables it to resist abrasion, indentation, machining and scratching.

1.3.8 Ductility

It is the property of a material which enables it to be drawn out or elongated to an appreciable extent before rupture occurs.

1.3.9 Brittleness

It is the property of a material, which is opposite to ductility. Material, having very little property of deformation, either elastic or plastic is called Brittle.

1.3.10 Creep

It is the property of the material which enables it under constant load to deform slowly but progressively over a certain period.

1.3.11 Stiffness

It is the property of a material which enables it to resist deformation.

1.3.12 Fatigue

The term fatigue is generally referred to the effect of cyclically repeated stress. Materials response and failure are in form of fracture under constant repeated loading. A material has a tendency to fail at lesser stress level when subjected to repeated loading.

1.3.13 Impact strength

The impact strength of a material is the quantity of work required to cause its failure per its unit volume. It thus indicates the toughness of a material.

1.3.14 Toughness

It is the property of a material which enables it to be twisted, bent or stretched under a high stress before rupture. Measure of how much force is required to deform the material to breaking point. The area under the stress strain diagram represents the toughness.

1.3.15 Hardness

Resistance of the material to penetration.

1.3.16 Resilience

The ability of the material to absorb energy

1.3.17 Thermal Conductivity

It is the property of a material which allows conduction of heat through its body. It is defined as the amount of heat in kilocalories that will flow through unit area of the material with unit thickness in unit time when difference of temperature on its faces is also unity.

1.3.18 Corrosion Resistance

It is the property of a material to withstand the action of acids, alkalis gases etc., which tend to corrode (or oxidize).

1.4   Cast Iron

Cast iron is a very popular material for machine construction because of its castability, machinability, cheapness and good damping capacity. These are alloys of iron and carbon but contain large proportions of silicon, sulphur, manganese, phosphorus. Iron-carbon alloys containing more than 2 percent carbon are called cast irons.

Composition: All cast irons contain iron, carbon, silicon, manganese, phosphorus and sulphur. Cast iron is iron containing 2 to 4 % carbon.

1.4.1 Properties of cast Iron

The properties of cast iron are given below:

- It is the least expensive casting material. All the raw materials used are relatively cheap – pig iron, cast iron scrap, steel scrap, lime stone, coke and iron ore.

- Cast iron has a lower melting point temperature (1140 to 1200) degrees Celsius than steel (1380-1500) degrees Celsius.

- It possesses high casting properties such as, high fluidity, low shrinkage, casting soundness, ease of production, and a higher yield.

- Cast iron can provide a very wide range of metallic properties ranging from a high yield strength to high ductility and toughness

- They possess a very high compressive strength, about 3 to 4 times that of its tensile strength.

- Cast irons can be machined easily.

- They provide high wear and hardness resistance.

- An important characteristic of cast iron is its high damping capacity. It is that property which permits a material to absorb vibrational stresses.

1.4.2 Applications

The few applications of cast iron are as:

- cast iron can be used for production of complex shapes i.e., machinery parts,

- piston rings,

- frames for electric motors,

- machine tool structure,

- cylinder blocks and head of I.C engine,

- household appliances,

- crank shaft,

- pumps and compressors,

- pipes,

- chain links,

- farm implements and tractors,

- used as excellent vibration damper in machines.

Depending upon the structure and form of the carbon present, cast-irons can be classified into different groups such, as white cast iron, grey iron, Nodular and malleable iron. The properties of the cast iron are affected by the following factors:

- Chemical composition of the iron.
- Rate of cooling of the casting in the mould (which depends on the section thickness in the casting).
- Type of graphite formed (if any).

1.4.3 Types of cast iron

Fig. 1.2 classifies the cast iron on the basis of structure and form of the carbon present.

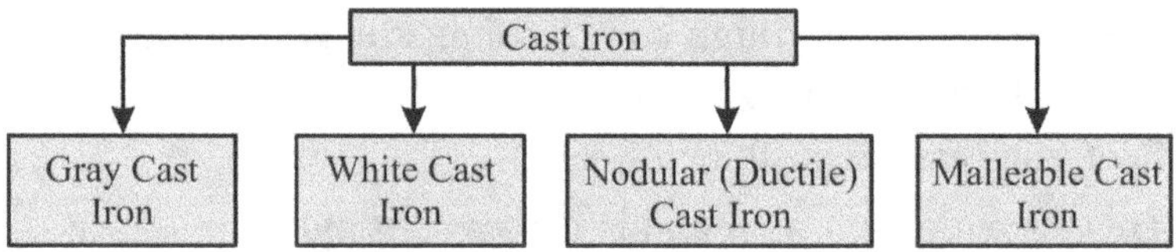

Fig. 1.2 Types of cast iron

(i)   Gray Cast Iron: Gray cast iron is by far the oldest and most common form of cast iron. The flake-like shape of graphite in Gray iron, see Fig. 1.3 and 1.4, exerts a dominant influence on its mechanical properties. The graphite flakes act as stress raisers which may prematurely cause localized plastic flow at low stresses, and initiate fracture in the matrix at higher stresses. As a result, Gray iron exhibits no elastic behavior but excellent damping characteristics, and fails in tension without significant plastic deformation. The presence of graphite flakes also gives gray iron excellent machinability and self-lubricating properties.

Properties of Gray Cast Iron:

- Graphite acts as a chip breaker and a tool lubricant.
- Very high damping capacity.
- Good dry bearing qualities due to graphite.
- After formation of protective scales, it resists corrosion in many common engineering environments.
- Brittle (low impact strength) which severely limits use for critical applications.

- Graphite acts as a void and reduces strength. Maximum recommended design stress is 1/4 of the ultimate tensile strength. Maximum fatigue loading limit is 1/3 of fatigue strength.

- Changes in section size will cause variations in machining characteristics due to variation in microstructure.

- Higher strength gray cast irons are more expensive to produce.

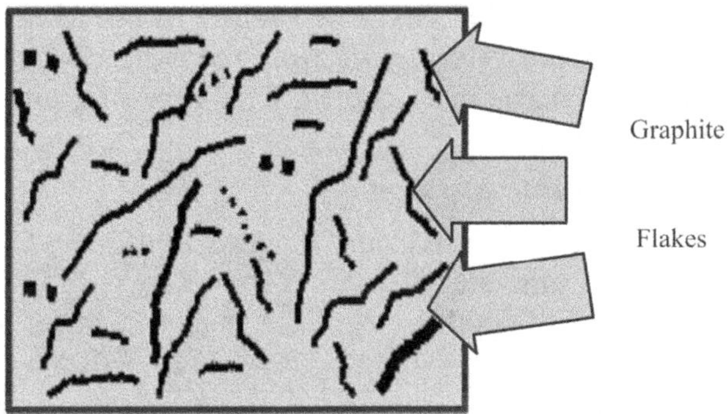

Fig. 1.3 Graphite flakes in gray cast iron

Fig. 1.4 Photomicrograph of gray cast iron

Applications:

- It is used in clutch plates and brake drums.
- It is widely used for bases or beds for machine tools.
- It is used in heat shock applications such as ingot or pig-moulds.
- It is used in elevator counterweights, industrial counter weights etc.

- It is used as cylinders, pistons and engine frames.
- It is used for gear housings, enclosures for electrical equipmets, pump housings etc.

(ii) White Cast Iron: White cast iron is unique in that it is the only member of the cast iron family in which carbon is present only as carbide. Due to the absence of graphite, it has a light appearance. The presence of different carbides, depending on the alloy content, makes white cast irons extremely hard and abrasion resistant but very brittle (Fig. 1.5). An improved form of white cast iron is the chilled cast iron. When localized area of a gray cast iron is cooled very rapidly from the melt, cast iron is formed at the place that has been cooled. This type of white cast iron is called chilled iron.

Properties of white cast iron:

- Chromium is used in small amounts to control chill depth.

- Because of the formation of chromium carbides, chromium is used in amount of 1 to 4 percent in chilled iron to increase hardness and improve abrasion resistance.

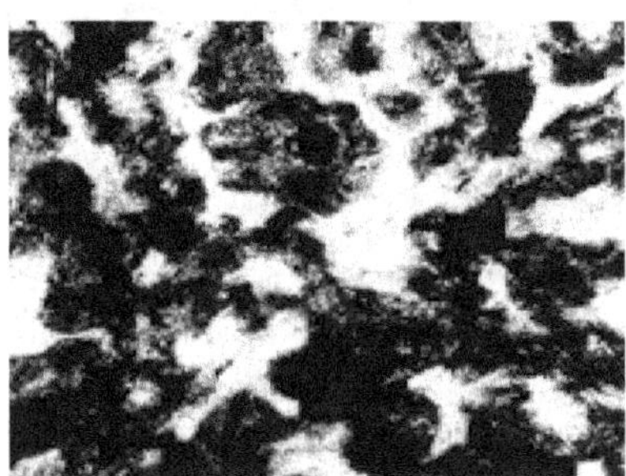

Fig. 1.5 White cast iron

- When added in amounts of 12 to 35 percent, chromium will impart resistance to corrosion and oxidation at elevated temperatures.

- If alloys such as nickel, chromium, or molybdenum are added, much of the austenite transforms to martensite instead of Pearlite.

- The hardness of chilled cast iron is generally due to the formation of martensite.

Applications:

- railway-car wheels,
- crushing rolls,
- stamp shoes and dies, and
- Heavy-duty machinery parts.

(iii) **Ductile Cast Iron (Nodular Cast Iron):** This structure is developed from the melt. The carbon forms into spheres when cerium, magnesium, sodium, or other elements are added to a melt of iron with a very low sulfur content that will inhibit carbon from forming. The control of the heat-treating process can yield pearlitic, ferritic, martensitic matrices into which the carbon spheres are embedded. Figure 1.6 and 1.7 shows the structure of ductile (modular) cast iron.

Properties:

- The advantages of ductile cast iron which have led to its success are numerous, but they can be summarized easily – versatility and high performance at low cost. In addition to cost advantages offered by all castings, ductile cast iron, when compared to steel and malleable cast iron, also offers further cost savings. Like most commercial cast metal, steel and malleable cast iron decrease in volume during solidification, and as a result, require feeders and risers to offset the shrinkage and prevent the formation of internal or external shrinkage defects.

- In other cases, it requires feeders that are much smaller than those used for malleable cast iron and steel. This reduced requirement for feed metal, increases the productivity of ductile cast iron and reduces its material and energy requirements, resulting in substantial cost savings.

- The use of the most common grades of ductile cast iron "as-cast" eliminates heat treatment costs, offering a further advantage.

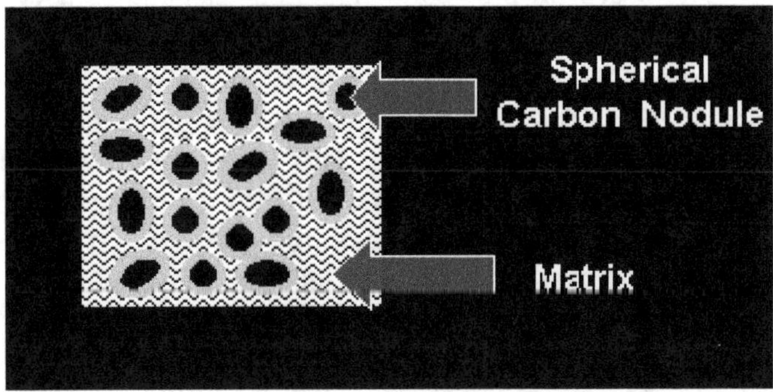

Fig. 1.6 Nodular (Ductile) cast iron and the spherical carbon embedded into the matrix

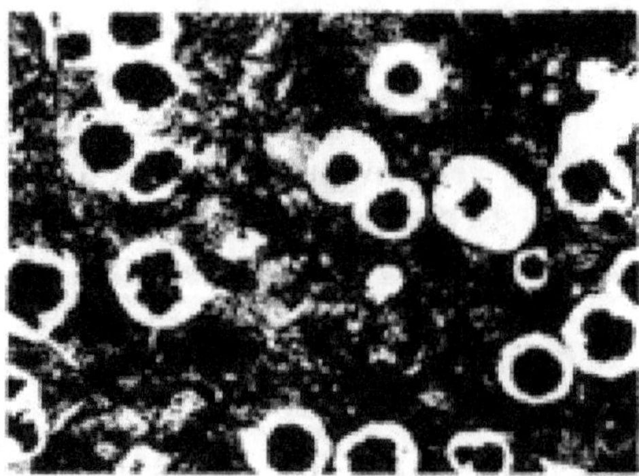

Fig. 1.7 Photomicrograph of nodular cast iron

Applications:

- Ductile cast iron is used for many structural applications, particularly those requiring strength and toughness combined with good machinability and low cost.

- The automotive and agricultural industries are the major users of ductile iron castings.

- Because of economic advantage and high reliability, ductile iron is used for such critical automotive parts as crankshafts, engine connecting rods, idler arms, wheel hubs, truck axles, front wheel spindle supports, disk brake calipers, Suspension system parts, power transmission yokes.

- High temperature applications for turbo housing and manifolds.

- High security valves for many applications.

- The cast iron pipe industry is another major user of ductile iron.

(iv) **Malleable Cast Iron:** If cast iron is cooled rapidly, the graphite flakes needed for gray cast iron do not get a chance to form. Instead, white cast iron forms. This white cast iron is reheated to about 1700°F for long periods of time in the presence of materials containing oxygen, such as iron oxide. At the elevated temperatures cementite (Fe_3C) decomposes into ferrite and free carbon. Upon cooling, the combined carbon further decomposes to small compact particles of graphite (instead of flake-like graphite seen in gray cast iron). If the cooling is very slow, more free carbon is released. This free carbon is referred to as temper carbon and the process is called malleableizing (Fig.1.8).

Properties:

- It has excellent machinability
- It has significant ductility
- It has good shock resistance properties.
- The major disadvantage is shrinkage. Malleable cast iron decreases in volume during solidification, and as a result, requires attached reservoirs (feeders and risers) of liquid metal to offset the shrinkage and prevent the formation of internal or external shrinkage defects.

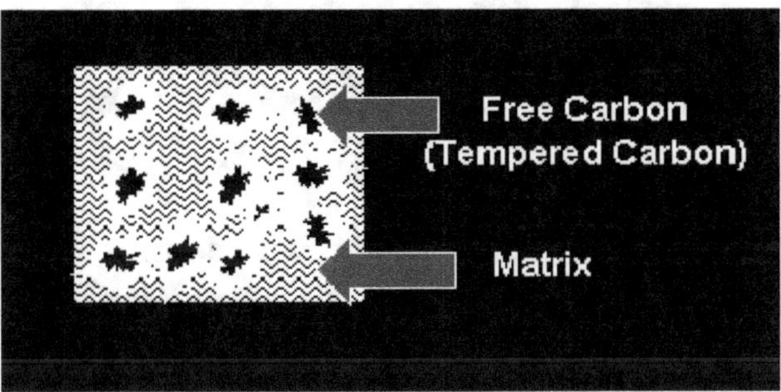

Fig. 1.8 Malleable cast iron

Figure 1.9 shows the ferritic malleable cast iron, which has a ferritic matrix and the tempered carbon particles are embedded into the matrix.

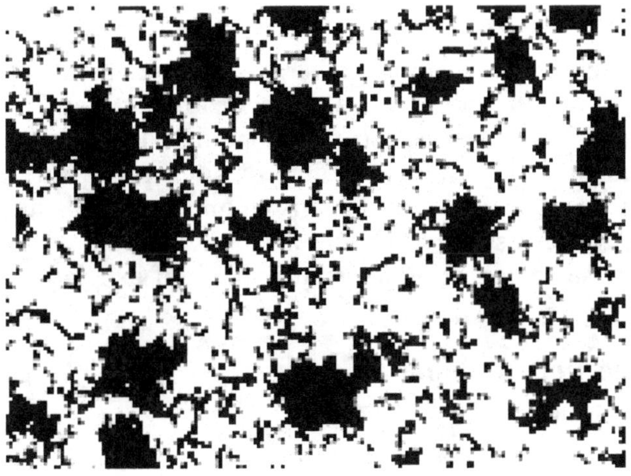

Fig. 1.9. Ferritic Malleable Cast Iron

Figure 1.10 shows the pearlitic malleable cast iron, which has a pearlitic matrix. By adding managanese to the structure, carbon is retained in the form of cementite.

Fig. 1.10 Pearlitic Malleable Cast Iron

Applications: Malleable cast iron is used for:

- Connecting rods and universal joint yokes.
- Transmission gears.
- Differential cases and certain gears.
- Compressor crankshafts and hubs, flanges.
- Pipe fittings and valve parts for railroad.
- Marine and other heavy-duty applications.

1.5 Carbon Steel

Steel is an alloy of iron, carbon and several other elements. It is an iron-carbon alloy which contains up to 2 percent carbon, the amount of carbon in steels has a marked effect on the properties. In addition to carbon, steels generally contain small amounts of silicon, sulphur, phosphorus and manganese. Manganese & silicon is an important constituent because of its ability to increase the hardness and strength of the steel. Steel is the most widely used engineering material because of its relative advantages over materials.

1.5.1 Properties of plain carbon steels

- It is cheap and readily available.
- It can provide a very wide range of mechanical properties, from a very soft condition to a very hard condition.

- Properties of steel can be improved easily by addition of small amounts of other elements known as alloying elements.
- Due to phase transformations which take place in steel, it responds to heat-treatment. Any desired combination of strength and ductility can be obtained by different heat treatments.
- Steel possesses good machinability and weldability.

1.5.2 Classification of steels

(i)	Low carbon steel (Mild steel)	less than 0.30% C
(ii)	Medium carbon steel	0.30 to 0.70% C
(iii)	High carbon steel	0.70 to 2.0% C

1.5.3 Low Carbon/Mild Steel

Mild steel is the commonest steel with a range of 0.05 to 0.30% carbon. Further it is classified into three types:

(i)   Dead mild steel- C 0.05 to 0.15%

(ii) Mild steel - C 0.15 to 0.20%

(iii) Mild steel – C 0.20 to 0.30%

Properties:

(i)   Bright fibrous structure.

(ii) Can be readily forged and welded.

(iii) Can be permanently magnetized.

(iv) Rusts rapidly.

(v)   Can be hardened and tempered but not easily.

Applications:

(i)   Wires.

(ii) Structural shapes.

(iii) Screw machine parts such as screws, nuts and bolts, storage tanks, automobile frames, galvanizes steel valves, sheet.

These steels are soft, very ductile, easily machined. Since the carbon content is low so heat treatment process is ineffective.

1.5.4 Medium Carbon Steel

These are used for rails, axels, gears and parts requiring high strength. Medium steel is containing carbon from 0.30 to 0.70%.

1.5.5 High Carbon Steels

Steels containing 0.70 to 2.0% carbon is called high carbon steel. Carbon is the most important alloying element in steel. Plain-carbon steels are very satisfactory where strength and other properties are not too severe. They are also used successfully at ordinary temperatures. It was also noticed that at high temperatures the plain carbon steel softness. Thus behaviour restricts the applicability. Alloying elements are added to overcome the limitations. They are mainly silicon, manganese, chromium and other elements added in very small proportions.

Composition:In high carbon steel the following elements are present in the following compositions:

Carbon	-	0.7 to 2%
Silicon	-	0.2 to 0.6%
Manganese	-	0.2 to 0.8%
Phosphorus	-	0.05 to 0.15%
Chromium	-	0.12 to 0.2%
Sulphur	-	0.005 to 0.05%

Properties:

(i)   it has fine granular structure,

(ii)   tough and more elastic than mild steel,

(iiii) can be hardened and tampered readily,

(iv)   cannot be easily forged and welded,

(v)   can be permanently magnetized,

(vi)   brittle and less ductile than mild steel,

(vii) rusts rapidly, absorbs shocks.

Applications: High carbon steels are widely used in:

(i)   Cutting tools such as: knives, drills, taps, hand tools, reamers and for abrasion-resisting properties.

(ii)   This steel is also used for making products having edges on cutlery, chisels, and shear blades and also for spring wire, wire rope.

1.5.6 Alloy steels

When elements like nickel, chromium, vanadium, molybdenum, tungsten are present in sufficient quantity, the steel is called **"alloy steel"**. In order to obtain special characteristics like high hardenability, high temperature properties, corrosion resistance etc., generally steel has to be alloyed with other alloying elements. Different alloying elements give rise to enhancing of different physical and mechanical properties.

Effect of alloying elements in steel:

Element	Effects
Carbon	Increases hardness, strength, decreases ductility
Nickel	Increases hardness, toughness, corrosion resistance
Chromium	Increases corrosion resistance and elastic limit
Tungsten and molybdenum	Increases strength and toughness for high speed cutting tools

Alloy steels are generally classified into two major types depending on the structural classification:

Low alloy steels: These possesses similar microstructure and heat treatment as that of plain carbon steels. These generally contain up to 3-4 % of one or more alloying elements for purpose of increasing strength, toughness and hardenability. The applications of low alloy steels are similar to those of plain carbon steels of similar carbon contents. Low alloy steels containing nickel are particularly suitable for applications requiring resistance to fatigue.

High alloy steels: Those steels that possess structures, and require heat treatments, that differ considerably from those of plain carbon steels. A few examples of high alloy steels are given below:

   (i)   *High-speed tool steels:* Tungsten and chromium form very hard and stable carbides. Both elements also raise the critical temperatures and, also, cause an increase in softening temperatures. High carbon steels rich in these elements provide hard wearing metal-cutting tools, which retain their high hardness at temperature up to 600°C. A widely used high-speed tool steel composition is containing 18% of tungsten, 4% of chromium, 1% of vanadium and 0.8% of carbon. This high alloy content enables them to be used as cutting tools at high cutting speed.

   (ii) *Stainless steels:* When chromium is present in amounts in excess of 12%, the steel becomes highly resistance to corrosion, owing to protective film of chromium oxide that forms on the metal surface. Chromium also raises the transformation temperature of iron, and tends to stabilize ferrite in the structure.

There are several types of stainless steels, and these are summarized below:

(a) Ferritic stainless steels contain between 12-25% of chromium and less than 0.1% of carbon.

(b) Martensitic stainless steels contain between 12-18 % of chromium, together with carbon contents ranging from 0.1 to 1.5 %.

(c) Austenitic steels contain both chromium and nickel. When nickel is present, the tendency of nickel to lower the critical temperatures over-rides the opposite effect of chromium, and the structure may become wholly austenitic.

(iii) *Maraging steels:* These are very high-strength materials that can be hardened to give tensile strengths up to 1900 MN/m^2. They contain 18% of nickel, 7% of cobalt and small amounts of other elements such as titanium, and the carbon content is low, generally less than 0.05%. A major advantage of marging steels is that after the solution treatment they are soft enough to be worked and machined with comparative ease.

Applications:

(i) General machinery parts

(ii) Aircraft

(iii) Taps

(iv) Hand tools, cutting tools

(v) Drills

(vi) Knives

So the most used elements with the alloy steel and with their amount as a percentage of:

- 2 % Manganese (Mn)
- 0.5 % Chrome (Cr) or Nickel (Ni)
- 0.3 % Tungsten (W) or Cobalt
- 0.1 % Molybdenum (Mo) or Vanadium
- Different amount of Aluminum (Al), Copper (Cu) and Silicon (Si).

Type 304 Stainless Steel: Type 304 stainless steel (containing 18%-20% chromium and 8%-10.5% nickel) is used in the tritium production reactor tanks, process water piping, and original process heat exchangers. This alloy resists most types of corrosion.

1.6   Steel and Iron Carbon Diagram

Steel is basically an alloy of iron and carbon, but several elements are used in various proportions and combinations to produce different types. Carbon is by far the most important element in steel: By definition, steel has a maximum carbon content of 2.0%. Structural steels normally contain less than 0.30% carbon, however, and in terms of chemical composition can be classified as either plain carbon or low-alloy steels. The chemical composition of steel is very important since it has a significant effect on the microstructure of the material and hence on its mechanical behavior and properties.

The changes in composition (cementite or iron carbide [Fe_3C] content), microstructure (pearlite content), and mechanical properties (strength and ductility, in particular) resulting from changes in the carbon content in annealed plain carbon steels are shown very clearly in Fig. 1.11. In the range of carbon contents shown in Fig. 1.11, a significant increase in strength and decrease in ductility are produced by an increase in the carbon content in the steel. In their simplest form, steels are alloys of Iron (Fe) and Carbon (C). The Fe-C phase diagram is a fairly complex one, but we will only consider the steel part of the diagram, up to around 7% Carbon.

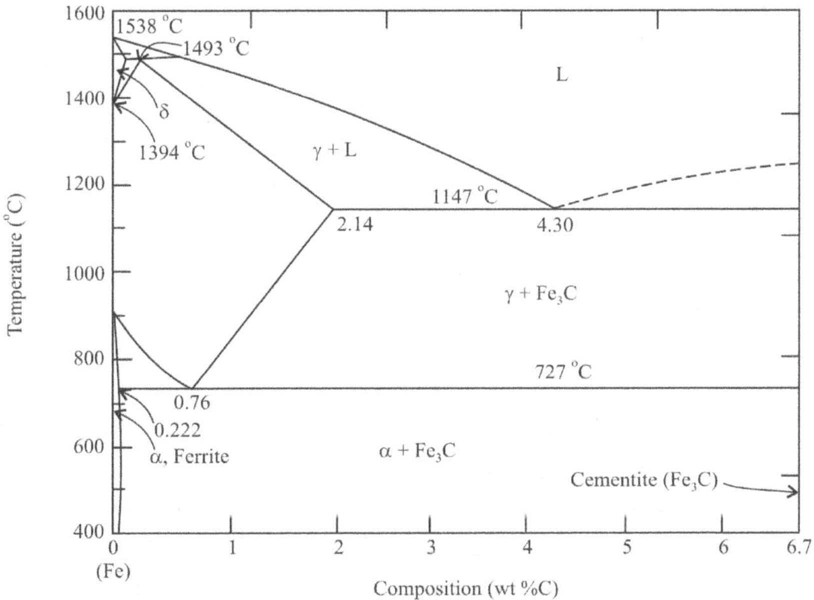

Fig. 1.11 Iron carbon diagram

The following features should be noted in the iron-cementite, Fe-Fe$_3$C, phase diagram

(i)   Iron goes through two allotropic transformations during heating or cooling: On continued cooling from a liquid melt, it first forms a body-centered cubic (BCC) structure, then a face-centered cubic (FCC) structure, and finally another BCC structure. All three allotropes will form interstitial solid solutions with carbon, identified as delta iron (δ), austenite (γ), and ferrite (α), respectively. A greater number of carbon atoms can be accommodated in the austenite than in the other two phases, since the interstitial holes in the FCC lattice are somewhat larger than those in the BCC lattices. Hence the maximum solubility of carbon in austenite is 2.0%, while it is much lower in delta iron (0.10%) and ferrite (0.025%). These solid solutions are relatively soft and ductile, but stronger than pure iron due to solid solution strengthening by the interstitial carbon atoms.

(ii) Cementite, or iron carbide, Fe$_3$C, is a stoichiometric intermetallic compound formed when the solubility of carbon in solid iron is exceeded. This compound contains 6.67% C, is extremely hard and brittle, and is present in all commercial steels. The degree of dispersion strengthening and the properties of the steel are hence controlled by properly regulating the amount, size, and shape of the Fe$_3$C phase.

(iii) A eutectoid reaction occurs as the austenite (γ) cools below 727°C. That is,

$$\gamma \rightarrow \alpha + Fe_3C$$

Since the two phases that form has different compositions, atoms must diffuse during the reaction: Most of the carbon in the austenite diffuses to the Fe$_3$C, and most of the iron to the ferrite (α). Since this redistribution of atoms is easiest if the diffusion distances are short, the α and Fe$_3$C grow as thin lamellae, or plates, forming a structure called pearlite. The various phases and its description are shown in Table 1.2.

Table 1.2 Phases in the Fe-Fe$_3$C System

Phase	Atomic packing	Description and comments
Liquid (iron)	d.r.p (densely random packing)	Liquid solution of C (carbon) in Fe
δ (also called delta iron)	b.c.c (body- centered cubic)	Random interstitial solid solution of C in b.c.c. Fe
γ (also called austenite)	f.c.c (face- centered cubic)	Random interstitial solid solution of C in f.c.c. Fe. Maximum solubility is 2 wt% at 1147^{0}C. It is normally not stable at room temperature.
α (also called ferrite)	b.c.c	Random interstitial solid solution of C in b.c.c. Fe. Maximum solubility is 0.02 wt% at 723^{0}C. It is the softest structure appears on the diagram.
Fe$_3$C (also called iron carbide or cementite)	orthorhombic	A hard and brittle interstitial compound of Fe and C containing 25 atomic % (6.7 wt%) C.
(α + Fe$_3$C) Pearlite	-	It is very fine platelike or lamellar eutectoid mixture of ferrite and cementite containing 0.8 wt% C and formed at 723^{0}C on very slow cooling

1.6.1 Martensite

It is a supersaturated solid solution of carbon in b.c.t (body-centered tetragonal) iron. This meta-stable phase is formed under very rapid cooling.

1.6.2 Ferrite

Ferrite is practically pure iron containing only 0.06 percent carbon at room temperature. The name "ferrite" comes from the Latin word, 'ferum', which means iron. It is also called as alpha iron. Ferrite is relatively soft and ductile phase. It can undergo extensive cold working. Ferrite is strongly magnetic at room temperature and becomes paramagnetic at 770 degrees Celsius.

1.6.3 Austenite

Austenite is a solid solution of carbon in gamma iron and, it can dissolve up to two percent of carbon at 1148 degrees Celsius. Austenite is also a soft and ductile phase but stronger and less ductile than ferrite. It is a non-magnetic (paramagnetic) phase and exists at temperature above 72.7 degrees Celsius.

1.6.4 Cementite

Cementite is an interstitial compound of iron containing 6.67 per cent of carbon. It is an extremely hard and brittle phase. Cementite is a magnetic phase at room temperature and becomes paramagnetic above 210 degrees Celsius. It is also known as iron-carbide.

1.6.5 Pearlite

Pearlite is an intimate mixture of ferrite and Cementite. It has a distinct lamellar structure and consists of alternate layers of ferrite and Cementite. Pearlite is so called because of the lustrous, mother-of-pearl appearance it presents in the microscope. Due to its distinct appearance under a microscope, Pearlite is called as micro constituent. Pearlite contains 88.5 percent ferrite and 11.5 percent Cementite. It is obtained from austenite, when the later is cooled slowly below 727 degrees Celsius. Pearlite has a variable hardness depending upon its fineness.

1.6.6 Ledeburite

Ledeburite is also a micro constituent consisting of a mixture of two phases, austenite and Cementite. It is obtained when the liquid alloy containing 4.3 percent carbon is cooled below 1148 degrees Celsius. Below 727 degrees Celsius austenite of Ledeburite changes to Pearlite giving it a characteristic appearance under microscope.

1.7   Stress vs Strain

1.7.1 Stress

The internal resistance of the material to counteract the applied load is called stress. Stress (σ) can be equated to the load per unit area or the force (F) applied per cross-sectional area (A) perpendicular to the force.

$$\text{Stress} = \sigma = \frac{F}{A}$$

where,

σ = stress (N/m^2);　　　　　　　　F = applied force (N);

A = cross-sectional area (m^2)

Stress intensity within the body of a component is expressed as one of three basic types of internal load. They are known as tensile, compressive, and shear. Fig. 1.12 illustrates the different types of stress. As illustrated in Fig. 1.12, the plane of a tensile or compressive stress lies perpendicular to the axis of operation of the force from which it originates. The plane of a shear stress lies in the plane of the force system from which it originates.

1.7.2 Types of stresses

Tensile stress: force acts to pull materials apart;

Compressive stress: the force squeezes material;

Shear stress: the force causes one part to slide on another part

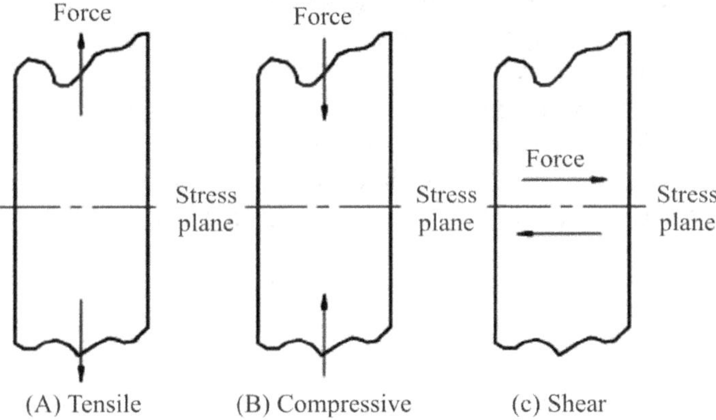

Fig. 1.12 Types of applied stress

  (i)   **Tensile stress:** Tensile stress is that type of stress in which the two sections of material on either side of a stress plane tend to pull apart or elongate as illustrated in Fig. 1.12(A).

  (ii) **Compressive stress:** Compressive stress is the reverse of tensile stress. Adjacent parts of the material tend to press against each other through a typical stress plane as illustrated in Fig. 1.12(B).

  (iii) **Shear stress:** Shear stress exists when two parts of a material tend to slide across each other in any typical plane of shear upon application of force parallel to that plane as illustrated in Fig. 1.12(C).

1.7.3 Strain

Strain is defined as change in length over original length of the member under tensile or compressive load. If L stands for original length, and ΔL represents change in length, then the change in length divided by the original length will give us the strain in the material.

ε = Lateral/Linear Strain = change in length/Original length

$\varepsilon = \Delta L/L$ [Unit of Length/Unit of Length] and therefore it is a dimensionless quantity.

1.7.4 Stress-Strain Diagram

The stress-strain diagram is a diagram with values of stress (load) as ordinate and strain (elongation, compression, deflection, twist etc.) as abscissa. Mechanical properties depend upon the crystal structure, its bonding forces, and the imperfections which exist within the crystal. The various mechanical properties can be defined or understand with the help of the above diagram. This is drawn with data obtained from a test on Universal Testing Machine (UTM). To establish a relationship between stress and strain, we can perform tests on a given specimen, one such test is the tensile test, where the specimen is loaded in tension in a machine (Tensile Test Machine) and the stress and strain are recorded.

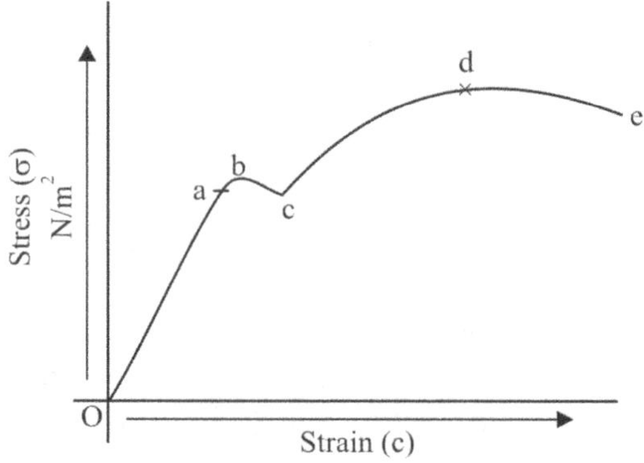

Fig. 1.13 Stress-strain diagram for steel

The stress-strain diagram for steel is shown in Fig. 1.13. The salient points are:

Point a: Limit of proportionality, O-a is a straight line and stress is proportional to strain. The slope of the line gives the value of Young's Modulus of Elasticity; E.

Point b: gives the yield point of the material and is called the elastic limit. This is the greatest stress that the material can endure without taking up permanent set after load is removed.

Point c: is called lower yield point.

Point d: gives the maximum or ultimate stress.

Point e: is called the breaking point and material fails.

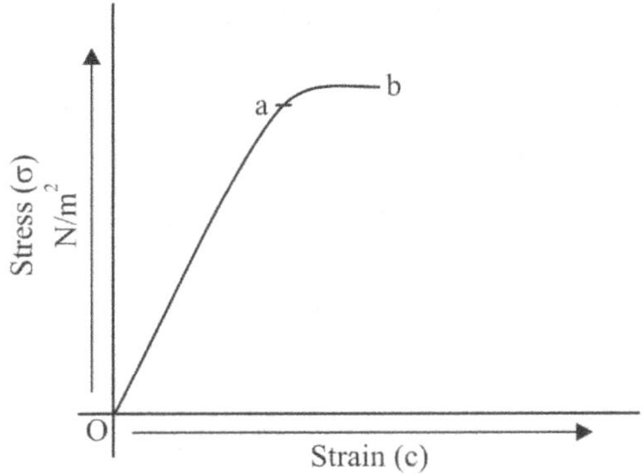

Fig. 1.14 Stress-strain diagram for brittle materials (Cast Iron)

The stress-strain ($\sigma - \varepsilon$) diagram for high carbon steel, cast iron and other brittle materials are shown in Fig.1.14.

Point a is the limit of proportionality and

Point b is the breaking point.

The curve does not have a yield point. Looking at a typical stress strain curve, when the specimen is loaded, it will deform and elongate in a linear fashion with a slope up to the proportional limit. At this point the material if loaded further will start to yield. If unloaded, the specimen will return to its origin O. (Approximately, with a 0.2% offset). After the yield the material is in plastic region and if unloaded the material will not return to its original length and will have what is called a permanent deformation or permanent set. The stronger the material the greater the slope of stress-strain curve.

1.7.5 Definition related to stress strain curve

(i)   **Ultimate Stress** (determined while testing): The highest point in stress strain curve before fracture. Maximum stress the specimen can be subjected to under testing. The point on the stress strain diagram after which the material elongates without further additional load and elongates until fracture.

(ii) **Ultimate Strength** determined by applying a factor of safety depends on the loading condition and type of member being loaded and importance of the member in structural stability, axial load on column, bending moment or shear in beam.

(iii) **Elastic Range:** Range on the stress strain curve up to the elastic or proportional limit.

(iv) **Elastic Limit/Proportional Limit:** Point on the stress strain curve up to which if load is removed, the material will return back to its original length and shape.

(v) **Yield Point:** The point on the stress strain curve after which the specimen/material will not return to its original length and shape. The crystal structure of the specimen/material begins to change up to plastic limit.

(vi) **Plastic-Strain hardening Range:** Range between Yield and Ultimate Stress.

(vii) **Plastic limit:** In the plastic range, the specimen will be able to take additional load – 1.2 times the yield strength.

(viii) **Young's Modulus (or) Modulus of Elasticity (E):** The slope of the stress strain curve is termed the Modulus of Elasticity or Young's Modulus.

(ix) **Hooke's Law:** There exists a linear relationship between stress and Strain (elongation) of a bar in tension within the elastic limit.

(x) **True Stress and Strain:** If the results of tensile testing are to be used to predict how a metal will behave under other forms of loading, it is desirable to plot the data in terms of true stress and true strain. True stress, r, is defined as

$$\sigma = F/A \qquad \qquad \text{.....(1.1)}$$

where A is the cross-sectional area at the time that the applied force is F. Up to the point at which necking starts, true strain, e, is defined as

$$e = \ln (L/L_0) \qquad \qquad \text{.....(1.2)}$$

This definition arises from taking an increment of true strain, $d\varepsilon$, as the incremental change in length, dL, divided by the length, L, at the time, $d\varepsilon = dL/L$, and integrating. As long as the deformation is uniform along the gage section, the true stress and strain can be calculated from the engineering quantities. With constant volume and uniform deformation, $LA = L_0A_0$:

$$A_0/A = L/L_0 \qquad \qquad \text{.....(1.3)}$$

Thus, according to, $A_0/A = 1 + e$

Eq. 1 can be rewritten as

$$\sigma = (F/A_0)(A_0/A)$$

and, with substitution for A_0/A and F/A_0, as

$$\sigma = s(1 + e) \qquad \qquad .....(1.4)$$

Substitution of $L/L_0 = 1 + e$ into the expression for true strain (Eq 2) gives

$$e = \ln(1 + e) \qquad \qquad .....(1.5)$$

At very low strains, the differences between true and engineering stress and strain are very small. It does not really matter whether Young's modulus is defined in terms of engineering or true stress strain.

It must be emphasized that these expressions are valid only as long as the deformation is uniform. Once necking starts, Eq. 1.1 for true stress is still valid, but the cross-sectional area at the base of the neck must be measured directly rather than being inferred from the length measurements. Because the true stress, thus calculated, is the true stress at the base of the neck, the corresponding true strain should also be at the base of the neck. Eq. 1.2 could still be used if the L and L_0 values were known for an extremely short gage section centered on the middle of the neck (one so short that variations of area along it would be negligible). Of course, there will be no such gage section, but if there were, Eq. 1.3 would be valid. Thus the true strain can be calculated as

$$\varepsilon = \ln(A_0/A) \qquad \qquad .....(1.6)$$

1.8 Testing of Materials

Materials are tested for one or more of the following purposes:

  (i)   To assure mechanical properties of materials like ductility, malleability, toughness etc.

  (ii) To determine suitability of a material for a particular application.

  (iii) To determine the surface or surface defects in raw materials or processed parts.

  (iv) To check chemical composition.

The various test and its applications are summarized in Table 1.3.

Table 1.3 Various Test and Application

S. No	Name of the test	Properties measured
1.	Tensile test	Tensile strength, yield point, elastic limit, Young's modulus, ductility, toughness, etc,
2.	Impact test	Toughness of a material under shock loading condition.
3.	Hardness test	Wear resistance, indentation resistance, scratch resistance or cutting ability of material.
4.	Fatigue test	Behaviour of material under repeated applied stress and its endurance limit.
5.	Creep test	Behaviour of material under a steady load over long period of time and creep limit of material.

1.8.1 Tensile Testing

Tensile testing involves the stretching of a piece of material, or putting it under strain, and examining how it behaves under test. The ductility of a metal can be found by tensile testing using an *extensometer*, which is an instrument for measuring the amount that a specimen stretches or extends during a test. These specimens may be rectangular or circular but must have a specified cross sectional area (CSA) as shown in Fig. 1.15.

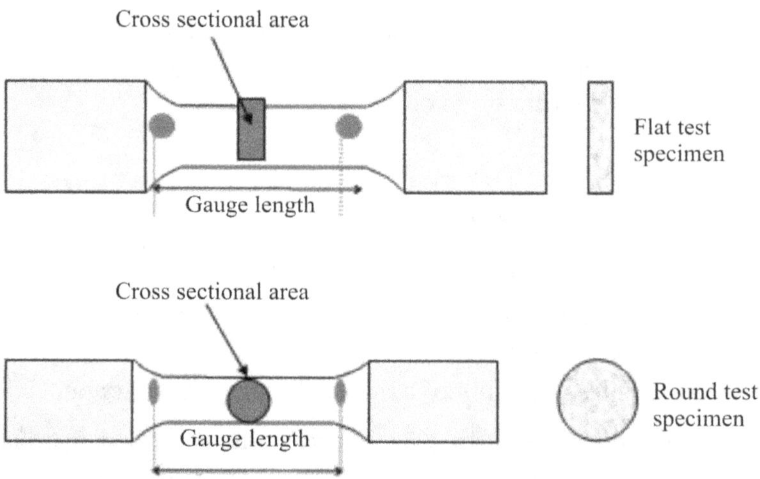

Fig. 1.15 Tensile test specimens

A material that returns to its original length has elasticity. Repeated tests are carried out making the load greater each time. While the material under test has elasticity, the extension or stretch is directly proportional to the load. This is known as *Hookes Law*. Metals are elastic in the early stage of tensile testing, plastics are not. When a material fails to return to its original length which means it has reached its *elastic limit* or the *limit of proportionality*. After the elastic limit the loads produce much larger extensions of the specimen. This is called the plastic region. At the end of this stage, the extension is even greater and a *yield point* is reached.

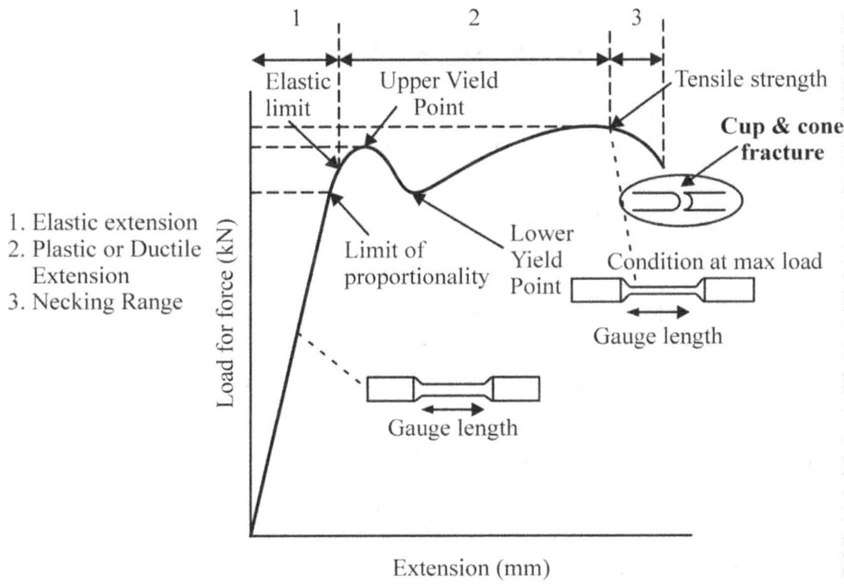

Fig. 1.16 Stress strain for material under tensile test

A further increase in the load causes the specimen to thin uniformly and then to *neck*. After necking the specimen will break or fracture. When the fracture occurs one side of the specimen has a rough cone shape and the other has a rough cup shape. This is known as a *cup and cone* fracture. Fig. 1.16 shows the stress-strain plot for a material under tensile test.

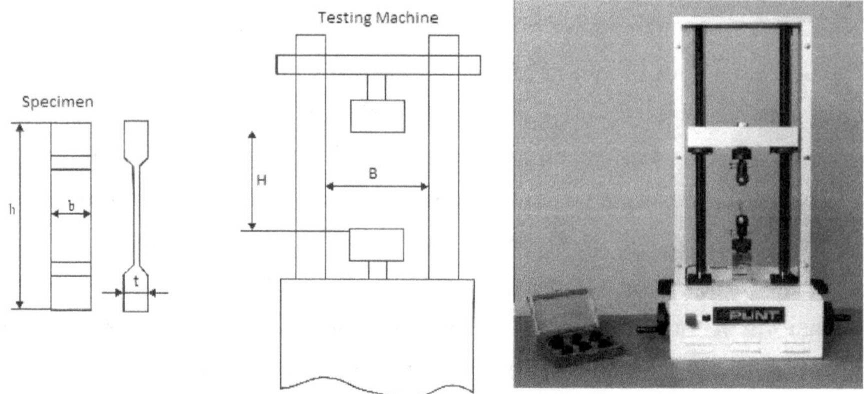

Fig. 1.17 Specimen and universal testing machine

The Test Procedure

- The specimen is set up in the tensile testing machine, the extensometer.
- One end is held in a vice and the other in a holding system.
- A controlled load, measured in newtons (N), is applied.

The amount of stretch or extension that this force causes is measured. Universal testing machine is used for performing shear, compression and tension. Fig. 1.17 shows the sketch and pictorial view of Universal Testing Machine (UTM)

There are two types of UTM.

(i)   Screw type   (ii)   Hydraulic type.

1.8.2 Shear testing

A shear force tends to shear the material into two parts. Shear force is used in cutting with scissors, tinsnips and guillotine.

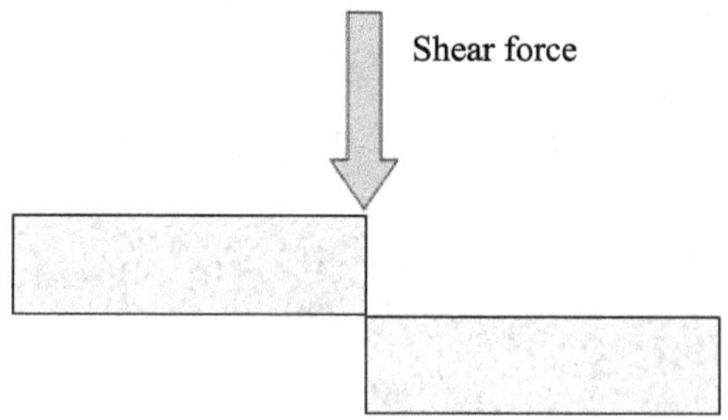

In actual practice when a beam is loaded the shear force at a section always comes to play along with bending moment. It has been observed that the effect of shearing stress as compared to bending stress is quite negligible. But sometimes, the shearing stress at a section assumes much importance in design calculations.

Details of UTM:

Capacity :   400 KN.

Range   :   0 – 400 KN.

Working Procedure:

1. Measure the diameter of the hole accurately.
2. Insert the specimen in position and grip one end of the attachment in the upper portion and the other end in the lower portion.
3. Switch on the main switch on the universal testing machine.
4. Bring the drag indicator in contact with the main indicator.
5. Gradually move the head control lever in left hand direction till the specimen shears.
6. Note down the load at which specimen shears.
7. Stop the machine and remove the specimen.

Model Observation:

Diameter of the specimen (d)= ----- mm

Cross sectional area in double shear , (A) = $2 \times \pi\, d^2/4$ mm^2

Shear Load taken by specimen at the time of failure (P) = ------ KN.

$$\text{Shear strength} = \frac{\text{Maximum shear force}}{\text{Area of the specimen}}$$

Shear strength of the given material = -------- N / mm^2

1.8.3 Parameters measured using UTM

(i) **Stress:** Stress can be defined as the amount of load or force carried by a unit area. It is normally written as Newtons per millimetre square, (N/mm^2).

$$\text{Stress} = \frac{\text{Load}}{\text{Cross sectional Area (CSA)}}$$

Example: If a bar, 15 mm wide × 8 mm deep, is pulled with a force of 360 kN, the stress is found as follows.

$$\text{Force} \quad = \quad 360 \text{ kN}$$

$$\text{Cross Sectional Area (CSA)} = \quad 120 \text{ mm}^2 \,(15 \times 8)$$

$$\textbf{Stress} \quad = \frac{360 \text{ kN}}{120 \text{ mm}^2} = \textbf{3 kN/mm}^2$$

(ii) **Strain:** Stress is the ratio of extension and original length. It is dimensionless quantity.

$$\text{Strain} = \frac{\text{Extension}}{\text{Original Length}}$$

Example: The extension is the amount by which the length changes and in a tensile tests the original length is the gauge length. If the bar in the previous example measures 2 metres before the force is applied, and 2.05 metres when the force is applied then the strain can be found as follows

$$\text{Strain} = \frac{05(2.05 - 2.0 \text{ extension})}{(\text{Original length})} = \textbf{0.025}$$

(iii) **Young's Modulus of Elasticity:** In the elastic range of a material, stress is directly proportional to strain, which is another way of stating *Hookes Law*. This is called Young's Modulus, is represented by 'E' and measured in kN/mm^2.

$$\text{Youngs Modulus of Elasticity (E)} = \frac{\text{Stress}}{\text{Strain}}$$

(iv) **Tensile Strength:** Tensile strength is the maximum force of load, in Newtons, applied to the specimen before it fractures divided by the original cross sectional area.

$$\text{Tensile Strength} = \frac{\text{Max Load}}{\text{Original CSA}}$$

(v) **Proof Stress:** Some materials do not have a well defined yield point or an indication at what stress, yield occurs. To overcome this, a value of stress, known as proof stress is used.

Reading 0.1% Proof Stress from a stress-strain graph

Proof stress is much easier to find from a stress-strain graph. Proof stress is read almost directly from this type of graph as shown in Fig.1.18. If the 0.1% proof stress is required, then a line is drawn from 0.001 strain, parallel to the straight part of the graph.

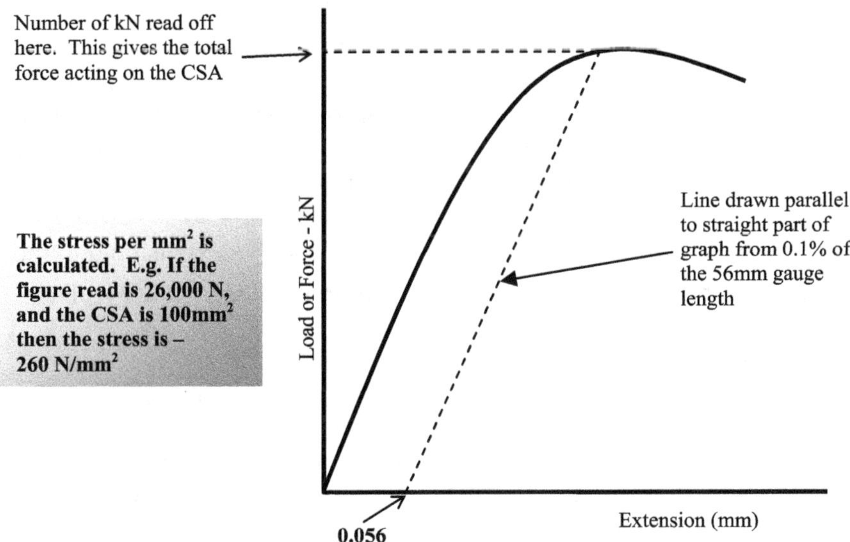

Fig. 1.18 Proof stress from a stress-strain graph

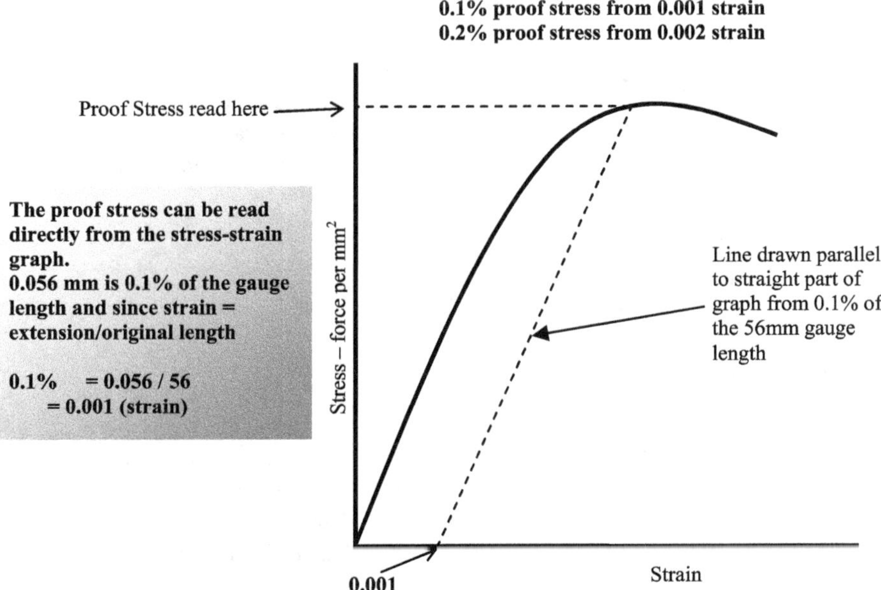

Fig. 1.19 Proof stress from a load vs extension

Reading 0.1% Proof Stress from a force-extension graph

A distance equal to 0.1% of the gauge length, from the origin, is located. If the gauge length is 56 mm, then the distance from the origin for 0.1% proof stress is 0.056 mm. A line, parallel to the straight part of the graph is drawn from the 0.056 mm mark until it cuts the graph line as shown in Fig. 1.19. The stress at that point is the 0.1% proof stress.

Load* vs *Extension Graph: The shape of the load vs extension diagram is useful in determining the properties of materials as shown in Fig.1.20.

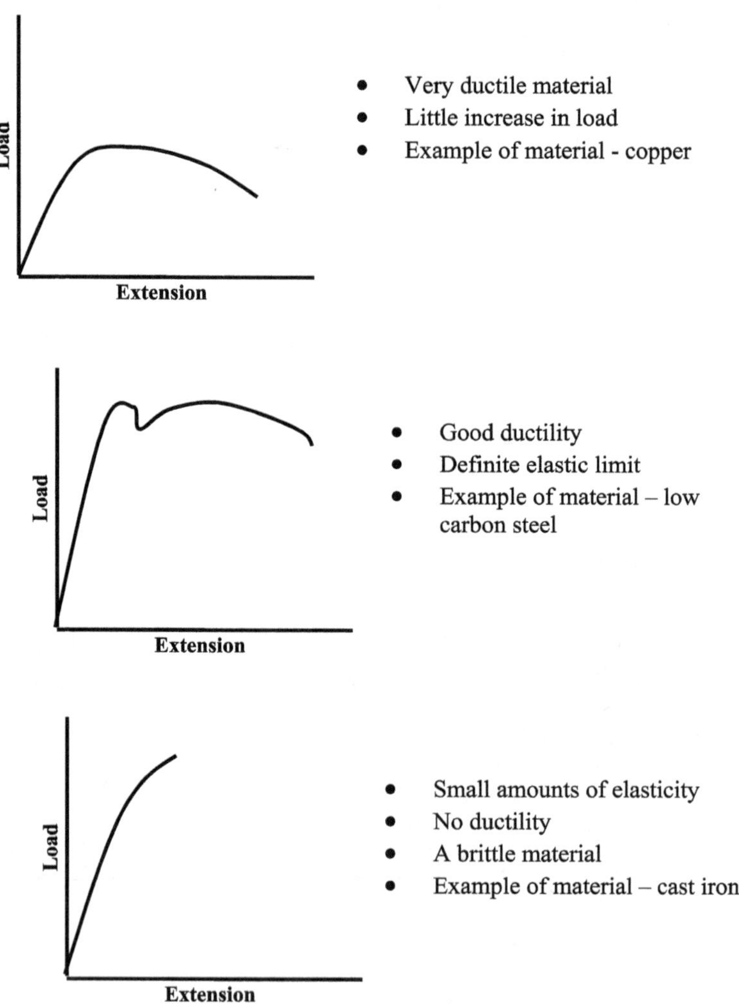

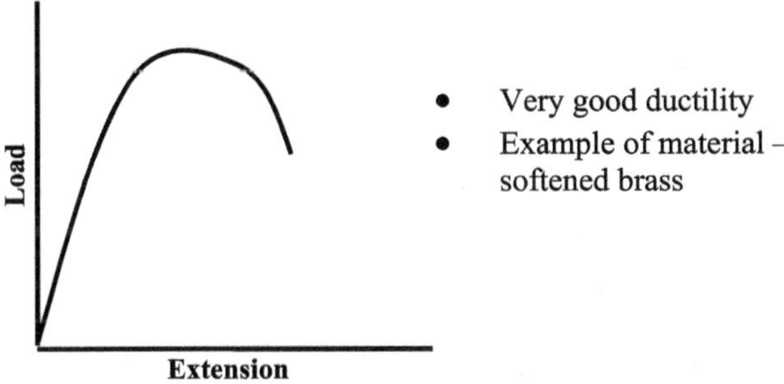

Fig. 1.20 Load Vs extension graph results

1.9   Hardness Testing

Hardness is tested by measuring the material's resistance to indentation or scratching of its surface. Hardness in metals may vary or can be changed by heat treatment.

A hard piece of metal, called the indenter, is pressed against the surface of the material being tested. The force used is measured and the indentation caused, is read directly. The softer the material, the deeper the indentation will be. The principle parts and picture of hardness testing machine is shown in Fig.1.21. There are three main types of hardness tests

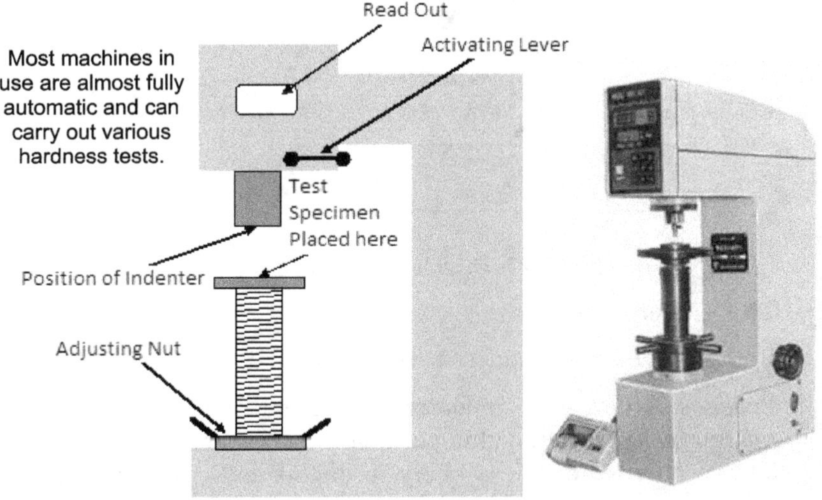

Fig. 1.21 The principal parts of a basic hardness testing machine

(i)   *The Brinell Test*, which uses a ball indenter.

(ii)   *The Vickers Test*, which uses a diamond, square based pyramid indenter.

(iii)   *The Rockwell Test*, which uses a steel ball for soft materials and a diamond cone for hard materials.

(i)   The Brinel Test [Refer Fig. 1.22]

Indenter: Hardened steel or Tungsten Carbide ball

Procedure: Ball indenter forced into the surface of the test piece by a suitable load. Diameter of indentation is measured and converted into a Brinell hardness number. The test piece must be 8 times thicker than the depth of the impression to prevent the table of the test machine absorbing the indenting force.

Uses: Used on iron castings and drop forgings.

Disadvantages: Cannot be used on thin material specimens.

Ball deforms with very hard material specimens.

Ball indents too much with soft material specimens.

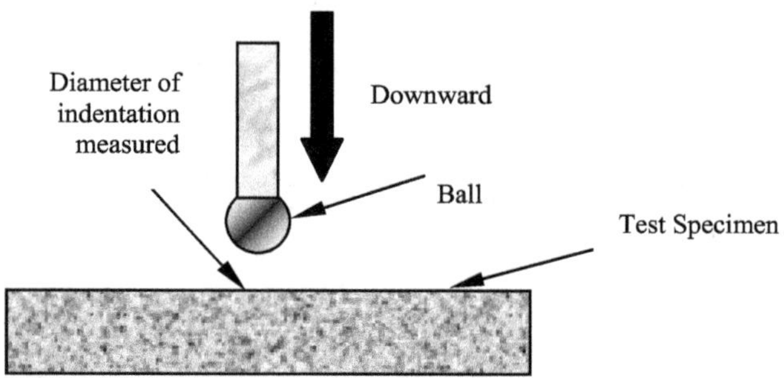

Fig. 1.22 Brinell test procedure

(ii)   The Vickers Test

Indenter: Diamond, square based pyramid – point angle of $136°$

Procedure: Diamond indenter forced into the surface of the material being tested making a square shaped impression. The diamond pyramid is less likely to distort under high forces than the steel ball.

The diagonal length of the impression is measured and this measurement is converted into a Vickers hardness number.

The test piece must be 5 times thicker than the depth of penetration required.

Uses: Used for very hard materials, gives a more accurate reading than Brinell. Easier to use. A smoother surface is required on test pieces which makes this method ideal for finished components.

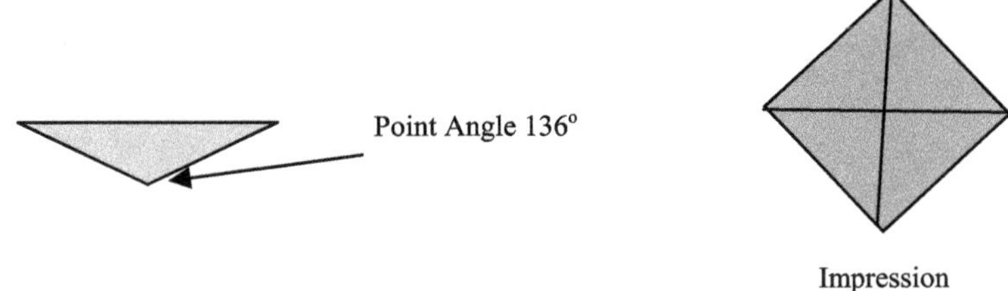

Diamond Indenter

(iii)　The Rockwell Test

Indenter: The two most common indenters are the Ball and Cone.

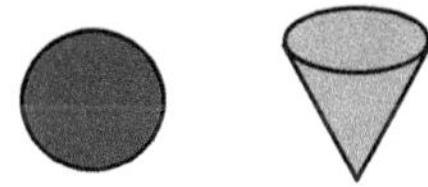

Ball and Cone Indenter

Procedure: Similar to the Vickers process.

Rockwell system has a large range of hardness scales.

Each scale has its own type and size of indenter, as well as a specified indenting force.

'C' scale indicates that a Cone was used and 'B' scale indicates that a Ball was used. The ball is used for soft materials and the cone is used for hard materials.

Uses: Particularly useful for rapid routine testing of finished components, as the hardness number is directly read from the dial. No preparation of the test piece is required.

Note: Both the Brinell and Vickers hardness tests measure the surface areas that the indenters make, to determine the hardness numbers but the Rockwell test measures how far the indenter moves into the material tested.

1.10 Toughness Testing

Toughness testing can also be called *Impact Testing* or *Notched Bar Testing*. The Fig. 1.23 shows the Impact testing machine. The two common methods of Impact Testing are

- **The Izod Test**

- **The Charpy Test**

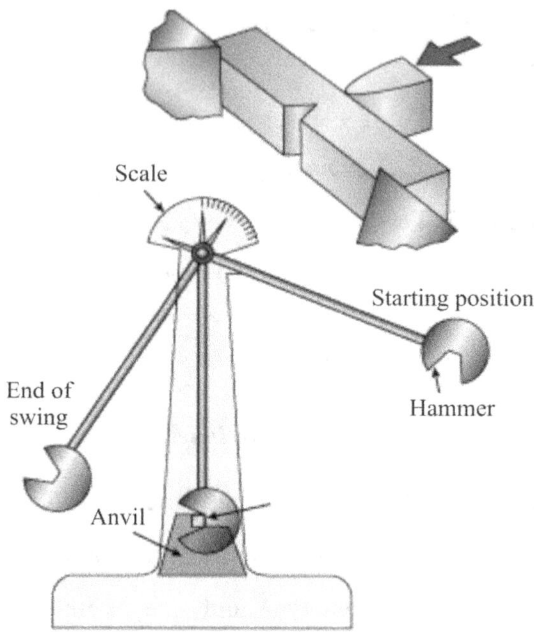

Fig. 1.23 Diagram of impact testing machine

Procedure: Notched specimens are held in a vice and are struck by a weighted pendulum. The energy absorbed in breaking the test piece is measured and a value for toughness is given. The height that the pendulum swings to after breaking the specimen indicates how much energy was absorbed.

The Izod Test [Refer Fig. 1.24]
- Specimen is held vertically
- Notch is facing the pendulum
- Striking energy of 167 Joules

'I' for Izod and the specimen stands in the vice like an 'I'.

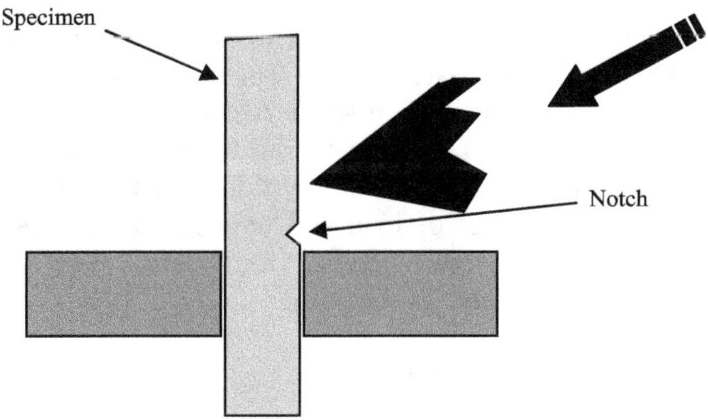

Fig. 1.24 Izod test

The Charpy Test [Refer Fig. 1.25]

- Specimen is held horizontally

- Notch is facing away from the pendulum

- Striking energy of 300 Joules – greater than the Izod because the pendulum is released from a higher position

- A different striker is fitted for this test – a knife edge striker

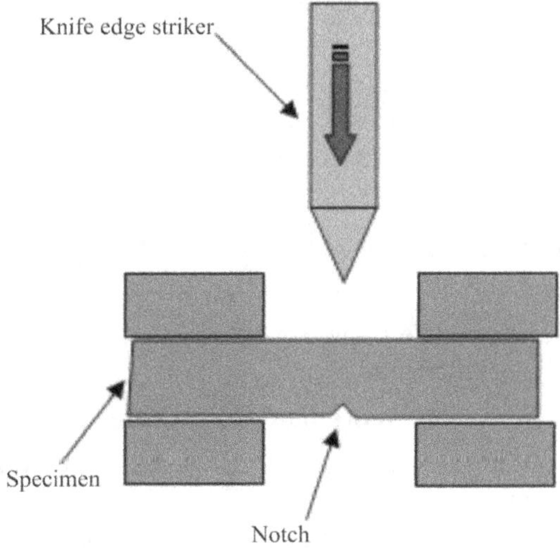

Fig. 1.25 Charpy test

1.11 Creep

Creep is the continuous slow deformation over time in materials subjected to steady persistent stress. The slow and continuous elongation of a material with time at constant stress and high temperature below elastic limit is called creep. In general, the higher the melting point of a metal, the higher its resistance to creep. Factors that affect creep in metals are length of time, size of load and temperature. At high temperatures, stresses even below the elastic limit can cause some permanent deformation on stress-strain diagram.

1.11.1   Stages of creep

Creep test is carried out at high temperature (Fig. 1.26). A creep curve (Fig. 1.27) is a plot of elongation of a tensile specimen versus time, for a given temperature and under constant stress. Tests are carried out for a period of a few days to many years. The test can be carried out on Universal Testing Machine with special attachments.

In the first stage the material elongates rapidly but at a decreasing rate. In the second stage, the rate of elongation is constant. In third stage, the rate of elongation increases rapidly until the material fails. The stress for a specified rate of strain at a constant temperature is called creep strength.

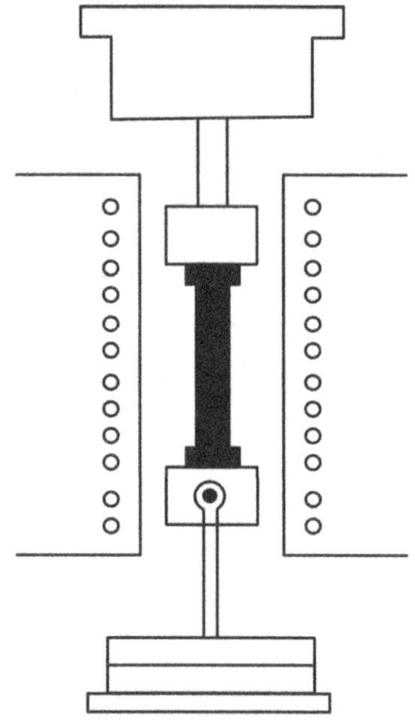

Fig. 1.26 Creep test design

Creep curve shows four stages of elongation:

(a)   Instantaneous elongation on application of load.

(b)   Primary creep: Work hardening decreases and recovery is slow.

(c)   Secondary creep: Rate of work hardening and recovery processes is equal.

(d)   Tertiary Creep: Grain boundary cracks. Necking reduces the cross-sectional area of the test specimen.

The creep strength is used for the design of blades and other parts of steam and gas turbines working at high temperatures.

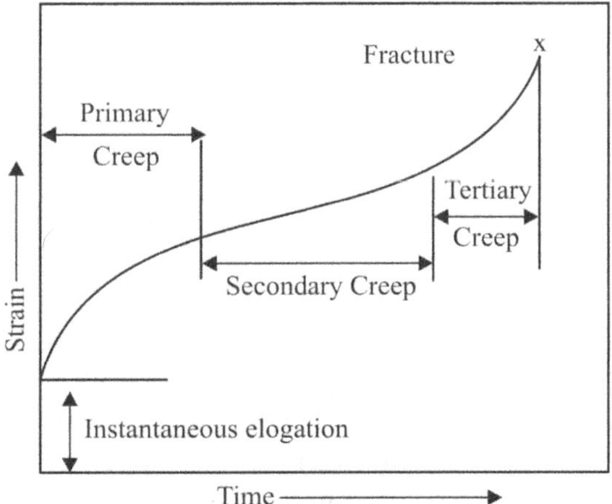

Fig. 1.27 Creep test curve

1.12 Fatigue and Fatigue Test

A material may become fatigued after it has been subjected to many small and alternating stresses over a period of time. Fatigue may start as a little crack on the surface of the component. Slowly the crack increases and moves across the section of the component until it fracture. Fatigue testing is often carried out by means of the Cantilever Beam Test. The fatigue strength of a material is the maximum stress at which failure may occur after a certain number of cyclic load applications. A component is designed to give a certain length of service under a specified loading cycle. Many components of high speed aero and turbine engines are designed for fatigue strength. The fatigue strength or endurance limit of material is used in the design of parts subjected to repeated alternating stresses over an extended period of time. Specimens are tested to failure using different loads. The number of cycles is noted for each load. The results of such tests are plotted as graphs of applied stress against the logarithm of the number of cycles of failure. The curve is known as S-N curve. The tests are carried out on special fatigue testing machines.

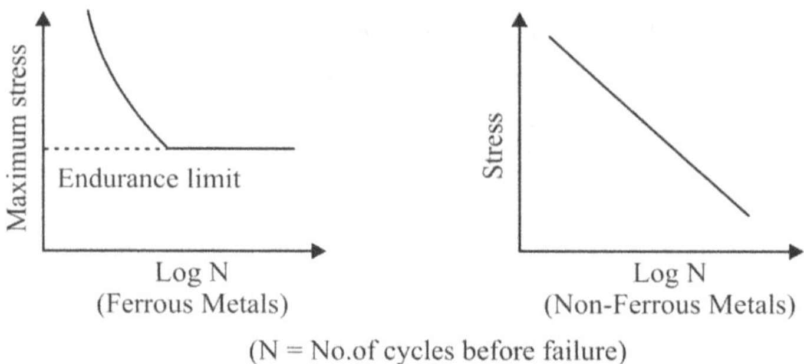

Fig. 1.28 Fatigue (S-N) curves

Many components are subjected to alternating or fluctuating loading cycles during service, and failure by fatigue is a fairly common occurrence. When a metal is tested to determine its fatigue characteristics, the test conditions usually involve the application of an alternating stress cycle with a mean stress value of zero. The results are plotted in the form of an S-N curve (Fig. 1.28), where S is the maximum stress in the cycle, and N is the number of cycles to failure. Most steels show an S-N curve of type (i), with a very definite fatigue limit, or endurance strength. This means that if the maximum stress in the stress cycles is less than this fatigue limit, fatigue failure should never occur. Many non-ferrous materials show S-N curves of type (ii) with no definite fatigue limit with these materials it is only possible to design for a limited life, and a limit of 106 or 107 cycles is often used.

Solved Numerical Examples

Example 1: A Force of 500 N acts on an area of 10 m^2. Determine the stress if the area is increased to 20 m^2

Solution:

Stress will be equal to 500/10 = 50 N/m^2

Increasing the area to 20 m^2 will decrease the stress to

500/20 = 25 N/m^2

If area is doubled the stress will be halved.

Example 2: A cylindrical rod of copper (E = 110 GPa) having a yield strength of 240 MPa is to be subjected to a load of 6660 N). If the length of the rod is 380 mm, what must be the diameter to allow an elongation of 0.50 mm

Solution:

Assuming that deformation is entirely elastic and employing the following equations

$$\sigma = \frac{F}{A_0} = \frac{F}{\pi\left(\dfrac{d_0^2}{4}\right)} = E\frac{\Delta l}{l_0}$$

Or, solving for d_0

$$d_0 = \sqrt{\frac{4\,l_0 F}{\pi\,E\,\Delta l}}$$

$$= \sqrt{\frac{(4)\,(380 \times 10^{-3}\,\text{m})\,(6660\ \text{N})}{(\pi)\,(110 \times 10^9\ \text{N/m}^2)\,(0.50 \times 10^{-3}\,\text{m})}}$$

$$= 7.65 \times 10^{-3}\ \text{m} = 7.65\ \text{mm}$$

Example 3: *A 3.0 m length of copper wire of diameter 0.4 mm is suspended from the ceiling. When a 0.5 kg mass is suspended from the bottom of the wire it extends by 0.9 mm. Calculate the strain and Stress in the wire. Also calculate the value of the Young modulus for copper.*

Solution:

$$\text{Strain} = \frac{\text{Extension}}{\text{Length}} = \frac{0.9 \times 10^{-3}\,\text{m}}{3.0\,\text{m}} = 3.0 \times 10^{-4}$$

$$\text{Stress} = \frac{\text{Load}}{\text{Cross sec tional area}} = \frac{5.0\ \text{N}}{\pi(0.2 \times 10^{-3}\,\text{m})} = 4.0 \times 10^7\,\text{Pa}$$

$$E = \frac{\text{Stress}}{\text{Strain}} = \frac{4.0 \times 10^7\,\text{Pa}}{3.0 \times 10^{-4}} = 1.3 \times 10^{11}\,\text{Pa}$$

Example 4: A long strip of rubber whose cross section measures 12 mm by 0.25 mm is pulled with a force of 3.0 N. What is the tensile stress in the rubber? Another strip of rubber originally 90 mm long is stretched until it is 120 mm long. What is the tensile strain?

Solution:

$$\text{Stress} = \frac{\text{Load}}{A} = \frac{3\text{N}}{12 \times 10^{-3}\,\text{mm} \times 0.25 \times 10^{-3}\,\text{mm}} = 1 \times 10^6\,\text{Pa}$$

$$\text{Strain} = \frac{\text{Extension}}{\text{Length}} = \frac{120\,\text{mm} - 90\,\text{mm}}{90\,\text{mm}} = 0.33$$

Review Questions

1. What is meant by mechanical properties of materials? State their importance in the design of a machine or structural element.

2. Explain the difference between malleability and ductility. Toughness, stiffness and strength.

3. Explain the term 'fatigue'. Also explain the term fatigue strength and fatigue limit related to fatigue.

4. Explain the difference between hardness and brittleness, strength and stiffness, elasticity and creep, malleability and ductility.

5. What do you understand from the term "Mechanical Properties of Materials"? On which factors does these properties mainly depend?

6. Explain the following in brief:

    (i) Impact strength   (ii) Plasticity

    (iii) Fatigue   (iv) Elasticity

7. Draw a typical "creep test" curve, showing different stages of elongation for a long time, high temperature creep test. State how the information is useful to the design engineers.

8. Differentiate between:

    (a) Hardness and impact resistance

    (b) Hardness and toughness

9. Explain: Brittleness, Stiffness and Ductility.

10. What property is dependent on time and temperature?

11. On what factors does the hardness of steel depend?

12. Briefly explain what do you understand by: Toughness, Fatigue, Creep, Hardness

13. Differentiate among strength, stiffness and toughness.

14. Explain (i) Elasticity (ii) Plasticity (iii) Toughness. (iv) Malleability

15. Discuss the factors that are taken into considerations in selecting materials for engineering design.

16. Mention the list of important physical properties of building material.

17. Describe the important mechanical properties of building materials.

18. Explain the following terms.

    (a)   Hardness   (b)   Chemical resistance

    (c)   Bulk density   (d)   Impact strength

    (e)   Thermal resistance   (f)   Elasticity

    (g)   Strength   (h)   Durability

    (i)   Creep   (j)   porosity

19. Differentiate between the following

    (a)   Density and bulk density

    (b)   Elasticity and plasticity

    (c)   Ductile materials and Brittle materials

    (d)   Hardness and impact strength

CHAPTER 2

Measurement

> **Measurement:** Temperature, Pressure Concept of Measurement Error and Uncertainly Analysis, measurement by Vernier Caliper, Micrometer, Dial gauges, Slip gauges, Sine-bar and Combination set; Introduction to Lathe, Drilling, Milling and Shaping machines., Velocity, Flow, Strain, Force and Torque measurement,

2.1 Mechanical Measurements

2.1.1 Measurement

Measurement is defined as "a set of operations having the object of determining the value of a quantity". In other words, a measurement is the evaluation of a quantity made after comparing it to a quantity of the same type which we use as a "unit".

2.1.2 Metrology

The science and grammar of measurement is defined as "the field of knowledge concerned with measurement".

46

Engineers design physical systems in the form of machines to serve some specified functions. The behavior of the parts of the machine during the operation of the machine needs to be examined or analyzed or designed such that it functions reliably. Such an activity needs data regarding the machine parts in terms of material properties. These are obtained by performing measurements in the laboratory.

2.1.3 General measurement scheme

Fig. 2.1 shows the schematic of a general measurement scheme. Not all the elements shown in the Fig. may be present in a particular case. The measurement process requires invariably a detector that responds to the measured quantity by producing a measurable change in some property of the detector. The change in the property of the detector is converted to a measurable output that may be either mechanical movement of a pointer over a scale or an electrical output that may be measured using an appropriate electrical circuit. This action of converting the measured quantity to a different form of output is done by a transducer. The output may be manipulated by a signal conditioner before it is recorded or stored in a computer. If the measurement process is part of a control application the computer can use a controller to control the measured quantity. The relationship that exists between the measured quantity and the output of the transducer may be obtained by calibration or by comparison with a reference value. The measurement system requires external power for its operation.

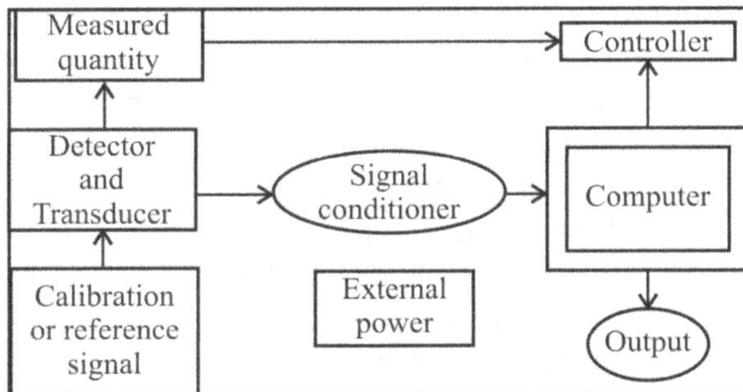

Fig. 2.1 Schematic of a General Measurement System

2.1.4 Measurement categories

Measurement is classified into four categories.

(i)   Primary quantity

(ii) Derived quantity

(iii)   Intrusive OR Probe method

(iv)   Non-intrusive

Measurement categories are described in detail now.

(i)   Primary quantity: It is possible that a single quantity that is directly measurable is of interest.   An example is the measurement of the diameter of a cylindrical specimen.   It is directly measured using an instrument such as vernier calipers.

We shall refer to such a quantity as a primary quantity.

(ii)   Derived quantity: There are occasions when a quantity of interest is not directly measurable by a single measurement process.   The quantity of interest needs to be estimated by using an appropriate relation involving several measured primary quantities.   The measured quantity is thus a derived quantity.   An example of a derived quantity is the determination of acceleration due to gravity (g) by finding the period (T) of a simple pendulum of length (L).   T and L are the measured primary quantities while g is the derived quantity.

(iii)   Intrusive or Probe method: Most of the time, the measurement of a physical quantity uses a probe that is placed inside the system.   Since a probe invariably affects the measured quantity the measurement process is referred to as an intrusive type of measurement.

(iv)   Non-intrusive method: When the measurement process does not involve insertion of a probe into the system the method is referred to as being non-intrusive.   Methods that use some naturally occurring process like radiation emitted by a body to measure a desired quantity relating to the system the method may be considered as non-intrusive.   The measurement process may be assumed to be non-intrusive when the probe has negligibly small interaction with the system. A typical example for such a process is the use of laser Doppler velocitymeter (LDV) to measure the velocity of a flowing fluid.

2.2   Temperature Measurement

In analyzing mechanical system, until now, only three basic quantities are required: displacement (L), time (T) and mass (M). Other mechanical quantities such as energy, force and momentum can be described by these basic quantities. However, heat or thermal phenomenon needs for one more basic quantity, which is temperature.

Temperature in the system is a characteristic that ensure whether it is in heat equilibrium with other systems. When two or more systems are in heat equilibrium, they have same temperature.

2.2.1 Temperature scale

The two temperature scales in use are the Fahrenheit and Celsius scales. These scales are based on a specification of the number of increments between the freezing point and boiling point of water at standard atmospheric pressure. The Celsius scale has 100 units between these points, while the Fahrenheit scale has 180 units. The absolute Celsius scale is called the Kelvin scale, while the absolute Fahrenheit scale is termed as the Rankine scale. Both absolute scales are so defined that they will correspond as closely as possible with the absolute scales represent the same physical state, and the ratio of two values is same, regardless of the absolute scale used; i.e.,

$$\left(\frac{T_2}{T_1} \right)_{Rankine} = \left(\frac{T_2}{T_1} \right)_{Kelvin}$$

The boiling point of water at 1 atm is arbitrarily taken as 100° on the Celsius scale and 212° on the Fahrenheit scale. The following relationship applies for conversion:

$$^\circ F = 32.0 + \frac{9}{5}\ ^\circ C$$

$$^\circ R = \frac{9}{5}\ K$$

2.2.2 Temperature Measurement Technique

There are several methods that usually used for temperature measurement:

   (i)   deformation in solid, liquid and gas

   (ii)   deformation in electrical resistance and voltage

   (iii)   deformation in radiation

   (i)   **Deformation in solid, liquid and gas:**

       Bi-metallic strip: A bi-metallic strip is used to convert a temperature change into mechanical displacement.

The strip consists of two layers, usually iron and copper. The two layers are joined together to form the strip. Owing to the difference in the constants of expansion of the two materials, a flat strip will bend one way (toward the iron part) if heated, and in the opposite direction if cooled below its normal temperature.

In some applications the bi-metal strip is used in the flat form. In others, it is wrapped into a coil, which gives greater sensitivity in a compact space.

Thermostats: In regulating thermostats that operate over a wide range of temperatures the bi-metal strip is mechanically fixed and attached to an electrical power source while the other (moving) end carries an electrical contact. In adjustable thermostats another contact is positioned with a regulating knob or lever. The position so set controls the regulated temperature, called the set point. Some thermostats use a mercury switch connected to both electrical leads. The angle of the entire mechanism is adjustable to control the set point of the thermostat. Depending upon the application, a higher temperature may open a contact (as in a heater control) or it may close a contact (as in a refrigerator or air conditioner).

The electrical contacts may control the power directly (as in a household iron) or indirectly, switching electrical power through a relay or the supply of natural gas or fuel oil through an electrically operated valve. In some natural gas heaters the power may be provided with a thermocouple that is heated by a pilot light (a small, continuously burning flame). In devices without pilot lights for ignition (as in most modern gas clothes dryers and some natural gas heaters and decorative fireplaces) the power for the contacts is provided by reduced household electrical power that operates a relay controlling an electronic ignitor, either a resistance heater or an electrically powered spark generating device.

Thermometers: A direct indicating dial thermometer (such as a patio thermometer or a meat thermometer) uses a bi-metallic strip wrapped into a coil, as does a common household thermostat. One end of the coil is fixed to the chassis of the device and the other is connected to an indicating needle.

Bimetallic thermometer: The sensitive element of which consists of two metal strips that have different coefficients of expansion and are brazed together. The distortions of the system in response to temperature variations are used as a measure of temperature. It is a type of deformation thermometer.

(ii) Deformation in electrical resistance and voltage

Thermocouple: In electronics, thermocouples are a widely used type of temperature sensor. They are cheap, interchangeable, have standard connectors and can measure a wide range of temperatures. The main limitation is accuracy; system errors of less than 1 °C can be difficult to achieve.

A thermopile is a group of thermocouples connected in series.

Principle of operation: In 1822, an Estonian physicist named Thomas Johann Seebeck discovered that when any conductor (such as a metal) is subjected to a thermal gradient, it will generate a small voltage. Thermocouples make use of this so-called Peltier-Seebeck effect.

Thermocouples produce an output voltage which depends on the temperature difference between the junctions of two dissimilar metal wires (WIRE A & WIRE B) as shown in Fig. 2.2. It is important to appreciate that thermocouples measure the temperature difference between two points, not absolute temperature.

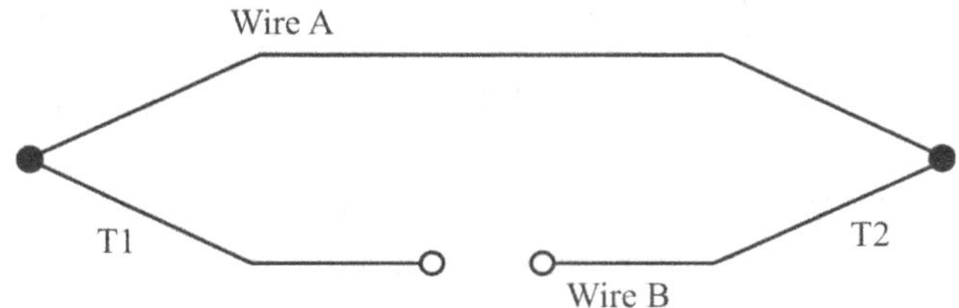

Fig. 2.2 Thermocouple

In most applications, one of the junctions — the "cold junction" — is maintained at a known (reference) temperature, whilst the other end is attached to a probe. For example, in Fig. 2.3, the cold junction will be at copper tracks on the circuit board. Another temperature sensor will measure the temperature at this point, so that the temperature at the probe tip can be calculated.

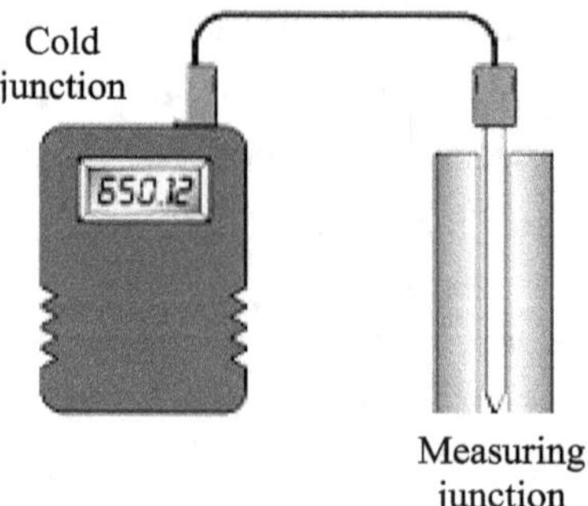

Fig. 2.3 Temperature Measurement with Probe

The relationship between the temperature difference and the output voltage of a thermocouple is nonlinear and is given by a complex polynomial equation (which is fifth to ninth order depending on thermocouple type). To achieve accurate measurements some type of linearisation must be carried out, either by a microprocessor or by analogue means.

Different types of thermocouples: A variety of thermocouples are available, suitable for different measuring applications (industrial, scientific, food temperature, medical research, etc.).

Type K (Chromel (Ni-Cr alloy) / Alumel (Ni-Al alloy)): The "general purpose" thermocouple. It is low cost and, owing to its popularity, it is available in a wide variety of probes. They are available in the −200 °C to +1200 °C range. Sensitivity is approximately 41 µV/°C.

Type E (Chromel / Constantan (Cu-Ni alloy)): Type E has a high output (68 µV/°C) which makes it well suited to low temperature (cryogenic) use. Another property is that it is non-magnetic.

Type J (Iron / Constantan): Limited range (− 40 to + 750 °C) makes type J less popular than type K. The main application is with old equipment that cannot accept "modern" thermocouples. J types cannot be used above 760 °C as an abrupt magnetic transformation causes permanent decalibration.

Type N (Nicrosil (Ni-Cr-Si alloy) / Nisil (Ni-Si alloy)): High stability and resistance to high temperature oxidation makes type N suitable for high temperature measurements without the cost of platinum (B, R, S) types. Designed to be an "improved" type K, it is becoming more popular.

Thermocouple types B, R, and S are all noble metal thermocouples and exhibit similar characteristics. They are the most stable of all thermocouples, but due to their low sensitivity (approximately 10 µV/°C) they are usually only used for high temperature measurement (>300 °C).

Type B (Platinum-Rhodium/Pt-Rh): Suited for high temperature measurements up to 1800 °C. Unusually type B thermocouples (due to the shape of their temperature-voltage curve) give the same output at 0 °C and 42 °C. This makes them useless below 50 °C.

Type R (Platinum / Rhodium): Suited for high temperature measurements up to 1600 °C. Low sensitivity (10 µV/°C) and high cost makes them unsuitable for general purpose use.

Type S (Platinum / Rhodium): Suited for high temperature measurements up to 1600 °C. Low sensitivity (10 µV/°C) and high cost makes them unsuitable for general purpose use. Due to its high stability type S is used as the standard of calibration for the melting point of gold (1064.43 °C).

Type T (Copper / Constantan): Suited for measurements in the −200 to 0 °C range. The positive conductor is made of copper, and the negative conductor is made of constantan.

Thermocouples are usually selected to ensure that the measuring equipment does not limit the range of temperatures that can be measured. Note that thermocouples with low sensitivity (B, R, and S) have a correspondingly lower resolution.

Resistance Temperature Detector (including compensated Wheatstone bridge): Resistance Temperature Detectors (RTD), as the name implies, are sensors used to measure temperature by correlating the resistance of the RTD element with temperature. Most RTD elements consist of a length of fine coiled wire wrapped around a ceramic or glass core. The element is usually quite fragile, so it is often placed inside a sheathed probe to protect it. The RTD element is made from a pure material whose resistance at various temperatures has been documented. A typical type of RTD design is shown in Fig. 2.4. The material has a predictable change in resistance as the temperature changes; it is this predictable change that is used to determine temperature. Common resistance materials for RTDs:

- Platinum (most popular and accurate)
- Nickel
- Copper
- Balco (rare)
- Tungsten (rare)

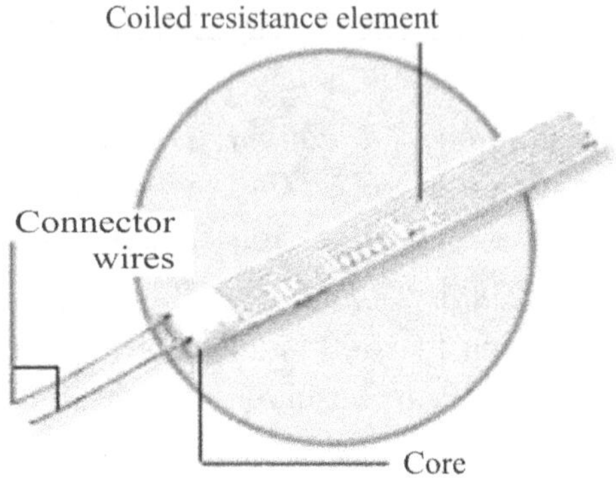

Fig. 2.4 Typical RTD Design

Benefit of using RTD: The RTD is one of the most accurate temperature sensors. Not only does it provide good accuracy, it also provides excellent stability and repeatability. RTDs are also relatively immune to electrical noise and therefore well suited for temperature measurement in industrial environments, especially around motors, generators and other high voltage equipment.

Wheatstone bridge: A Wheatstone bridge is a measuring instrument invented by Samuel Hunter Christie in 1833 and improved and popularized by Sir Charles Wheatstone in 1843. It is used to measure an unknown electrical resistance by balancing two legs of a bridge circuit, one leg of which includes the unknown component. Its operation is similar to the original potentiometer except that in potentiometer circuits the meter used is a sensitive galvanometer.

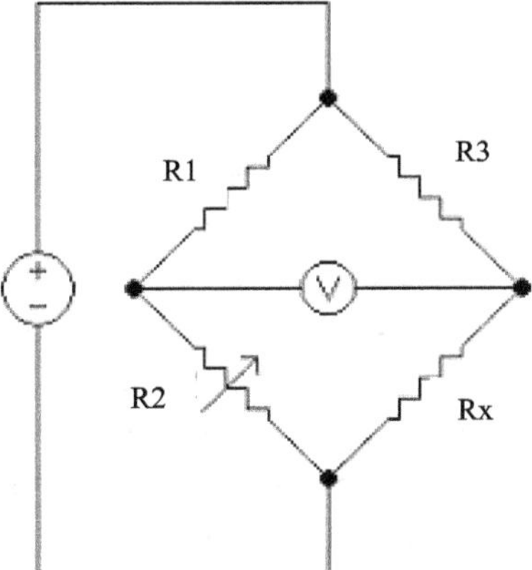

Fig. 2.5 Wheatstone Bridge

Here, Rx is the unknown resistance to be measured; R1, R2 and R3 are resistors of known resistance and the resistance of R_2 is adjustable. If the ratio of the two resistances in the known leg (R2/R1) is equal to the ratio of the two in the unknown leg (Rx/R3), then the voltage between the two midpoints will be zero and no current will flow between the midpoints. R_2 is varied until this condition is reached. The current direction indicates if R_2 is too high or too low.

Detecting zero current can be done to extremely high accuracy Therefore, if R1, R2 and R3 are known to high precision, then Rx can be measured to high precision. Very small changes in Rx disrupt the balance and are readily detected.

Alternatively, if R1, R2, and R3 are known, but R2 is not adjustable, the voltage or current flow through the meter can be used to calculate the value of Rx. This setup is frequently used in strain gauge measurements, as it is usually faster to read a voltage level off a meter than to adjust a resistance to zero the voltage.

The Wheatstone bridge illustrates in Fig. 2.5, the concept of a difference measurement, which can be extremely accurate. Variations on the Wheatstone bridge can be used to measure capacitance, inductance, impedance and other quantities, such as the amount of combustible gases in a sample, with an explosimeter. The Kelvin bridge was one specially adapted for measuring very low resistances. This was invented in 1861 by William Thomson, Lord Kelvin. The concept was extended to alternating current measurements by James Clerk Maxwell in 1865, and further improved by Alan Blumlein in about 1926.

(iii)   Deformation in radiation

Pyrometer: A pyrometer is non-contact temperature measuring device; generally the term is applied to instruments measuring temperatures above 600 degrees Celsius. It is typically used to measure temperatures of glowing hot metals in a steel mill or foundry. It is also known as an optical pyrometer (Fig.2.6). One of the most common pyrometers is the absorption-emission pyrometer which is a thermometer for determining gas temperature from measurement of the radiation emitted by a calibrated reference source before and after this radiation has passed through and been partially absorbed by the gas. Both measurements are made over the same wavelength interval. To measure the temperature of incandescent metals, you look through the pyrometer at the glowing metal, and turn a knob or ring which adjusts the temperature of a glowing filament projected into your field of view. When the color of the filament matches the color of the metal, you can read the temperature from a scale on the filament color adjusting knob/ring.

Optical Pyrometer: Optical pyrometers measure the radiation from the target in a narrow band of wavelengths of the thermal spectrum. The oldest devices use the principle of optical brightness in the visible red spectrum around 0.65 microns (Fig.2.6). These instruments are also called single color pyrometers. Optical pyrometers are now available for measuring energy wavelengths that extend into the infrared region. The term single color pyrometer has been broadened by some authors to include narrow band radiation thermometers as well. Some optical designs are manually operated. The operator sights the pyrometer on target. At the same time he/she can see the image of an internal lamp filament in the eyepiece. In one design, the operator adjusts the power to the filament, changing its color, until it matches the color of the target. The temperature of the target is measured based upon power being used by the internal filament. Another design maintains a constant current to the filament and changes the brightness of the target by means of a rotatable energy-absorbing optical wedge. The object temperature is related to the amount of energy absorbed by the wedge, which is a function of its annular position.

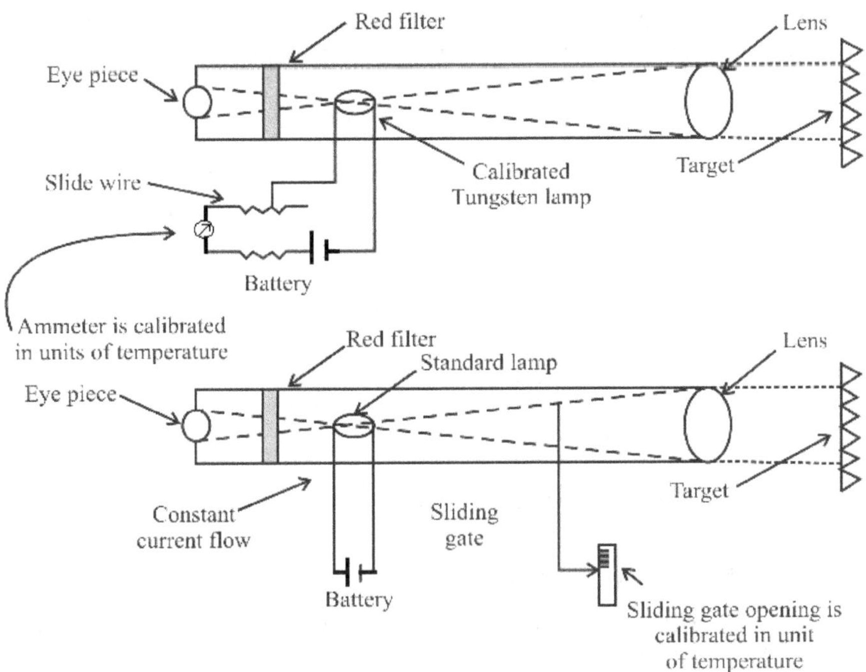

Fig. 2.6 Optical Pyrometry by Visual comparison

Automatic optical pyrometers, sensitized to measure in the infrared region, also are available. These instruments use an electrical radiation detector, rather than the human eye. This device operates by comparing the amount of radiation emitted by the target with that emitted by an internally controlled reference source. The instrument output is proportional to the difference in radiation between the target and the reference. A chopper, driven by a motor, is used to alternately expose the detector to incoming radiation and reference radiation. In some models, the human eye is used to adjust the focus. Fig. 2.7 is a schematic of an automatic optical pyrometer with a dichroic mirror.

Radiant energy passes through the lens into the mirror, which reflects infrared radiation to the detector, but allows visible light to pass through to an adjustable eyepiece. The calibrate flap is solenoid-operated from the amplifier, and when actuated, cuts off the radiation coming through the lens, and focuses the calibrate lamp on to the detector. The instrument may have a wide or narrow field of view. All the components can be packaged into a gun-shaped, hand-held instrument. Activating the trigger energizes the reference standard and read-out indicator. Optical pyrometers have typical accuracy in 1% to 2% of full-scale range.

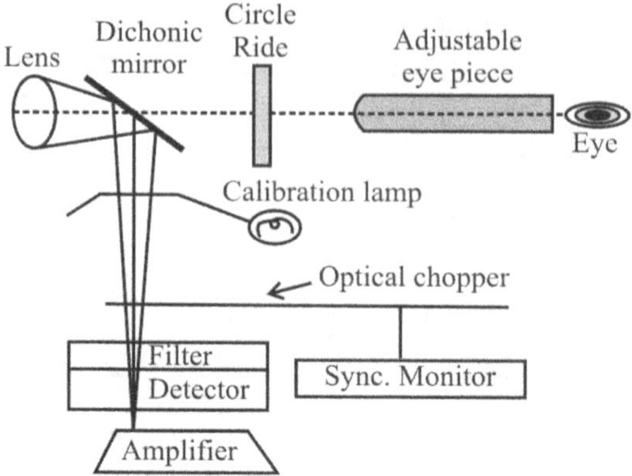

Fig. 2.7 An Automatic Optical Pyrometer

2.3   Pressure Measurement

Pressure is defined as a force per unit area and the most accurate way to measure low air pressure is to balance a column of liquid of known weight against it and measure the height of the liquid column so balanced. The units of measure commonly used are inches of mercury (in. Hg), using mercury as the fluid and inches of water (in. w.c.), using water or oil as the fluid.

Unit Conversion:

$1\ lbf/in^2 = 0.07031\ kgf/cm^2 = 6865\ N/m^2$

$1\ tonf/in^2 = 157.5\ kgf/cm^2 = 15.44\ MN/m^2$

$1\ kgf/cm^2 = 0.09807\ MN/m^2 = 0.9807\ bar$

$1\ lbf/ft^2 = 47.88\ N/m^2$

$1\ ft\ H_2O = 62.43\ lbf/ft^2 = 2989\ N/m^2$

$1\ in\ Hg = 70.73\ lbf/ft^2 = 3386\ N/m^2$

$1\ mmHg = 1\ torr = 133.3\ N/m^2$

$1\ bar = 14.50\ lbf/in^2 = 105\ N/m^2$

$1\ Atm = 14.70\ lbf/in^2 = 10.3\ mH_2O = 1.013 \times 105\ N/m^2$

2.3.1   Pressure Gauge and Absolute Pressure

Many techniques have been developed for the measurement of reduced pressures.

Gauges are either direct- or indirect-reading. Those that measure pressure by calculating the force exerted on the surface by incident particle flux are called direct reading gauges. Indirect gauges record the pressure by measuring a gas property that changes in a predictable manner with gas density.

Absolute Pressure: Pressure measurement relative to atmospheric pressure.

Gauge Pressure: Gauge pressure is the pressure indicated on a gauge. As pressure is normally measured as a difference from another pressure (typically atmospheric pressure); the gauge will only read the pressure difference -not the absolute pressure (Fig. 2.8).

Vacuum Pressure: occurs when pressure is negative in gauge pressure.

Absolute Pressure = Atmospheric Pressure ± Pressure Gauge

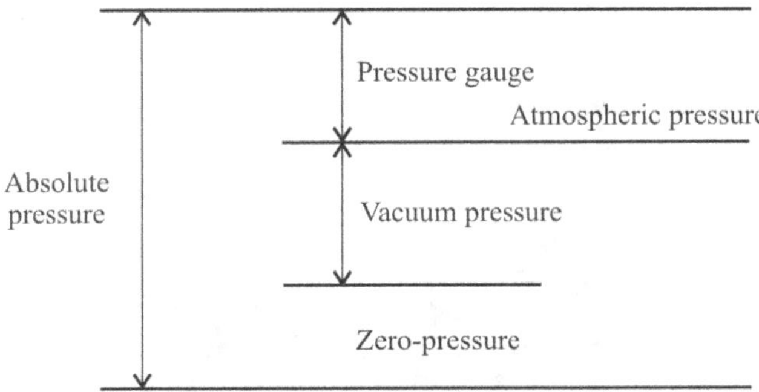

Fig. 2.8 Atmospheric, Absolute, Vacuum Pressure

2.3.2 Pressure Measurement Technique
(i)   Elastic deformation method
(ii) Liquid columns method

(i)   Elastic deformation method

Bourdon tube: In 1849 the Bourdon tube pressure gauge was patented in France by Eugene Bourdon.

A pressure or vacuum gauge usually consists of a closed coiled tube connected to the chamber or pipe in which pressure is to be sensed. As the pressure increases the tube will tend to uncoil, while a reduced pressure will cause the tube to coil more tightly. This motion is transferred through a link to a gear train connected to an indicating needle. The needle is presented in front of a card face inscribed with the pressure indications associated with particular needle deflections as shown in Fig. 2. 9.

In the following pictures the transparent cover face has been removed and the mechanism removed from the case. This particular gauge is a combination of vacuum and pressure gauge used for automotive diagnosis. The left side of the face, used for measuring manifold vacuum, is calibrated in centimeters of mercury vacuum on its inner scale and inches of mercury vacuum in its outer scale. The right portion of the face is used to measure fuel pump pressure and is calibrated in kilograms per square centimeter on its inner scale and pounds per square inch on its outer scale.

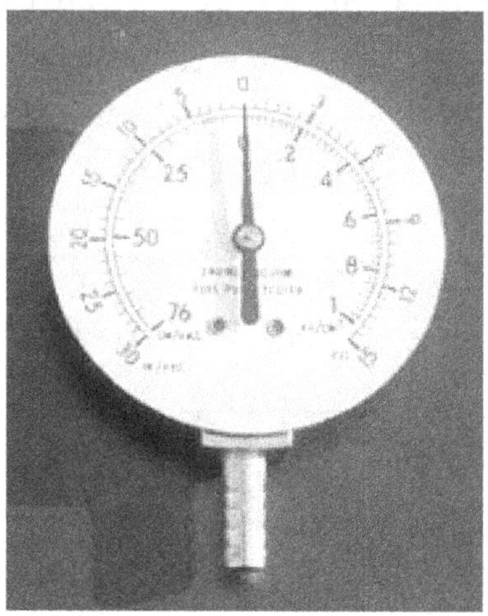

Fig. 2.9 Indicator Side

Mechanical details are shown in Fig. 2.10 and Fig. 2.11.

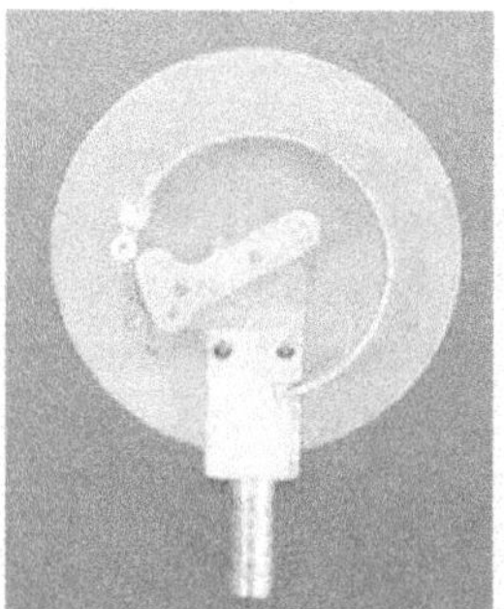

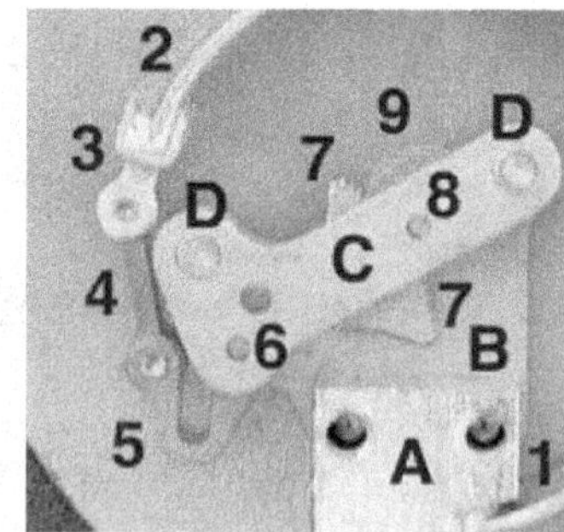

Fig. 2.10 Mechanical Side   **Fig. 2.11:** Mechanical Details

Stationary parts:

A: Receiver block. This joins the inlet pipe to the fixed end of the Bourdon tube (1) and secures the chassis plate (B). The two holes receive screws that secure the case.

B: Chassis Plate. The face card is attached to this. It contains bearing holes for the axles.

C: Secondary Chassis Plate. It supports the outer ends of the axles.

D: Posts to join and space the two chassis plates.

Moving Parts:

1.   Stationary end of Bourdon tube. This communicates with the inlet pipe through the receiver block.

2.   Moving end of Bourdon tube. This end is sealed.

3.   Pivot and pivot pin.

4.   Link joining pivot pin to lever (5) with pins to allow joint rotation.

5.   Lever. This is an extension of the sector gear (7).

6.   Sector gear axle pin.

7.   Sector gear.

8.   Indicator needle axle. This has a spur gear that engages the sector gear (7) and extends through the face to drive the indicator needle. Due to the short distance between the lever arm link boss and the pivot pin and the difference between the effective radius of the sector gear and that of the spur gear, any motion of the Bourdon tube is greatly amplified. A small motion of the tube results in a large motion of the indicator needle.

9.   Hair spring to preload the gear train to eliminate gear lash and hysteresis.

Diaphragms and Bellows: One common element used to convert pressure information into a physical displacement is the diaphragm. A diaphragm is like a spring, construction is shown in Fig. 2.12.

A diaphragm (membrane or corrugated) is a thin elastic circular plate supported about its circumference. Membranes are made of metal or nonmetallic material, such as plastic. The material chosen depends on the pressure range expected and the fluid in contact with it.

Diaphragm type pressure transducers are well suited for both static and dynamic pressure measurements. They have good linearity and resolution, as well as fast speed of response compared with the tube and bellows types.

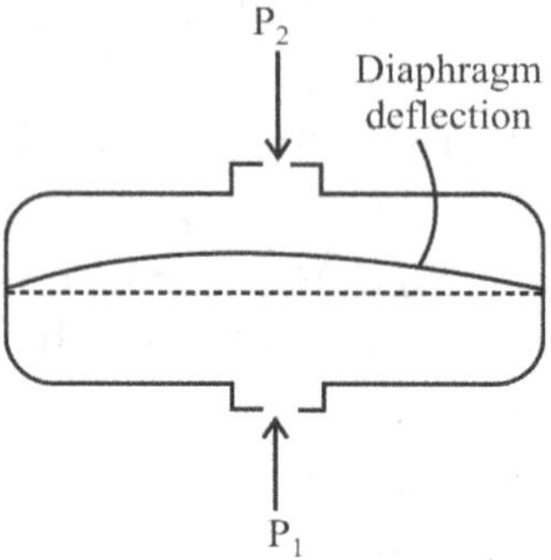

Fig. 2.12 Diaphragm

While a bellows (see Fig.2.13) is another device much like the diaphragm that converts a pressure differential into a physical displacement, except that here the displacement is much more a straight-line expansion. Fig.2.13 also shows how an LVDT can be connected to the bellows so that the pressure measurement is converted directly from displacement to voltage. In addition, the displacement and pressure are nearly linearly related, and because the LVDT voltage is linear with displacement, the voltage and pressure are also linearly related.

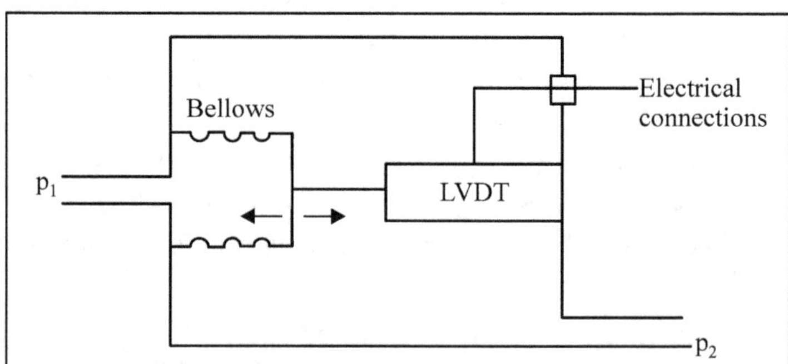

Fig. 2.13 Bellows

(ii)　Liquid columns method

Manometer: A manometer is a pressure measuring instrument, often also called pressure gauge. Liquid manometers, because of inherent accuracy and simplicity, have applications in every industry and laboratory. They are unique in being both basic pressure measurement instruments and standards for calibration of other instruments. Despite their simplicity and wide usage, their principle of operation, the advantages of various types and the basic accuracy factors associated with them are not known by all.

The manometer has many advantages in this age of technology. Containing no mechanical moving parts, needing nothing but the simplest of measurements, the primary standard manometer is readily available at modest cost. The principle of the manometer has not changed since its inception, however great strides have been made in its arrangement and the application of the instrument to various industrial measurement requirements. Whereas formerly the manometer was considered as a laboratory instrument, today we find the manometer commonly used to measure pressures ranging from as high as 600 inches of mercury to space vacuums. The manometer utilizes the hydrostatic (standing liquid) balance principle wherein a pressure is measured by the height of the liquid it will support. For example, the weight of a column of mercury at 0 deg C that is one inch high and one inch in cross sectional area is 0.4892 pounds. Thus we can say that a column of mercury one inch high imposes a force of 0.4892 pounds per square inch or 0.4892 PSI.

Description: The oldest type is the liquid-column manometer. A very simple version is a U-shaped tube half-full of liquid where the measured pressure is applied to one side of the tube whilst the reference pressure (which might be that of the atmosphere) is applied to the other. The difference in liquid level represents the applied pressure. It is quite easy to make a homemade manometer. A single-limb liquid-column manometer has a larger reservoir instead of one side of the U-tube and has a scale beside the narrower column. The column may be inclined to further amplify the liquid movement. Liquid-column manometers can be used to measure small differences between great pressures.

A second type uses the deflection of a flexible membrane that seals a fixed pressure reference volume to determine the pressure. The amount of deflection is repeatable for known pressures so the pressure can be determined using a lookup table. A third variant (Bourdon gauge) uses a coiled tube which as it expands due to pressure increase causes a rotation of an arm connected to the tube. One use of manometers is to measure vacuum pressures, especially in the range from 0.001 atmospheres to 1 atm. They are helpful because the deflection of the manometer is not dependent upon the type of gas being measured, unlike other types of vacuum gauges in this pressure range. The deflection of the piston is often one half of a capacitor, so that when the piston moves, the capacitance of the device changes. This is a common way (with proper calibrations) to get a very precise, electronic reading from a manometer, and this configuration is called a capacitive manometer vacuum gauge.

U-type manometer: The manometer principle is most easily demonstrated in the U-type manometer illustrated in Fig. 2.14. Here, with both legs of the instrument open to atmosphere or subjected to the same pressure, gravity forces the surfaces of the liquid to be at exactly the same level or reference zero.

As illustrated in Fig. 2.14, if a pressure is applied to the left side of the instrument the applied pressure depresses the fluid in the left leg and raises it in the right until the unit weight of the fluid as indicated by "H" exactly balances the pressure. Thus, if the fluid in one side stands two inches higher than the fluid in the other, and the fluid is mercury, the pressure balanced is 2 × 0.4892 or 0.9784 PSI.

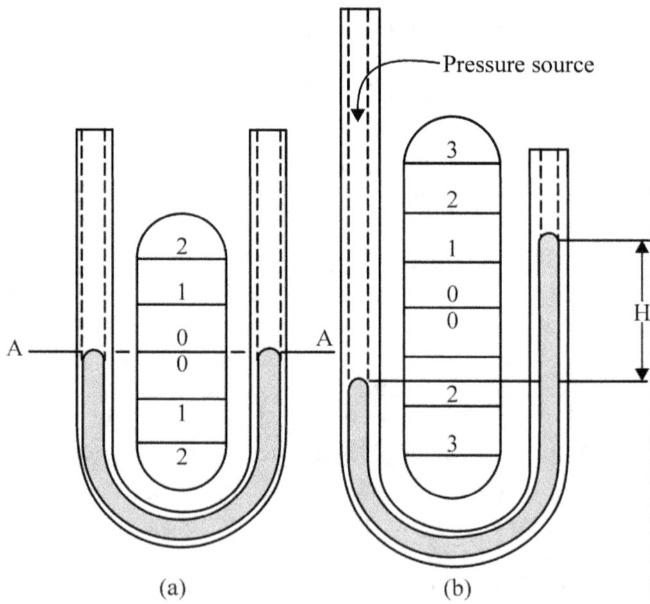

Fig. 2.14 U-type manometer

Note that only the height of the fluid from the surface in one tube to the surface in the other is the actual height fluid opposing and balancing the pressure. This is true regardless of the shape or size of the tubes. If the tubes are unsymmetrical, more or less fluid may be moved from one to the other, but the height of fluid required to obtain equilibrium is independent of all but the fluid density and vertical height.

Types of Pressure Measurement: There are three types of pressure measurement – Positive pressure or gauge pressure are those greater than atmospheric; negative pressures or vacuums are pressures less than atmospheric; differential pressure is the difference between two pressures.

All three types are readily measured with the manometer, which in itself is somewhat unusual. Connecting one leg of the U-tube to a source of gauge pressure, as illustrated in Fig. 2,14, depresses the fluid in the connected leg and raises it in the vented leg. If, however, our air supply line should be changed to a vacuum line, the only effect would be to reverse the fluid movement and it would rise in the connected leg and recede in the open leg.

Differential pressures are measured by connecting one leg to each of the two pressures.

Then the higher pressure depresses the fluid in one leg while the lower pressure allows the fluid to rise in the other: Again the true differential is measured by the difference in height of fluid in the two legs. At the same time the manometer indicates which of the two pressures is higher, since the higher pressure depresses the fluid column.

Well-type manometer: The principles of manometric measurements have been discussed in reference to the U-type manometer. However the manometer has been arranged in other forms to provide greater convenience and to meet varying service requirements. The well type manometer is one of these variations. As illustrated in Fig.2.15, if one leg of the manometer is increased many times in area to that of the other, the volume of fluid displaced will represent very little change of height in the smaller area leg. This condition results in an ideal arrangement whereby it is necessary to read only one convenient scale adjacent to a single indicating tube rather than two in the U-type. The larger area leg is called the "well".

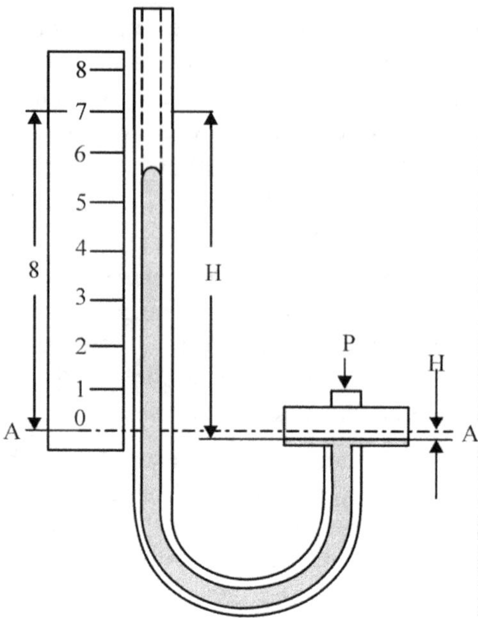

Fig. 2.15 Well-type manometer

For this reason, the well type lends itself to use of direct reading scales graduated in meaningful units for the process or test variable involved. It does, however, place certain operational requirements not found with the U-type. The higher pressure source being measured must always be connected to the well connection "P". A lower pressure source must always be connected to the top of the tube, and a differential pressure must always have the higher pressure source connected at the well connection "P". In any measurement the source of pressure must be connected in a manner that will cause the indicating fluid to rise in the indicating tube.

The true pressure still follows the principles previously outlined and is measured by the difference between the fluid surfaces. It is apparent that there must be some drop in the well level. This is readily compensated for by spacing the scale graduations in the exact amount required to reflect and correct for this "well drop". By carefully controlling tolerances of the well area and internal diameter of the indicating tube, Meriam well type manometers and scales are manufactured to a high degree of accuracy.

Inclined manometer: Many applications require accurate measurement of low pressure such as drafts and very low differentials, primarily in air and gas installations. In these applications the manometer is arranged with the indicating tube inclined, as in Fig.2.16, therefore providing an expanded scale. This arrangement can allow 12" of scale length to represent 1" of vertical liquid height. With scale subdivisions to 0.01 inches of liquid height, the equivalent pressure of 0.000360 PSI per division can be read using water as the indicating fluid.

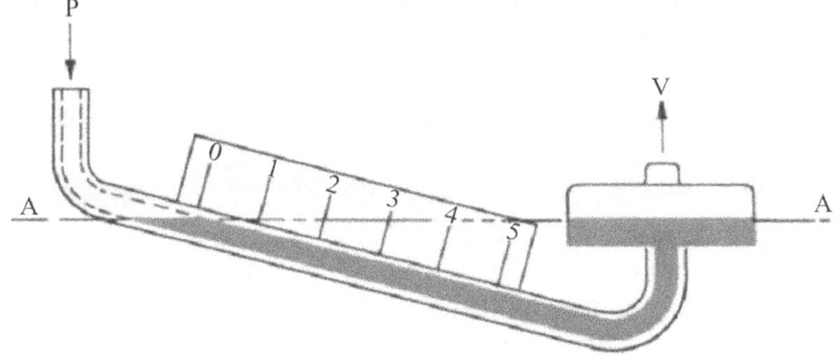

Fig. 2.16 Inclined manometer

Barometer: A barometer is an instrument used to measure atmospheric pressure. Schematic drawing of a simple mercury barometer with vertical mercury column and reservoir at base is shown in Fig. 2.17.

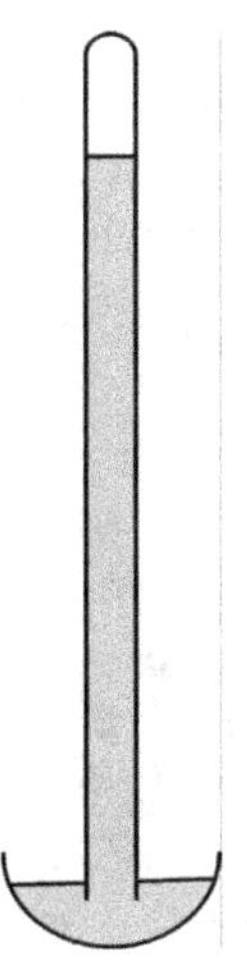

Fig. 2.17 Barometer

Water-based barometers: This concept of "decreasing pressure means bad weather" is the basis for a primitive weather prediction device called a weather glass or thunder glass. It can also be called a "storm glass" or a "Goethe thermometer".

It consists of a glass container with a spout. The container is filled with water up to about the middle of the spout; some air is left in the main body of the container. The design is such that when the air pressure decreases, the pressure of the air pocket inside the device will push some of the water up the spout. If the air pressure is low enough, some of the water may even drip out of the spout. These devices are essentially a water-based version of the mercury barometer. The "Thunder Glass" is extremely susceptible to the ambient temperature which will alter the height of the water column in the spout.

Mercury barometers: A standard mercury barometer has a glass column of about 30 inches (about 76 cm) in height, closed at one end, with an open mercury-filled reservoir at the base. Mercury in the tube adjusts until the weight of the mercury column balances the atmospheric force exerted on the reservoir. High atmospheric pressure places more downward force on the reservoir, forcing mercury higher in the column. Low pressure allows the mercury to drop to a lower level in the column by lowering the downward force placed on the reservoir.

The first barometer of this type was devised by Evangelista Torricelli, a student of Galileo Galilei, in 1643. Torricelli had set out to create a perfect vacuum, and an instrument to measure air pressure. He succeeded in creating a vacuum in the top of a tube of mercury. Torricelli also noticed that the level of the fluid in the tube changed slightly each day and concluded that this was due to the changing pressure in the atmosphere. He wrote: "We live submerged at the bottom of an ocean of elementary air, which is known by incontestable experiments to have weight".

The mercury barometer's design gives rise to the expression of atmospheric pressure in inches or millibars: the pressure is quoted as the level of the mercury's height in the vertical column. 1 atmosphere is equivalent to about 29.9 inches of mercury. The use of this unit is still popular in the United States, although it has been disused in favor of SI or metric units in other parts of the world. Barometers of this type can usually measure atmospheric pressures in the range between 28 and 31 inches of mercury.

2.4 Velocity Measurements

2.4.1 Tachometer

A Tachometer is used for measuring rotational speed. It can be used to measure speed of a rotating shaft. It can also be used to measure flow of liquid by attaching a wheel with inclined vanes. Tachometers can be classified on different basis. (i) Data acquisition – contact or non contact types. (ii) Measurement technique – time based or frequency based technique of measurement. (iii) Analog or digital type.

Comparison between analog and digital type techometer:

Analog Tachometer	Digital Tachometer
It has a needle and dial type of interface	It has a LCD or LED readout
There is no provision for storage of readings	There is Memory provided for storage
It cannot compute average, deviation, etc	It can perform statistical functions like averaging, etc

Classification Based on Data Acquisition Technique
- *Contact type*: The wheel of the tachometer needs to be brought into contact with the rotating object.
- *Non Contact type*: The measurement can be made without having to attach the tachometer to the rotating object.

Classification Based on Measurement Technique
- *Time Measurement*: The tachometer calculates speed by measuring the time interval between the incoming pulses.
- *Frequency Measurement*: The tachometer calculates speed by measuring the frequency of the incoming pulses.

Comparison between Contact and Non Contact Tachometers

Contact type	Non contact type
The tachometer has to be in physical contact with the rotating shaft	The tachometer does not need to be in physical contact with the rotating shaft
Preferred where the tachometer is generally fixed to the machine	Preferred where the tachometer needs to be mobile
Generally, optical encoder / magnetic sensor is attached to shaft of tachometer	Generally, laser is used or an optical disk is attached to rotating shaft and read by a IR beam or laser

2.4.2 Stroboscope

A stroboscope, also known as a strobe, is an instrument used to make a cyclically moving object appear to be slow-moving, or stationary. The principle is used for the study of rotating, reciprocating, oscillating or vibrating objects. Machine parts and vibrating strings are common examples. In its simplest form, a rotating disc with evenly-spaced holes is placed in the line of sight between the observer and the moving object. The rotational speed of the disc is adjusted so that it becomes synchronized with the movement of the observed system, which seems to slow and stop. The illusion is caused by temporal aliasing, commonly known as the stroboscopic effect. In electronic versions, the perforated disc is replaced by a lamp capable of emitting brief and rapid flashes of light. The frequency of the flash is adjusted so that it is an equal to, or a unit fraction below or above the object's cyclic speed, at which point the object is seen to be either stationary or moving backward or forward, depending on the flash frequency.

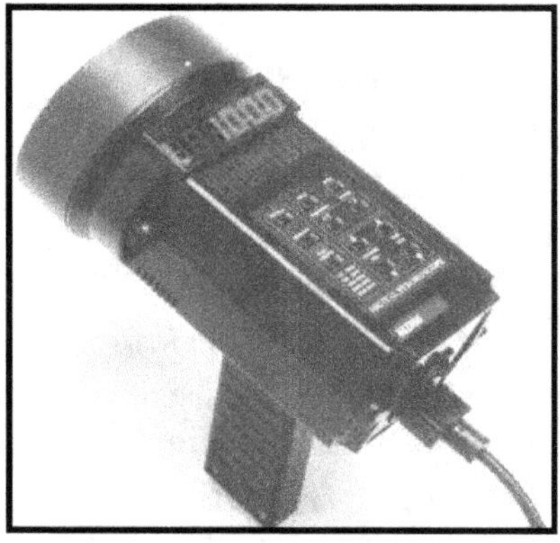

Fig. 2.18 Stroboscope

Other applications of stroboscope: Stroboscopes play an important role in the study of stresses on machinery in motion, and in many other forms of research. They are also used as measuring instruments for determining cyclic speed. As a timing light they are used to set the ignition timing of internal combustion engines. In medicine, stroboscopes are used to view the vocal cords for diagnosis of conditions that have produced (hoarseness). The patient hums or speaks into a microphone which in turn activates the stroboscope at either the same or a slightly different frequency. The light source and a camera are positioned by endoscopy. Another application of the stroboscope can be seen on many gramophone turntables. The edge of the platter has marks at specific intervals so that when viewed under fluorescent lighting powered at mains frequency, provided the platter is rotating at the correct speed, the marks appear to be stationary. This will not work under incandescent lighting, as incandescent bulbs don't strobe. For this reason, some turntables have a neon bulb next to the platter.

A strobe light is also used in some alarm systems to give a visual warning for people who may be hard of hearing and cannot hear the alarm bell.

2.4.3 Venturi Meter

The Venturi meter is a device for measuring flow rate in a pipe. It consists of a rapidly converging section which increases the velocity of flow and hence reduces the pressure. It then returns to the original dimensions of the pipe by a gently diverging 'diffuser' section. The angle of the diffuser is usually between 6 and 8 degrees. The efficiency of the diffuser of increasing pressure back to the original is rarely greater than 80%. By measuring the pressure differences the flow rate can be calculated. This is a particularly accurate method of flow measurement as energy losses are very small.

The venturi meter consists of two conical pipes connected as shown in the Fig. 2.19. The minimum cross section diameter is called throat. The angles of the conical pipes are established to limit the energy losses due to flow separation.

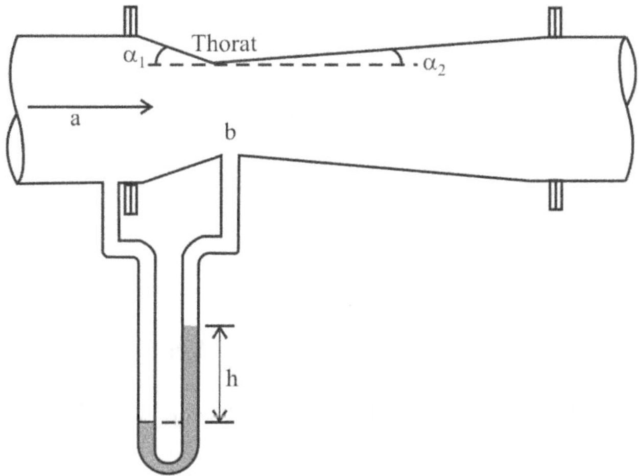

Fig. 2.19 Venturi Meter

The flow obstruction produced by the venturi meter produces a local loss that is proportional to the flow discharge.

Pressure taps are located upstream and downstream of the venturi meter, immediately outside the variable diameter areas, to measure the losses produced through the meter.

Flow rate measurements are obtained using **Bernoulli equation** and the **continuity equation** (see below the derivation). An experimental coefficient is used to account for the losses occurring in the meter (V_a and V_b are the upstream and downstream velocities and ρ is the density. (A_a and A_b are the cross sectional areas).

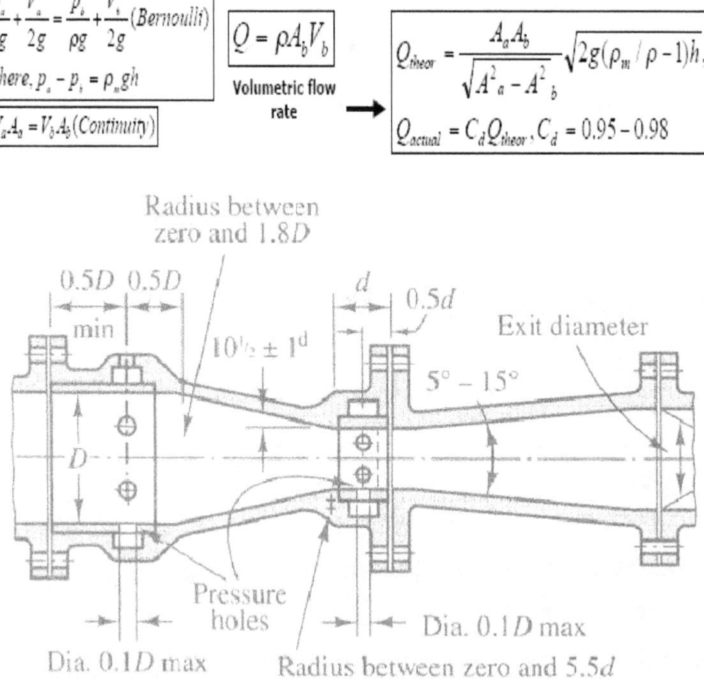

Fig. 2.20 Constructional detail of Venturi Meter

The construction of a venturimeter is shown in Fig. 2.20. Here it is so designed that the change in the flow path is gradual. As a result, there is no permanent pressure drop in the flow path. The discharge coefficient C_d varies between 0.95 and 0.98. The construction also provides high mechanical strength for the meter. However, the major disadvantage and disadvantages compare to orifice are:

Main advantages are:

- Low head loss. Around 90% of the pressure is recovered.
- Less affected by upstream flow disturbance
- Even more robust
- The venturi tube is suitable for clean, dirty and viscous liquid and some slurry services self-cleaning
- Less affected by erosion

The disadvantages compared to the orifice are:

- Occupies longer length of pipe
- More expensive (manufacture and installation)

2.4.4 Orifice Meter

An orifice is a pipeline with a manometer for measuring the drop in pressure (differential) as the fluid passes through the orifice. The minimum cross sectional area of the jet is known as the "vena contracta". As the fluid approaches the orifice the pressure increases slightly and then drops suddenly as the orifice is passed. It continues to drop until the "vena contracta" is reached and then gradually increases until at approximately 5 to 8 diameters downstream a maximum pressure point is reached that will be lower than the pressure upstream of the orifice. The decrease in pressure as the fluid passes through the orifice is a result of the increased velocity of the gas passing through the reduced area of the orifice. When the velocity decreases as the fluid leaves the orifice the pressure increases and tends to return to its original level. All of the pressure loss is not recovered because of friction and turbulence losses in the stream. The pressure drop across the orifice increases when the rate of flow increases. When there is no flow there is no differential. The differential pressure is proportional to the square of the velocity, it therefore follows that if all other factors remain constant, then the differential is proportional to the square of the rate of flow.

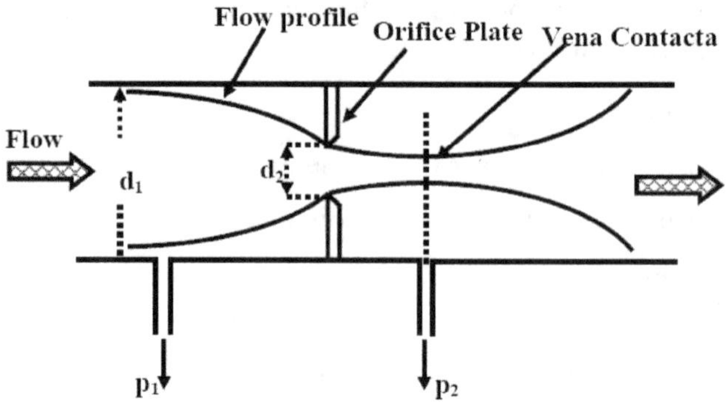

Fig. 2.21 Orifice type Flowmeter

Depending on the type of obstruction, we can have different types of flow meters. Most common among them is the orifice type flowmeter, where an orifice plate is placed in the pipe line, as shown in Fig. 2.21. If d_1 and d_2 are the diameters of the pipe line and the orifice opening, then the flow rate can be obtained using expression given below by measuring the pressure difference $(p_1 - p_2)$.

Orifice meter is similar to venturi meter. Hence, the equation for the orifice meter is similar to that of venturi.

C_d is defined as the ratio of the actual flow and the ideal flow and is always less than one. There are in fact two main reasons due to which the actual flow rate is less than the ideal one. The first is that the assumption of frictionless flow is not always valid. The amount of friction depends on the Reynold's number (Re). The more important point is that, the minimum flow area is not the orifice area A_2, but is somewhat less and it occurs at a distance from the orifice plate, known as the Vena Contracta, and we are taking a pressure tapping around that point in order to obtain the maximum pressure drop. As a result, the correction factor $C_d < 1$. The typical value of C_d for orifice plate varies between 0.6 and 0.7.

The major advantages of orifice plate are that it is low cost device, simple in construction and easy to install in the pipeline. The orifice plate is a circular plate with a hole in the center. Pressure tappings are normally taken distances D and 0.5D upstream and downstream the orifice respectively (D is the internal diameter of the pipe). But there are many more types of pressure tappings which are in use.

The major disadvantage of using orifice plate is the permanent pressure drop that is normally experienced in the orifice plate. The pressure drops significantly after the orifice and can be recovered only partially. The magnitude of the permanent pressure drop is around 40%, which is sometimes objectionable. It requires more pressure to pump the liquid. This problem can be overcome by improving the design of the restrictions.

2.4.5 Pitot Tube

The Pitot tube is a simple and inexpensive instrument for the measurement of fluid velocity often taken for granted today. Pitot tube is widely used for velocity measurement in aircraft. Its basic principle can be understood from Fig. 2.22(a). If a blunt object is placed in the flow channel, the velocity of fluid at the point just before it, will be zero. Then considering the fluid to be incompressible, from Bernloulli's equation, we have,

$$\frac{p_1}{\gamma} + \frac{v_1^2}{2g} = \frac{p_2}{\gamma} + \frac{v_2^2}{2g}$$

where γ is the specific weight of the fluid.

Now $v_2 = 0$.

Therefore,

$$\frac{v_1^2}{2g} = \frac{p_2 - p_1}{\gamma}$$

or
$$v_1 = \sqrt{\frac{2g}{\gamma}(p_2 - p_1)}$$

after correction,

$$v_1 = C_p \sqrt{\frac{2g}{\gamma}(p_2 - p_1)}$$

C_p is a dimensionless coefficient to correct deviation from the Bernoulli eqn. it is usually between 1.0 and 0.98. The typical construction of a Pitot tube is shown in Fig. 2.22(b).

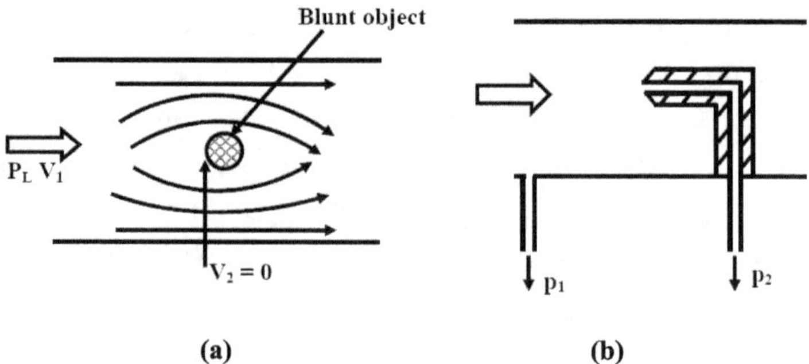

Fig. 2.22 Basic Principle of Pitot Tube (a) Construction Pitot Tube (b)

A pitot tube should have a number of characteristics:

(i)   It should draw no more current than necessary to do the job (It should be thermostatic.).

(ii) It should pretty much take care of itself, that is, it should remember to turn itself on and off (It should be automatic.).

(iii) It should be built of a low thermal-conductivity strong stuff (It should be of fiber glass.).

(iv) It should be low drag. On a regular pitot tube, the perpendicular part of the L-shaped tube exerts a drag about equal to half-a-foot of wing strut. It all adds up. Likewise the junction with the wing must be faired too (It should be aerodynamic.).

(v)   It should protect itself against mud wasps and various bugs attracted to its hole.

(vi) It should be rugged.

(vii) It should work well enough. (Perfect airspeed indicator shouldn't be done with a pitot tube at all.)

2.4.6 Rotameter

The orificemeter, Venturimeter and flow nozzle work on the principle of constant area variable pressure drop. Here the area of obstruction is constant, and the pressure drop changes with flow rate. On the other hand Rotameter works as a constant pressure drop variable area meter. It can only be used in a vertical pipeline. Its accuracy is also less (2%) compared to other types of flow meters. But the major advantages of rotameter are, it is simple in construction, ready to install and the flow rate can be directly seen on a calibrated scale, without the help of any other device, e.g. differential pressure sensor etc. Moreover, it is useful for a wide range of variation of flow rates (10:1).

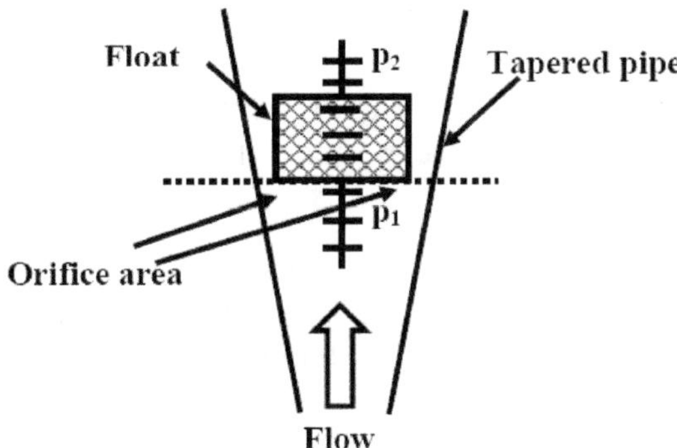

Fig. 2.23 Basic Construction of a Rotameter.

The basic construction of a rotameter is shown in Fig. 2.23. It consists of a vertical pipe, tapered downward. The flow passes from the bottom to the top. There is cylindrical type metallic float inside the tube. The fluid flows upward through the gap between the tube and the float. As the float moves up or down there is a change in the gap, as a result changing the area of the orifice. In fact, the float settles down at a position, where the pressure drop across the orifice will create an upward thrust that will balance the downward force due to the gravity. The position of the float is calibrated with the flow rate.

Let us consider,

γ_1 = Specific weight of the float

γ_2 = Specific weight of the fluid

v_f = Volume of the float

A_f = Area of the float.

A_t = Area of the tube at equilibrium (corresponding to the dotted line)

For incompressible fluid, we have, for the orifice,

$$Q = \frac{C_d A_2}{\sqrt{1 - \left(\frac{A_2}{A_1}\right)^2}} \sqrt{\frac{2g}{\gamma_2}(p_1 - p_2)}$$

Advantages

(i)   A rotameter requires no external power or fuel, it uses only the inherent properties of the fluid, along with gravity, to measure flow rate.

(ii) A rotameter is also a relatively simple device that can be mass manufactured out of cheap materials, allowing for its widespread use.

Disadvantages

(i)   Due to its use of gravity, a rotameter must always be vertically oriented and right way up, with the fluid flowing upward.

(ii) Due to its reliance on the ability of the fluid or gas to displace the float, graduations on a given rotameter will only be accurate for a given substance at a given temperature. The main property of importance is the density of the fluid; however, viscosity may also be significant. Floats are ideally designed to be insensitive to viscosity; however, this is seldom verifiable from manufacturers specifications. Either separate rotameters for different densities and viscosities may be used, or multiple scales on the same rotameter can be used.

(iii) Rotameters normally require the use of glass (or other transparent material), otherwise the user cannot see the float. This limits their use in many industries to benign fluids, such as water.

(iv) Rotameters are not easily adapted for reading by machine; although magnetic floats that drive a follower outside the tube are available.

2.5   Strain Measurement

2.5.1 Strain gauge

A strain gauge is a device used to measure deformation (strain) of an object (Fig. 2.24). The most common type of strain gauge consists of a flexible backing which supports a metallic foil pattern etched onto the backing. As the object is deformed, the foil pattern is deformed, causing its electrical resistance to change. This resistance change, usually measured using a Wheatstone bridge circuit, can be used to calculate the exact amount of deformation by means of the quantity known as the gauge factor. The gauge factor of a strain gauge relates strain to change in electrical resistance. The gauge factor GF is defined by the formula

$$GF = \frac{\Delta R / R_G}{\epsilon}$$

where R_G is the resistance of the undeformed gauge, ΔR is the change in resistance caused by strain, and ϵ is strain.

For measurements of small strain, semiconductor strain gauges are often preferred over foil gauges. A semiconductor gauge usually has a larger gauge factor than a foil gauge.

Semiconductor gauges tend to be more expensive, more sensitive to temperature changes, and are more fragile than foil gauges. The strain gauge has been in use for many years and is the fundamental sensing element for many types of sensors, including pressure sensors, load cells, torque sensors, position sensors, etc.

The majority of strain gauges are foil types, available in a wide choice of shapes and sizes to suit a variety of applications. They consist of a pattern of resistive foil, which is mounted on a backing material. They operate on the principle that as the foil is subjected to stress, the resistance of the foil changes in a defined way.

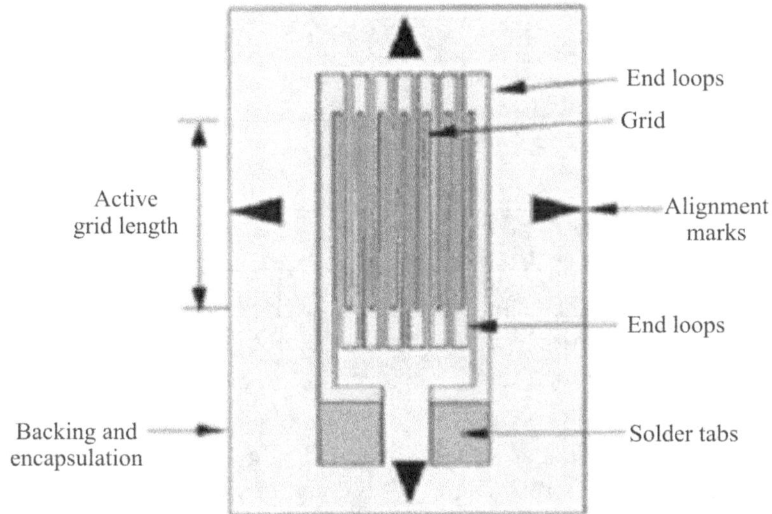

Fig. 2.24 Strain Gauge

Extensive ranges of differing strain gauges pattern are available for wide variety of application as shown in Fig. 2.25.

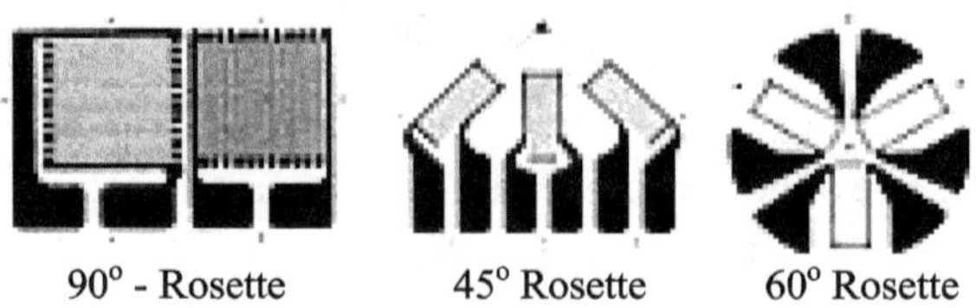

Fig. 2.25 Some of the available Gauge Patterns

The strain gauge is connected into a Wheatstone bridge circuit with a combination of four active gauges (full bridge), two gauges (half bridge), or, less commonly, a single gauge (quarter bridge) (Fig. 2.26). In the half and quarter circuits, the bridge is completed with precision resistors.

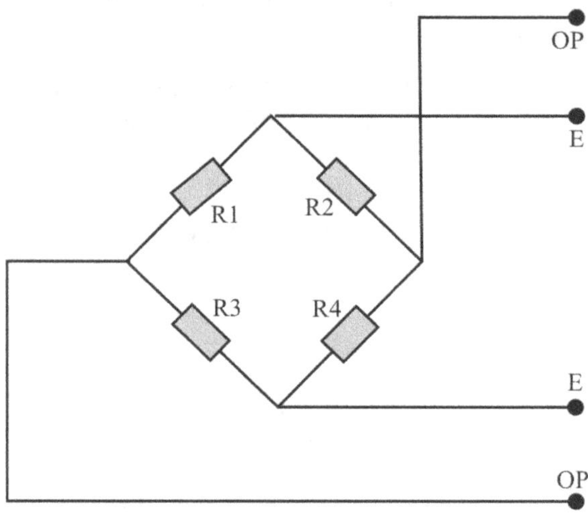

Fig. 2.26 Wheatstone Bridge Circuit

The complete Wheatstone bridge is excited with a stabilized DC supply and with additional conditioning electronics, can be zeroed at the null point of measurement.

As stress is applied to the bonded strain gauge, a resistive change takes place and unbalances the Wheatstone bridge. This results in a signal output, related to the stress value. As the signal value is small, (typically a few millivolts) the signal conditioning electronics provides amplification to increase the signal level to 5 to 10 volts, a suitable level for application to external data collection systems such as recorders or PC Data Acquisition and Analysis Systems.

Load cell: A load cell is a transducer that converts force into a measurable electrical output. Although there are many varieties of load cells, strain gage based load cells are the most commonly used type.

2.5.2 Working principles

Load cell designs can be distinguished according to the type of output signal generated (pneumatic, hydraulic, electric) or according to the way they detect weight (bending, shear, compression, tension, etc.) Hydraulic load cells are force-balance devices, measuring weight as a change in pressure of the internal filling fluid. In a rolling diaphragm type hydraulic load cell, a load or force acting on a loading head is transferred to a piston that in turn compresses a filling fluid confined within an elastomeric diaphragm chamber. As force increases, the pressure of the hydraulic fluid rises. This pressure can be locally indicated or transmitted for remote indication or control. Output is linear and relatively unaffected by the amount of the filling fluid or by its temperature. If the load cells have been properly installed and calibrated, accuracy can be within 0.25% full scale or better, acceptable for most process weighing applications. Because this sensor has no electric components, it is ideal for use in hazardous areas. Typical hydraulic load cell applications include tank, bin, and hopper weighing. For maximum accuracy, the weight of the tank should be obtained by locating one load cell at each point of support and summing their outputs. Pneumatic load cells also operate on the force-balance principle. These devices use multiple dampener chambers to provide higher accuracy than can a hydraulic device. In some designs, the first dampener chamber is used as a tare weight chamber. Pneumatic load cells are often used to measure relatively small weights in industries where cleanliness and safety are of prime concern. The advantages of this type of load cell include their being inherently explosion proof and insensitive to temperature variations. Additionally, they contain no fluids that might contaminate the process if the diaphragm ruptures.

Disadvantages include relatively slow speed of response and the need for clean, dry, regulated air or nitrogen. Strain-gage load cells convert the load acting on them into electrical signals. The gauges themselves are bonded onto a beam or structural member that deforms when weight is applied. In most cases, four strain guages are used to obtain maximum sensitivity and temperature compensation. Two of the gauges are usually in tension, and two in compression, and are wired with compensation. When weight is applied, the strain changes the electrical resistance of the gauges in proportion to the load. Other load cells are fading into obscurity, as strain gauge load cells continue to increase their accuracy and lower their unit costs.

2.6   Force Measurement

2.6.1 Analytical balance

A balance (also beam balance or laboratory balance) is used to accurately measure the mass of an object. This class of measuring instrument uses a comparison technique in its conventional form of a beam from which a weighing pan and scale pan are suspended. To weigh an object, it is placed on the measuring pan, and standard weights are added to the scale pan until the beam is in equilibrium (Fig.2.27).

Fig. 2.27 Weighing Balance

When the weights on the plates of this balance are equal, the needle mid-rod points straight up. While the word "weigh" or "weight" is often used, any balance scale actually measures mass, which is not dependent upon the force of gravity, as opposed to a scale with a spring, which measures weight. Mass is properly measured in grams, kilograms, pounds, ounces, or slugs; while weight is in newtons or pound force. An analytical balance is an instrument used to measure mass to a very high degree of precision. The weighing pan(s) of a high accuracy (0.1 mg or better) analytical balance are inside a see-through enclosure with doors so dust does not collect and so any air currents in the room do not affect the delicate balance. Also, the sample must be at room temperature to prevent natural convection from forming air currents inside the enclosure, affecting the weighing. Very precise measurements are achieved by ensuring that the fulcrum of the beam is friction-free (a knife edge is the traditional solution), by attaching a pointer to the beam which amplifies any deviation from a balance position; and finally by using the lever principle, which allows fractional weights to be applied by movement of a small weight along the measuring arm of the beam.

2.6.2 Platform scale

Fig. 2.28 shows the internal arrangement of platform scales. For platform scales, major scale adjustment can be achieved through adding one from several standard masses that available; minor adjustment can be achieved with adjusting the sliding mass along the measurement scale until the balance is achieved. The purpose to use this device is to do measurement on large force with small standard masses. Sliding mass is used to do comparison procedure as fast as possible, and in the same time, to avoid from using too much standard masses. In terms of application, it can be extended in measuring large force with doing a modification on the device. One of the devices that apply the concept used in platform scale is Material Testing Machine (or Universal Testing Machine). Usually, this machine is used to find the strength of material and its mechanical characteristics through the test on the materials.

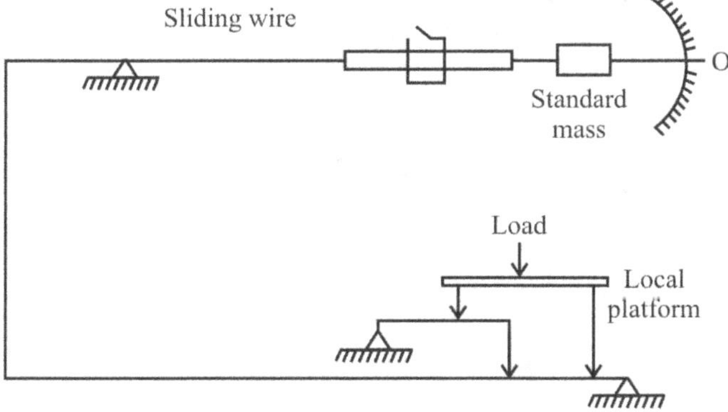

Fig. 2.28 Platform Scales

2.6.3 Mechanical balance

Fig. 2.29 shows the schematic view of mechanical balance. The concept of mechanical balance is as follows:

- Consider unknown force (in this case, W for mass A), where this force can be measured by moving the mass B until equilibrium is achieved.

- At this moment, where B in location Bo, using moment principle,

$$Wa = wx,$$

where w is weight of B.

- Because weight is multiplication between mass and gravitational acceleration, the above relation can be written as

$$Ma = mx \qquad \text{or}$$

$$M = (m/a)x$$

where M: mass of A; m: mass of B

- From this, it can be understood that M is proportional with x.

- The scale, S can be calibrated directily in M and W, if the value of gravity is assumed to remain unchanged.

- The range in force measurement can be increased through adding the mass C at the end of the beam of that balance. If m' is a mass for C, an equilibrium can be achieved when

$$Ma = mx + m'c$$

$$M = (mx)/a + (m'c)/a$$

- Because m' is constant, calibration mark at scale can be used for range which is higher than the previous one, in condition that fixed quantity m' is added to each reading.

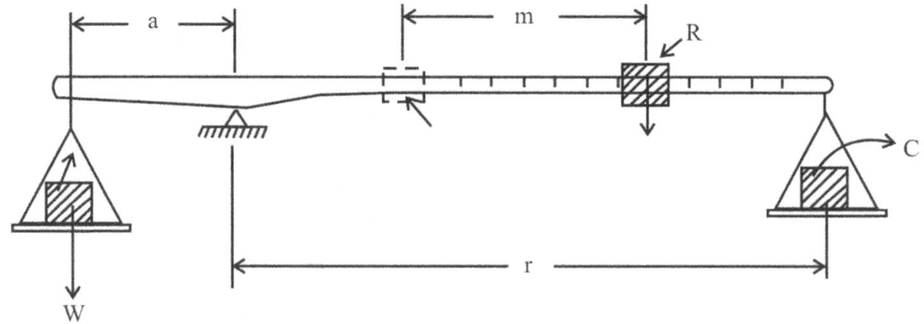

Fig. 2.29 Mechanical Balance

2.6.4 Elastic displacement method

(i) **Spring balance:** Some weighing scales such as a Jolly balance (named after Phillipp Gustav von Jolly who invented the balance about 1874) use a spring with a known spring constant (see Hooke's law) and measure the displacement of the spring by any variety of mechanisms to produce an estimate of the gravitational force applied by the object, which can be simply hung from the spring or set on a pivot and bearing platform. Rack and pinion mechanisms are used to convert the linear spring motion to a dial reading (Fig.2.30). A spring weighing scale can measure forces transmitted through the scale in any direction.

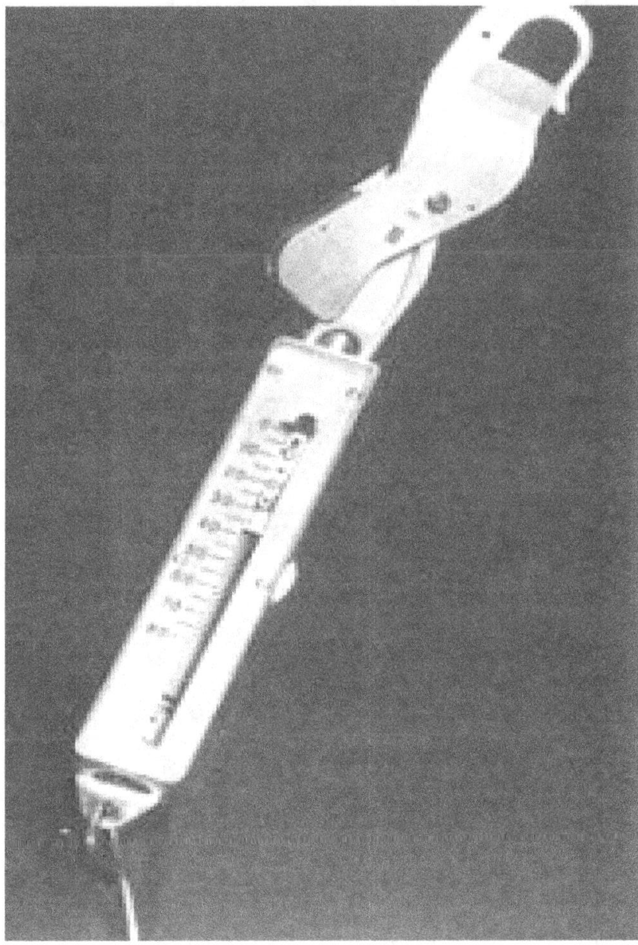

Fig. 2.30 Spring Weighing Scale

Spring scales typically measure force, which can be measured in units of force such as newtons or pounds-force. Spring scales typically cannot be used for commercial applications unless their springs are temperature compensated or used at a fairly constant temperature. The spring scales which are legal for commerce can be calibrated for the accurate measurement of mass (the quantity measured for weight in commerce) in the location in which they are used.

Fig. 2.31 Proving Ring

They can give an accurate measurement in kilograms or pounds for this purpose.

(ii) **Steel/proving ring:** The proving ring is a device used to measure force. As seen in Fig. 2.31, it consists of an elastic ring of known diameter with a measuring device located in the center of the ring. Proving rings come in a variety of sizes. They are made of a steel alloy. Manufacturing consists of rough machining from annealed forgings, heat treatment, and precision grinding to final size and finish.

The concept behind the proving ring is illustrated in the Fig. 2.32.

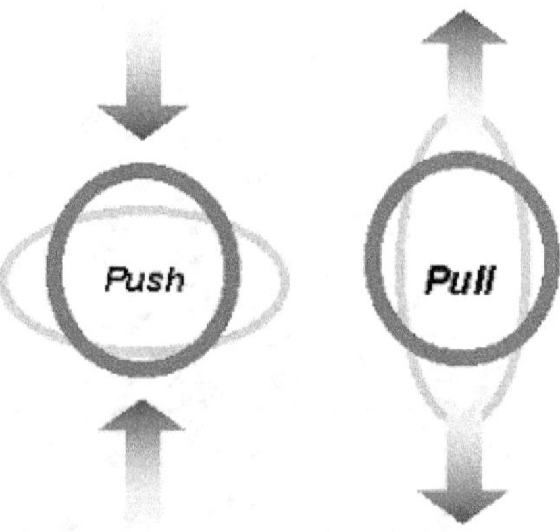

Fig. 2.32 Schematic diagram of changes in the ring diameter as compression (push) and tension (pull) forces are applied.

Proving rings can be designed to measure either compression or tension forces. Some are designed to measure both. The basic operation of the proving ring in tension is the same as in compression. However, tension rings are provided with threaded bosses and supplied with pulling rods that are screwed onto the bosses. The proving ring consists of two main elements, the ring itself and the diameter-measuring system, shown on the right in the exploded view of a proving ring (Fig. 2.33). Forces are applied to the ring through the external bosses. The resulting change in diameter, referred to as the deflection of the ring, is measured with a micrometer screw and the vibrating reed mounted diametrically within the ring. The micrometer screw and the vibrating reed are attached to the internal bosses of the ring. In modern rings, the upper and lower internal and external bosses are machined as an integral part of the ring to avoid mechanical interferences during the application of force.

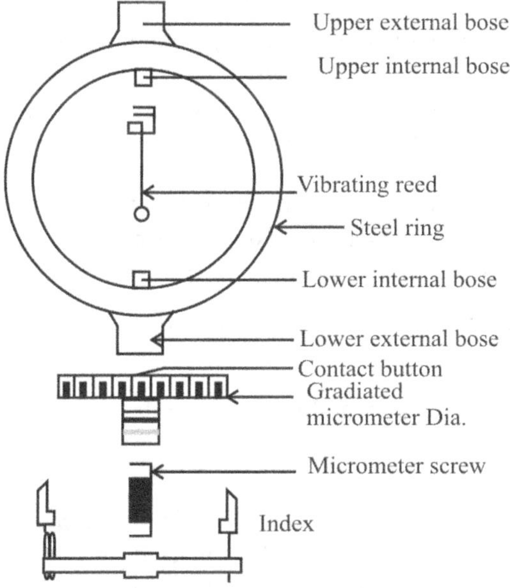

Fig. 2.33 Design of Proving Rings

To read the diameter of ring, the vibrating reed is set in motion by gently tapping it with a pencil (Fig. 2.34). As the reed is vibrating, the micrometer screw on the spindle is adjusted until the button on the spindle just contacts the vibrating reed, dampening out its vibrations. When this occurs a characteristic buzzing sound is produced. At this point a reading of the micrometer dial indicates the diameter of the ring. The number of divisions on the micrometer dial and the graduation of the vernier index vary by type of proving ring. Typically, proving rings are designed to have a deflection of about 0.84 mm (0.033 in) to 4.24 mm (0.167 in). The relative measurement uncertainty can vary from 0.075 % to 0.0125 %.

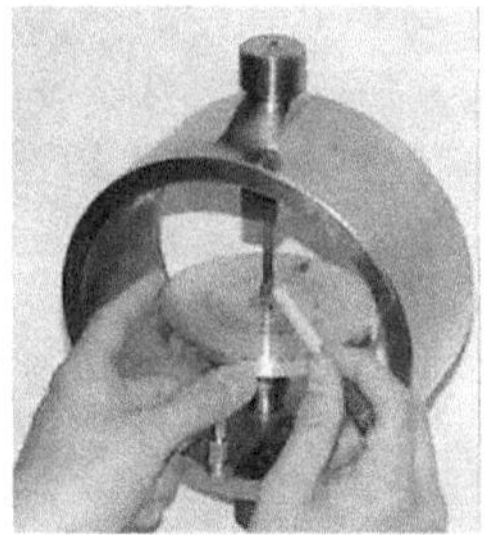

Fig. 2.34 Read the Diameter of the Ring

2.7   Torque Measurement

Torque is one of the mechanical quantity, which is also called "moment". It is measured by unit Nm. Generally, Nm is same with Joule but Joule usually used for describing a unit for energy. Torque measurement is related to find generated or absorbed power at the rotating machine.

There are several machines that apply and measure a torque. Usually these types of machine is used exclusively for power measurement, which is known as brakes or dynamometer.

Other method for measuring torque includes:

- direct strain method
- angular displacement method

2.7.1 Power measurement

A dynamometer or "dyno" for short is a device used to measure power and torque produced by an engine. It is said to be created by the "Father of Computing" Charles Babbage. There are two types of dynos; One that gets bolted directly to an engine, known as an engine dyno, or a dyno that can measure power and torque without removing the engine from the frame of the vehicle, this is known as a chassis dyno. In general, Dynamometers are useful in the development and refinement of modern day engine technology. The concept is to use a dyno to measure and compare power transfer at different points on a vehicle, thus allowing the engine or drive train to be modified to get more efficient power transfer.

For example, if an engine dynamo shows that a particular engine achieves 400 N·m (300 lbf·ft) of torque, and a chassis dynamo shows only 350 N·m (260 lbf·ft), one would know to look to the drive train for the major improvements. Dynamometers are typically very expensive pieces of equipment, reserved for certain fields that rely on them for a particular purpose.

2.8 Uncertainty Analysis

Uncertainty can be described as that portion of the measurement beyond which we are not sure of its true value. Each time a measurement is taken (mass, volume, length etc.) we rely upon a mechanical or visual point of reference in order to assign the appropriate value. These values, no matter how carefully they are obtained contain some degree of what is referred to as **uncertainty**. The uncertainty of a measurement tells us something about its quality. Uncertainty of measurement is the doubt that exists about the result of any measurement. You might think that well-made rulers, clocks and thermometers should be trustworthy, and give the right answers. But for every measurement – even the most careful – there is always a margin of doubt. In everyday speech, this might be expressed as 'give or take' ... e.g. a stick might be two metres long 'give or take a centimetre'.

A weighting balance used to obtain data for experiment reads to 0.001 g (1 mg). The true value could be as low as 0.0005 g or as high as 0.0014 g. The fourth decimal place (which cannot be read on this balance) causes us to be uncertain about the true value of the third decimal place. To account for this we assign an uncertainty to the balance reading obtained as ± 0.001g (e.g. 158.709 ± 0.001g). Similarly a balance which reads 0.01g would have an assigned uncertainty of ± 0.01g.

For volumetric (graduated cylinders, Mohr pipettes and burettes) and length measurements a general rule is to assign an uncertainty of one half of the smallest scale division. For example, the smallest division on a 100 mL graduated cylinder is 1 mL; a prudent uncertainty would be ± 0.5 mL. This may vary for each person and some what depends on the person's eye sight, some will be able to estimate 0.2 mL others will do well to estimate 1.0 ml. Be consistent, use values you believe are appropriate for your ability.

A measurement is a process whereby the value of a quantity is estimated. All measurements are accompanied by error. Our lack of knowledge about the sign and magnitude of measurement error is called measurement uncertainty. A measurement uncertainty estimate is the characterization of what we know statistically about the measurement error. Therefore, a measurement result is only complete when accompanied by a statement of the uncertainty in that result.

The general uncertainty analysis procedure consists of the following steps:

(i)   Define the Measurement Process

(ii)   Develop the Error Model

(iii) Identify the Error Sources and Distributions

(iv)   Estimate Uncertainties

(v)   Combine Uncertainties

(vi)   Report the Analysis Results

2.8.1 Uncertainty Calculations

Examples are given to illustrate how the High-Low method is used to estimate errors. The following is example applied to the Table 2.1.

Note: While the calculations shown here are dealt with as a separate topic, they may be included in the **Calculations** section of the lab report. Only one example of each calculation is illustrated.

Table 2.1 Data for Unknown

Volume		Mass	
(mL $\pm$ 0.2 mL)		(g $\pm$ 0.001 g)	
V_0	49.8	M_0	158.709
V_1	56.8	M_1	179.089
V_2	63.2	M_2	199.679
V_3	68.5	M_3	222.135

Using the High-Low Method: Look at the data for V_0. V_0 can be as high as 50.0 mL or as low as 48.6 mL, similarly the masses M_0 can also be as high as 158.710 g or as low as 159.708 g. Using the high-low method we obtain the results in Table 2.2

Table 2.2 Mass and Volume Uncertainty

ΔM_1		ΔM_2
High 179.090 – 158.708 = 20.382 g		High 57.0 mL – 49.6 mL = 7.4 mL
Mid 179.089 – 158.709 = 20.380 g		Mid 56.8 mL – 49.8 mL = 7.0 mL
Low 179.088 158.0710 = 20378 g		Low 56.6 mL – 50.0 mL = 6.6 mL
ΔM_1 = 20.380 g + 0.002 g		ΔV_1 = 7.0 mL $\pm$ 0.4 mL

Once the uncertainity is calculated for each mass and volume change, a similar set of calculations may be performed when determining the density, shown in Table 2.3.

Table 2.3: Density Uncertainty

d_1
High 20.382 g / 6.6 mL = 3.088 g/mL
Mid 20.380 g / 7.0 mL = 2.911 g/mL
Low 20.378 g / 7.4 mL = 2.754 g/mL
d_1 = 2.9 g/mL $\pm$ 0.17 g/mL

I have used separate tables to illustrate the calculations and results. You should put these all together in just one (1) table which will allow all the data to be viewed. Look at Table 2.4.

Table 2.4 Uncertainty Summery for all the Experimental Data

	ΔM_X (g)		ΔM_X (mL)		d_X (g/mL)
ΔM_1 (g)	20.380 $\pm$ 0.002	ΔV_1	7.0 $\pm$ 0.4	d_1	2.9 $\pm$ 0.2
ΔM_2 (g)	40.97 $\pm$ 0.002	ΔV_2	13.4 $\pm$ 0.4	d_2	3.1 $\pm$ 0.1
ΔM_3 (g)	63.326 $\pm$ 0.002	ΔV_3	18.7 $\pm$ 0.4	d_3	3.4 $\pm$ 0.1

2.8.2 Error Calculations using Standard Deviation

A better method used to estimate the spread of the data is called standard deviation and denoted by the Greek Letter σ (lower case, sigma). It is calculated using the formula shown to the right. Fortunately many scientific calculators have this function built in.

$$\sigma = \frac{\sqrt{\left(x_1 - \bar{x}\right) + \left(x_2 - \bar{x}\right)^2 + \left(x_1 - \bar{x}\right)......\left(x_n - \bar{x}\right)}}{N-1}$$

$$\sigma = \frac{\sqrt{\left(X_1 - \bar{X}\right)^2 + \left(X_2 - \bar{X}\right)^2 + \left(X_3 - \bar{X}\right)^2 \ldots \ldots \left(X_n - \bar{X}\right)^2}}{(N-1)}$$

where

$X_1, X_2, X_3, ..X_n$ are the individual values

$$\bar{X} \ (\text{average value of } X) = \frac{X_1 + X_2 + X_3 + ..X_n}{N}$$

N = the number of experimental trials

The standard deviation is calculated using the density values 2.9 g/mL, 3.1 g/mL and 3.4 g/mL and found to be 0.3 g/mL. The average density is then expressed as 3.1 g/mL ± 0.3g/mL. (Although I have used values rounded off to 2 significant figures, care must be taken, round off final answers, only after all chain calculations are done.)

2.8.3 Calculation of Confidence Limits

A more advanced approach is to calculate the data spread based on a confidence limit. The confidence limit is a statistical evaluation of the data spread and is calculated using the following formula:

$$\textit{\textbf{Confidence Limit}} = \pm \frac{t\sigma}{\sqrt{N}}$$

The std. dev. (σ) described in the previous is the same value used to determine the confidence limit for the specific data set. The Factor t values are obtained from a statistical table, note that the values of t vary for not only the number of observations but also to the level we wish to express our confidence. N is the number of observations made. The normally accepted confidence level is 95%.

Thus for the density data std. dev. = 0.25, for the 95% confidence level and 3 observations (the number of times the density was determined) the Factor t = 4.30. This leads to a confidence limit = 0.62. The result is then expressed as 3.1 ± 0.6 which means we are 95% confident the true value of the density lies some where between 3.7 and 2.5.

For example: We might say that the length of a certain stick measures 20 centimetres plus or minus 1 centimetre, at the 95 percent confidence level. This result could be written:

20 cm ±1 cm, at a level of confidence of 95%.

The statement says that we are 95 percent sure that the stick is between 19 centimetres and 21 centimetres long.

2.9   Concept of Error of a Measurement

The error of a measurement is defined as the difference between the measured value and the true physical value of the quantity. The error can not be calculated exactly unless the true value of the quantity is known. As there is some error is always present in every measurement so it is very difficult to find out true value of quantity. Hence the exact error can not be found out.

2.9.1   Types of error

Error can be classified into two general classes as:

(i)   Systematic / Bias Error

(ii)   Random / Precision Error

(i)   ***Systematic / Bias Error:*** An error due to some known physical law by which it might be predicted, these errors produced by the same cause affect the mean in the same sense, and do not tend to balance each other but rather give a definite bias to the mean. An error which results from some bias in the measurement process and is not due to chance, in contrast to random error. Since such errors alter the instrument reading with a fixed magnitude and with same sign from one reading to another, therefore, the error is commonly termed as *instrumental bias* (Fig.2.35). These types of error are due to following reasons:

Instrument error: Certain errors are inherent in the instrument system. These may be caused due to poor design/ construction of the instruments. Error in the divisions of graduated scales, inequality of the balance arms, irregular springs tension, etc, cause such error. Instrument errors can be avoided by

(i)   Selecting a suitable instrument for a given application.

(ii)   Applying suitable correction after determining the amount of instrument error.

(iii)   Calibrating the instrument against a suitable standard.

Environmental errors: These types of errors are caused due to variation of conditions external to the measuring device, including the conditions in the area surrounding the instrument. Commonly occurring changes in the environmental conditions that may affect the instrument characteristics are the effects of changes in temperature, barometric pressure, humidity, wind forces, magnetic or electrostatic fields etc. e.g. change in ambient temperature causes errors due to expansion of the measuring tape. Similarly, buoyant effect of the wind causes errors on weights of a chemical balance.

Loading errors: Such errors are caused by the act of measurement on the physical system being tested. Common examples of this type are:

(i)   Introduction of additional resistance in the circuit by the measuring milliammeter which may alter the circuit current by significant amount.

(ii)   An obstruction type flow meter may partially block or disturb the flow conditions and consequently the flow rate shown by the meter may not be same as before the meter installation.

It may be noted that systematic errors can be eliminated or alternatively, a suitable correction can be applied by properly calibrating the instruments. Therefore, instrument calibrations become very necessary as the measurements play a dominant role in any experiment. Further, systematic errors can also be sub divided into static and dynamic error. Static errors are caused due to limitations of the instruments as well as due to certain shortcomings in the measurement process. e.g., a static error is introduced in the micrometer reading when excessive torque is applied on the shaft of the micrometer screw. Dynamic errors are caused in the measurement if the instrument is not responding fast enough to follow the changes in the measured variable.

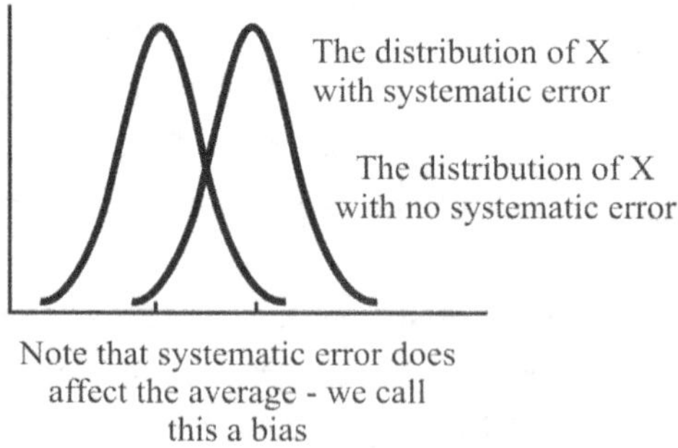

Fig. 2.35 Systematic Errors

(ii) Random errors: Random error is caused by any factors that randomly affect measurement of the variable across the sample. For instance, each person's mood can inflate or deflate their performance on any occasion. In a particular testing, some children may be feeling in a good mood and others may be depressed. If mood affects their performance on the measure, it may artificially inflate the observed scores for some children and artificially deflate them for others. The important thing about random error is that it does not have any consistent effects across the entire sample. Instead, it pushes observed scores up or down randomly. This means that if we could see all of the random errors in a distribution they would have to sum to 0 – there would be as many negative errors as positive ones. The important property of random error is that it adds variability to the data but does not affect average performance for the group (Fig. 2.36). Because of this, random error is sometimes considered *noise*.

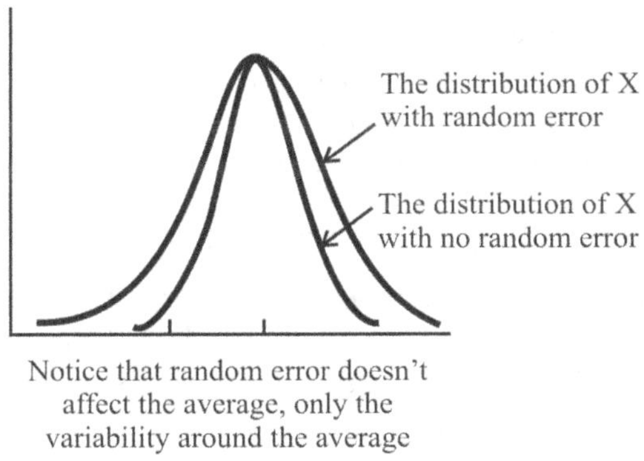

Fig. 2.36 Random Errors

Following are some of the main contributing factor to random error.

Inconsistencies associated with accurate measurement of small quantities: The outputs of the instruments become inconsistent when very accurate measurements are being made. This is because when the instruments are built or adjusted to measure small quantities, the random errors (which are of the order of the measured quantities) become noticeable. For example, if one wishes to measure the weight of a bucket of water to the nearest kilogram, there would be no excuse of one observation differing from another, no matter how many times the measurements are made. However, if one attempts to determine the same weight to the nearest milligram, individual observation are sure to differ unless these are performed with extreme care and without prejudice.

Presence of certain system defects: System defects such as large dimensional tolerances in mating parts and the presence of friction contribute to errors that are either positive or negative depending on the direction of motion, the former causes backlash error and the latter causes slackness in the meter bearings. One way of detecting and correcting such errors is to measure the quantity first while increasing and then decreasing the magnitude. This procedure is based on the *method of symmetry.*

Effect of unrestrained and randomly varying parameters: Chance errors are also caused due to effect of certain uncontrolled disturbances which influence the instrument output. Line voltage fluctuations, vibrations of the instrument supports, etc. are common examples of this type. The experimenter should try to reduce the influence of such randomly varying parameters to the minimum. But, in spite of this, there are always certain residual contributions due to these random perturbations.

2.9.2 Miscellaneous type of gross errors

There are certain errors that cannot be strictly classifies as either systematic or random as they are partly random. Therefore, such errors are termed miscellaneous type of gross errors. This class is mainly caused by the following:

  (i) ***Personal or human error***: These are caused due to the limitations in, the human senses. For example, one may sometimes consistently read the observed value either high or low and thus introduce systematic errors in the results. While at another time one may record the observed value slightly differently than the actual reading and consequently introduce random error in the data. Therefore, it becomes necessary to exercise extreme care with mature and considered judgement in recording the observations so as to reduce such errors.

  (ii) ***Errors due to faulty components/adjustments***: Sometimes there is misalignment of moving parts, electrical leakage, poor optics, etc. in measuring systems. These may simultaneously cause, for example, zero shift coupled with zero drift which are systematic and random errors, respectively. In such cases, it becomes necessary to determine the magnitude of the systematic and random errors from the overall gross error. The procedure is usually consists of repeating a measurement for a sufficiently large number of times by feeding a 'standard' signal to the instrument. The difference between the mean value of the signal and the standard signal gives the best estimate of systemic error. Further, the estimate of uncertainty which represent the random error in measurement in evaluated from the dispersion of data.

(iii) *Improper application of instrument:* Errors of this type of error are caused due to the use of instrument in condition which do not conform to the desired design /operating conditions. For example, extreme vibrations mechanical shock or pick up due to electrical noise could introduce so much gross error as to mask the test information. In such cases it is necessary to stop the experiment until the disturbing elements causing the chaotic type of gross errors are eliminated.

2.9.3 Reducing measurement error

So, how can we reduce measurement errors, random or systematic? One thing you can do is to pilot test your instruments, getting feedback from your respondents regarding how easy or hard the measure was and information about how the testing environment affected their performance. Second, if you are gathering measures using people to collect the data (as interviewers or observers) you should make sure you train them thoroughly so that they aren't inadvertently introducing error. Third, when you collect the data for your study you should double-check the data thoroughly. All data entry for computer analysis should be "double-punched" and verified. This means that you enter the data twice, the second time having your data entry machine check that you are typing the exact same data you did the first time. Fourth, you can use statistical procedures to adjust for measurement error. These range from rather simple formulas you can apply directly to your data to very complex modeling procedures for modeling the error and its effects. Finally, one of the best things you can do to deal with measurement errors, especially systematic errors, is to use multiple measures of the same construct. Especially if the different measures don't share the same systematic errors, you will be able to triangulate across the multiple measures and get a more accurate sense of what's going on.

2.9.4 Error versus uncertainty

It is important not to confuse the terms 'error' and 'uncertainty'.

Error is the difference between the measured value and the 'true value' of the thing being measured.

Uncertainty is a quantification of the doubt about the measurement result.

Whenever possible we try to correct for any known errors: for example, by applying corrections from calibration certificates. But any error whose value we do not know is a source of uncertainty.

2.9.5 Where do errors and uncertainties come from?

Many things can undermine a measurement. Flaws in the measurement may be visible or invisible. Because real measurements are never made under perfect conditions, errors and uncertainties can come from:

The measuring instrument: instruments can suffer from errors including bias, changes due to ageing, wear, or other kinds of drift, poor readability, noise (for electrical instruments) and many other problems.

The item being measured: which may not be stable (Imagine trying to measure the size of an ice cube in a warm room.).

The measurement process: the measurement itself may be difficult to make. For example measuring the weight of small but lively animals presents particular difficulties in getting the subjects to co-operate.

'Imported' uncertainties: calibration of your instrument has an uncertainty which is then built into the uncertainty of the measurements you make (But remember that the uncertainty due to not calibrating would be much worse.).

Visual alignment is an operator skill. A movement of the observer can make an object appear to move. 'Parallax errors' of this kind can occur when reading a scale with a pointer.

Operator skill: some measurements depend on the skill and judgement of the operator.

One person may be better than another at the delicate work of setting up a measurement, or at reading fine detail by eye. The use of an instrument such as a stopwatch depends on the reaction time of the operator. (But gross mistakes are a different matter and are not to be accounted for as uncertainties.)

Sampling issues: the measurements you make must be properly representative of the process you are trying to assess. If you want to know the temperature at the work-bench, don't measure it with a thermometer placed on the wall near an air conditioning outlet. If you are choosing samples from a production line for measurement, don't always take the first ten made on a Monday morning.

The environment: temperature, air pressure, humidity and many other conditions can affect the measuring instrument or the item being measured.

Where the size and effect of an error are known (e.g. from a calibration certificate) a correction can be applied to the measurement result. But, in general, uncertainties from each of these sources, and from other sources, would be individual 'inputs' contributing to the overall uncertainty in the measurement.

2.9.6 Distribution - the 'shape' of the errors

The spread of a set of values can take different forms, or probability distributions.

(i)　*Normal distribution:* In a set of readings, sometimes the values are more likely to fall near the average than further away. This is typical of a normal or Gaussian distribution. You might see this type of distribution if you examined the heights of individuals in a large group of men. Most men are close to average height; few are extremely tall or short.

Fig. 2.37 shows a set of 10 'random' values in an approximately normal distribution. A sketch of a normal distribution is shown in Fig. 2.38.

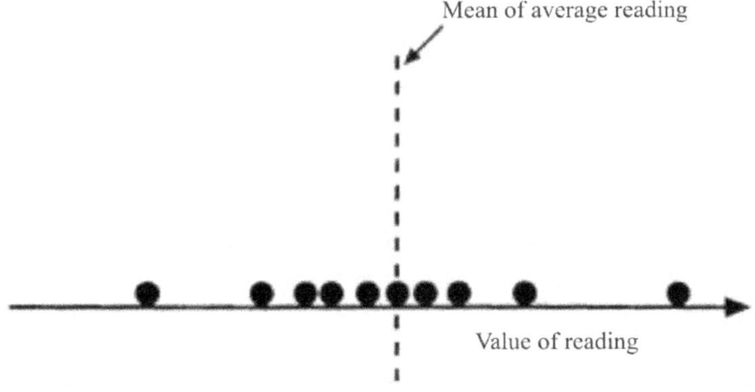

Fig. 2.37 'Blob Plot' of a Set of Values Lying in a Normal Distribution

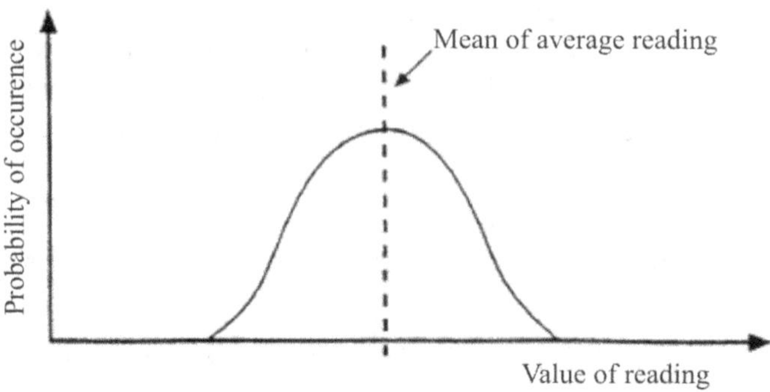

Fig. 2.38 Sketch of a 'Normal' Distribution

(ii) **Uniform or rectangular distribution:** When the measurements are quite evenly spread between the highest and the lowest values, a rectangular or uniform distribution is produced. This would be seen if you examined how rain drops fall on a thin, straight telephone wire, for example. They would be as likely to fall on any one part as on another. Fig. 2.39 shows a set of 10 'random' values in an approximately rectangular distribution. A sketch of a rectangular distribution is shown in Fig. 2.40.

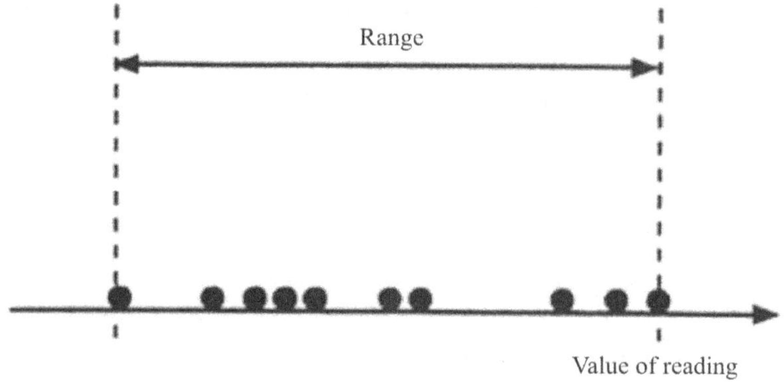

Fig. 2.39 'Blob Plot' of a Set of Values Lying in a Rectangular Distribution.

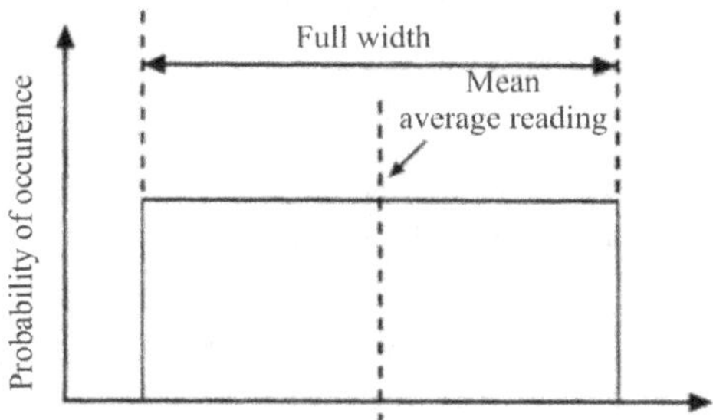

Fig. 2.40 Sketch of a Rectangular Distribution

(iii) **Other distributions:** More rarely, distributions can have other shapes, for example, triangular, M-shaped (bimodal or two-peaked), or lop-sided (skew).

2.9.7 Calibration

The relationship between the value of the input to the measurement system and the system's indicated output value is established during calibration of the measurement system. The quantity to be measured being the measure and, which we call m, the sensor must convert m into an electrical variable called s. The expression $s = F(m)$ is established by calibration. By using a standard or unit of measurement, we discover for these values of m (m_1, m_2 ....m_i) electrical signals sent by the sensor (s_1, s_2 ... s_i) and we trace the curve s(m), called the sensor calibration curve (Fig. 2.41).

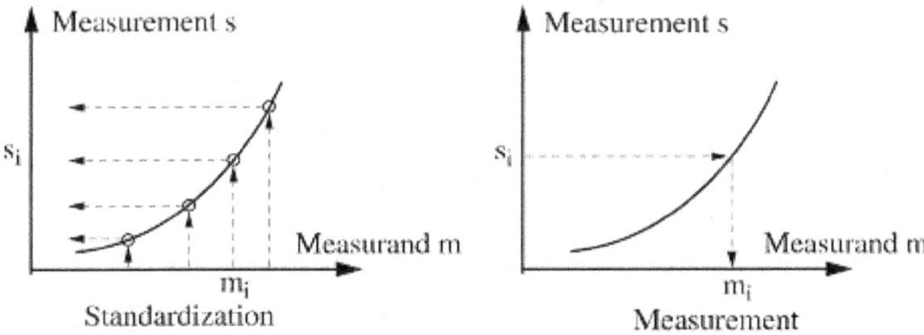

Fig. 2.41 Sensor Calibration Curve

2.9.8 Static Performance parameters

The definitions of various static performance parameters of the instruments are given below:

(i) *Resolution:* The smallest increment of change in the measured value that can be determined from the instrument's readout scale.

The resolution is often on the same order as the precision; sometimes it is smaller.

(ii) *Sensitivity:* The change of an instrument's output per unit change in the measured quantity. Typically, an instrument with higher sensitivity will have also finer resolution, better precision, and higher accuracy.

(iii) *Range:* The proper procedure for calibration is to apply known inputs ranging from the minimum to the maximum values for which the measurement system is to be used. These limit the operating range of the system.

(iv) **Hysteresis:** An instrument is said to exhibit hysteresis when there is a difference in reading on whether the value of the measured quantity is approached from above or below.

(v) **Accuracy and Precision:** Accuracy of a system can be estimated during calibration. If the input value of calibration is known exactly, then it can be called the true value. The accuracy of a measurement system refers to its ability to indicate a true value exactly. Accuracy is related to absolute error, ε:

ε = true value-indicated value from which the percent accuracy is found by

$$A = \left(1 - \frac{|\varepsilon|}{\text{true value}}\right) \times 100$$

(vi) *Precision:* Or repeatability of a measuring system refers to the ability of the system to indicate a particular value upon repeated but independent applications of a specific value input. Precision of a measurement describes the units used to measure something

Precision Example: How long is the pencil?

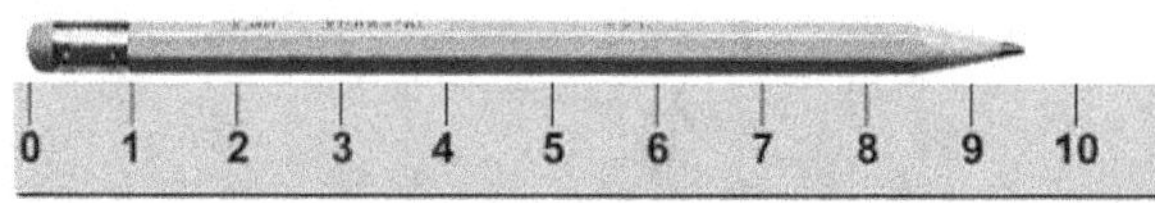

The best you can say is 'about 9 centimetres'

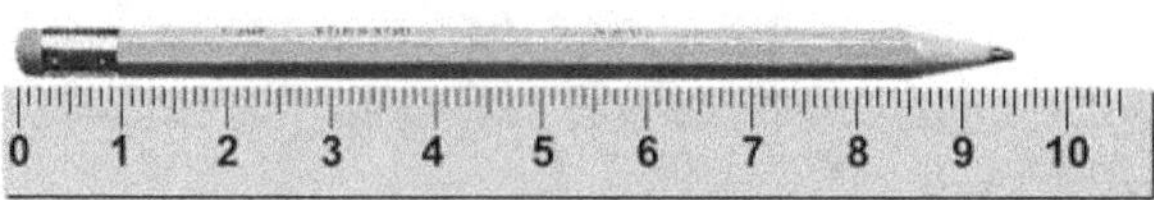

The best you can say is 'about 9.5 centimetres'. This second measurement is more precise, because you used a smaller unit to measure with.

- It is impossible to make a perfectly precise measurement.

- Accuracy can be improved up to but not beyond the precision of the instrument by calibration.

2.10 Measuring Instruments

Basic quantities for measurement in mechanical field are measurement for length, mass, time, temperature and angle. These quantities can be obtained using certain measurement devices. Several measurement devices can measure some of the quantities, while the remains need to use specific measurement devices. The measuring instruments, tools and gauges are used to measure, check and set the dimension specified. They are used to detect inaccuracy in machining so as to eliminate defects and rejections. Their applications are numerous. The measuring instruments are dimensional control instruments used to measure the exact size of object. These are adjustable devices and can measure with an accuracy of 0.001mm or better.

Classification of mechanical measurement and its measurement devices that will be discussed are as follows:

No.	Mechanical Qunatities	Measurement devices
1.	Linear dimension	<ul><li>Ruler</li><li>Caliper</li><li>Micrometer caliper</li><li>Vernier caliper</li><li>Dial guage</li></ul>
2.	Mass	<ul><li>Mechanica balance</li><li>Spring balance</li></ul>
3.	Force	<ul><li>Spring coil</li><li>Proving coil</li><li>Strain gauge</li><li>Load cell</li></ul>
4.	Torque	<ul><li>Brake</li><li>Dynamomemter</li><li>Strain gauge</li><li>Transducer</li></ul>

The measuring instruments may also be classified as:

(i) ***Precision or non-precision measuring instruments:*** The precision instruments are those which have ability to measure parts with accuracy of 0.001mm or better.

(ii) The non-precision instruments are limited to the measurements of parts to a visible line graduation on the instrument used, such as graduated rule or scale etc.

(iii) ***Line measuring or end measuring instruments:*** In line measuring instruments, the end of a dimension being measured is aligned with graduation of the scale from which the length is read directly such as a steel rule. In end measuring instruments, the measurements are taken between the two ends as in micrometers, calipers, gauge block etc. These instruments are more useful and important for precision measuring work.

(iv) ***Direct or indirect type measuring instruments:*** The direct type measuring instruments are used to determine the actual size and dimension of a work piece.

The indirect type measuring instruments are used to compare measurements of work piece with the direct measuring instrument.

2.10.1 Types of measuring instruments

Every measuring instrument has limited use and can be used for one or more purposes, i.e., for measuring linear dimensions; angular dimensions are for measuring surface finish. The various measuring instruments, depending upon their use, are as follows;

(i) ***Linear Measurements:*** The following are the various instruments used for linear measurement:

Non-precision instruments

Steel rule

Calipers

Dividers

Depth gauge

Telescopic gauge

Precision instruments

    Micrometers

    Vernier calipers

    Vernier height gauge

    Vernier depth gauge

    Vernier gear caliper

    Slip gauge

(ii) **Angular measurements:** The following are the various instruments used for angular measurements:

Non-precision instruments:

    Protractor

    Adjustable bevel

    Combination set

    T-square

Precision instruments:

    Universal bevel protractor

    Sine bar

    Dividing head (indexing head)

    Angle gauge blocks

    Spirit level

    Auto-collimator

(iii) **Surface measurements:** The following are the various instruments used for surface measurements:

    Straight edge

    Surface plate

    Surface gauge

    Optical flat

    Interferometer

    Profilometer

    Profilogram

2.10.2 Vernier caliper or scale

A vernier scale lets one read more precisely from an evenly divided straight or circular measurement scale. It is fitted with a sliding secondary scale that is used to indicate where the measurement lies when it is in-between two of the marks on the main scale.

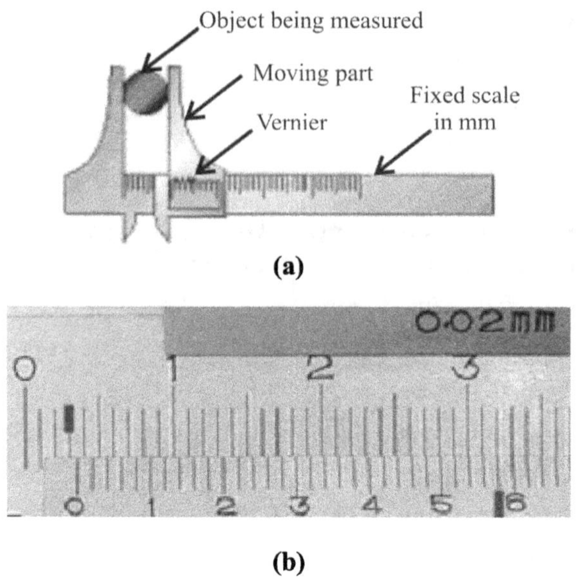

(a)

(b)

Fig. 2.42 (a) Vernier Calipers (b) Large View of Vernier Caliper

When a measurement is taken by mechanical means using vernier caliper, the measure is read off a finely marked data scale (the "fixed" scale, in the diagram). The measure taken will usually be between two of the smallest gradations on this scale. The indicating scale ("vernier" in the Fig. 2.42a) is used to provide an even finer additional level of precision without resorting to estimation. An enlarged view of the calipers, they have an accuracy of 0.02 mm as shown in Fig.2.42 b. The reading is 3.58 mm. The 3 mm is read off from the upper (fixed) data scale. The 0.58 mm is obtained from the lower (sliding) indicating scale at the point of closest alignment between the two scales. The superimposed markings show where the readings are taken (Fig. 2.42 b).

Construction of Vernier Caliper: The indicating scale is constructed so that when its zero point is coincident with the start of the data scale, its gradations are at a slightly smaller spacing than those on the data scale and so do not coincide with any on the data scale. N gradations of the indicating scale would cover N-1 gradations of the data scale (where N is the number of divisions the maker wishes to show at the finer level).

Use of vernier scale: When a length is measured the zero point on the indicating scale is the actual point of measurement, however this is likely to be between two data scale points. The indicator scale measurement which corresponds to the best-aligned pair of indicator & data gradations yields the value of the finer additional precision digit.

Examples: On instruments using decimal measure, as shown in the Fig. 2.43, the indicating scale would have 10 gradations covering the same length as 9 on the data scale. Note that the vernier's 10th gradation is omitted. On an instrument providing angular measure, the data scale could be in half-degrees with an indicator scale providing 30 1-minute gradations (spanning 29 of the half-degree gradations).

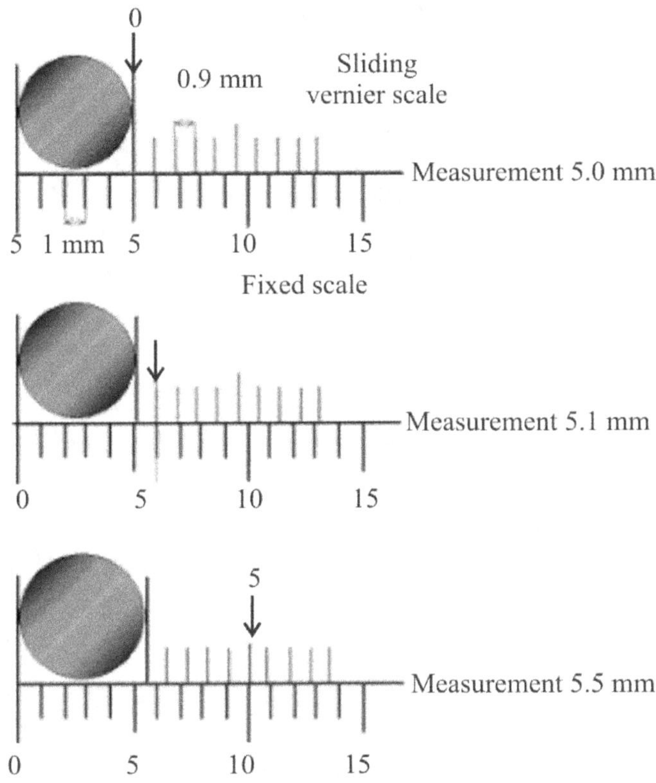

Fig. 2.43 Measurement process

2.10.3 Micrometer

A micrometer is a widely used device in mechanical engineering for precisely measuring thickness of blocks, outer and inner diameters of shafts and depths of slots. Appearing frequently in metrology, the study of measurement, micrometers have several advantages over other types of measuring instruments like the Vernier caliper – they are easy to use and their readouts are consistent. There are three types of micrometers based on their application (Fig. 2.44):

- External micrometer
- Internal micrometer
- Depth micrometer

An external micrometer is typically used to measure wires, spheres, shafts and blocks. An internal micrometer is used to measure the opening of holes, and a depth micrometer typically measures depths of slots and steps. The precision of a micrometer is achieved by a using a fine pitch screw mechanism.

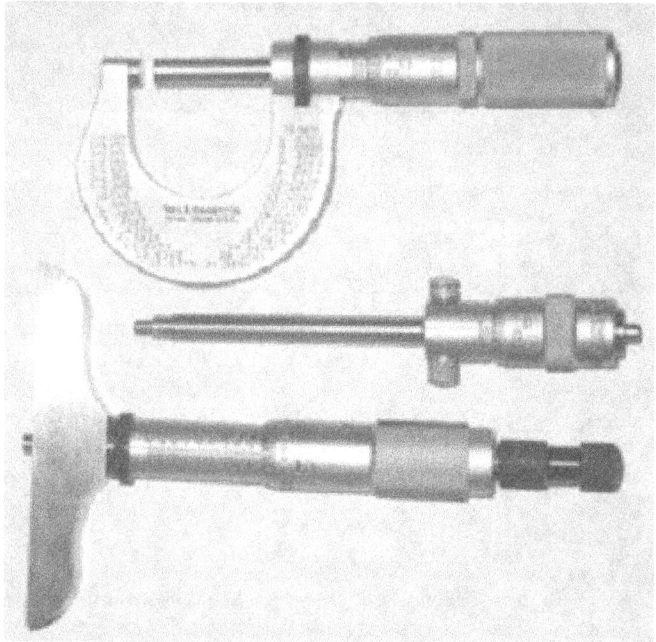

Fig. 2.44 External, internal, and depth micrometers

Reading a Metric Micrometer: The spindle of an ordinary metric micrometer has 2 threads per millimeter, and thus one complete revolution moves the spindle through a distance of 0.5 millimeter. The longitudinal line on the frame is graduated with 1 millimeter divisions and 0.5 millimeter subdivisions. The thimble has 50 graduations, each being 0.01 millimeter (one-hundredth of a millimeter). To read a metric micrometer, note the number of millimeter divisions visible on the scale of the sleeve, and add the total to the particular division on the thimble which coincides with the axial line on the sleeve (Fig. 2.45).

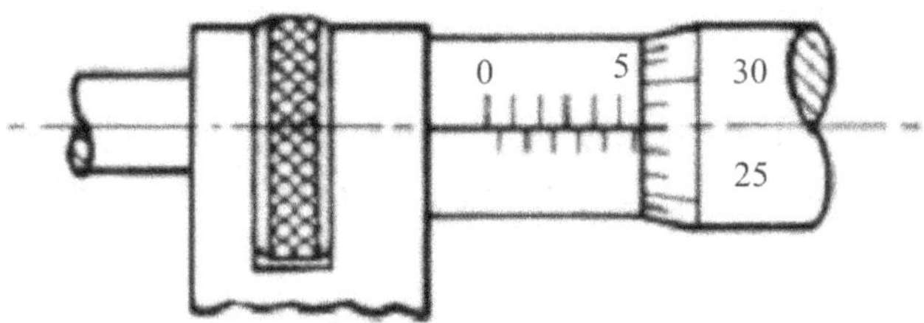

Fig. 2.45 Metric Micrometer

Suppose that the thimble were screwed out so that graduation 5, and one additional 0.5 subdivision were visible (as shown in Fig. 2.45), and that graduation 28 on the thimble coincided with the axial line on the sleeve. The reading then would be 5.00 +0.5 +0.28 = 5.78 mm. Some micrometers are provided with a vernier scale on the sleeve in addition to the regular graduations to permit measurements within 0.002 millimeter to be made.

Micrometers of this type are read as follows: First determine the number of whole millimeters (if any) and the number of hundredths of a millimeter, as with an ordinary micrometer, and then find a line on the sleeve vernier scale which exactly coincides with one on the thimble. The number of this coinciding vernier line represents the number of two-thousandths of a millimeter to be added to the reading already obtained. Thus, for example, a measurement of 2.958 millimeters would be obtained by reading 2.5 millimeters on the sleeve, adding 0.45 millimeter read from the thimble, and then adding 0.008 millimeter as determined by the vernier. Note: 0.01 millimeter = 0.000393 inch, and 0.002 millimeter = 0.000078 inch (78 millionths). Therefore, metric micrometers provide smaller measuring increments than comparable inch unit micrometers— the smallest graduation of an ordinary inch reading micrometer is 0.001 inch; the vernier type has graduations down to 0.0001 inch. When using eithera metric or inch micrometer, without a vernier, smaller readings than those graduated may of course be obtained by visual interpolation between graduations.

Use of Micrometer: The micrometers are precision measuring instruments. These are available in English system or in metric system, but the metric system micromenters are mostly used.

Outside Micrometer: It is mainly used to measrue the outside diameter of a job or length of a small part. It can measure the dimension to an accuracy of 0.01mm.

Screw Thread Micrometer: It is designed to measure the pitch diameter of screw threads to an accuracy of 0.01mm. In the construction, the screw thread, is similar to outside micrometer with the following differences.

- The movable spindle is pointed

- The end of the anvil is of the same form as the screw thread to be measured.

- The different pairs of interchangeable vee-anvils and spindle points are used with this micrometer. In order to measure the pitch diameter; the pointed end of the spindle and the sides of the vee-anvil should contact the surfaces of the thread. The reading on the micrometer is read in the similar way as in outside micrometer.

Depth Gauge Micrometer: The depth gauge micrometer (also known as depth micrometer) is used to measure the depth of holes, slots and recessed areas to an accuracy of 0.01mm.

The micrometer head acts as a reference surface and is held firmly and perpendicular to the centre line of the hole. The screwed spindle does the actual measuring. The graduated sleeve has the reference line and fixed graduations in the reverse direction i.e., in the order 0, 9, 8, 7, 6 etc. The thimble is a tabular cover fastened to the spindle and moves with the spindle. The bevelled edge of the thimble has graduations from 0 to 50 and every fifth is numbered. The locking ring is provided to lock the micrometer at any desired reading.

The ratchet stop is a small extension to the thimble. The extension rods are used for larger ranges of measurement, as in inside micrometer. It is inserted through the top of the micrometer.

In using this instrument, it should be first ensured that the edge of the hole is free from burrs. Care should be taken when the spindle touched the bottom of the hole and micrometer head rests on the surface. The accuracy in reading depends upon the sense of touch. The reading of this micrometer is taken in the similar way as outside micrometer.

Inside Micrometer: The inside micrometer is used for measuring large internal diameters (over 50 mm) to an accuracy of 0.01mm. It works on the same principle as that of outside micrometer.

The inside micrometer consists of the following parts:

- Micrometer unit
- Extension rod
- Spacing collor
- Handle

2.10.4 Dial indicator or dial gauges

Dial indicators are instruments used to accurately measure a small distance. They may also be known as a Dial gauge, Dial Test Indicator (DTI) or as a "clock" (Fig. 2.46).

The definition of small obviously depends on the observer however a range between 1 mm (0.040") and 50 mm (2") may be thought of as typical with a travel of 10 mm (approx 0.5") being perhaps the most common. The dial gauge is a common type of mechanical comparator used on lathe. It is mainly used for checking parallelism and concerntricity of rods, holes and flatness of surfaces to an accuracy of 0.01mm.

Probe type gauges: They typically consist of a graduated dial and needle (thus the clock terminology) to record the minor increments, with a smaller embedded clock face and needle to record the number of needle rotations on the main dial. They may be graduated to record measurements of between 0.01mm (0.001"), which is not a direct unit conversion) down to 0.001mm (0.001") for more accurate usage. The probe (or plunger) moves perpendicular to the object being tested by either retracting or extending from the indicators body.

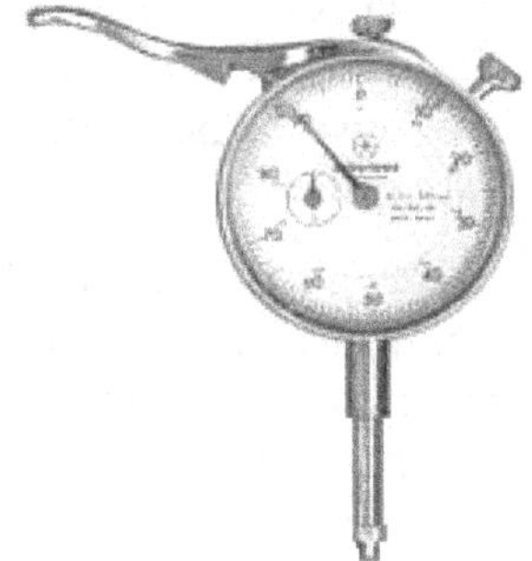

Fig. 2.46 0.1 mm-20 mm dial test indicator

The dial face can be rotated to any position, this is used to orient the face towards the user as well as set the zero point, there will also be some means of incorporating limit indicators (the two metallic tabs visible in the right image, at 90 and 10 respectively), these limit tabs may be rotated around the dial face to any required position. There may also be a lever arm available that will allow the indicator's probe to be retracted easily.

A situation that utilizes all these features would be in the inspection department of a manufacturing facility. The inspection department would have the Dial Test indicator (DTI) set up in a fixture (possibly a magnetic base) which would secure the DTI and allow its adjustment to read zero at the optimal size of a sample part, the two limit tabs would be set to the extremes of the parts tolerance, finally the lever arm would allow the probe to be retracted when loading the sample to be tested, between the DTI and base. This lever arm reduces the chance of applying undue force to the probe or DTI, possibly upsetting its accuracy. The tip of the probe may be interchanged with a range of shapes and sizes depending on application.

Lever type: A Lever arm test indicator or Finger indicator has a smaller movement range (perhaps 1 mm depending on the model) and measures the deflection of the arm, the probe does not retract but swings in an arc around its hinge point.

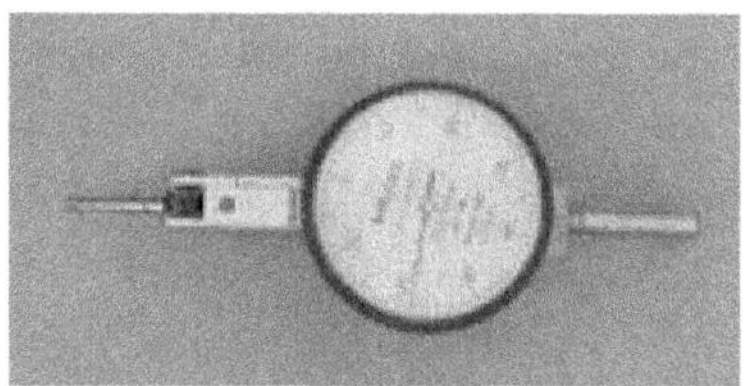

Fig. 2.47 Lever Arm Test Indicator

The lever may be interchanged for length or ball diameter and permits measurements to be taken in narrow grooves or small bores where the body of a probe type may not reach. The model shown in Fig. 2.47 is bidirectional, some types may have to be switched via a side lever to be able to measure in the opposite direction.

Digital type: With the advent of electronics and LCDs the clock face and analog display has been replaced with digital displays, these have the added advantage of being able to record and transmit the data electronically to a computer. This process is known as Statistical process control (SPC) and involves a computer recording and interpreting the results, this also reduces the risk of the operator introducing recording errors. Digital indicators can also be switched between imperial and metric units with the press of a button, thereby increasing the DTI's versatility.

Applications

- As outlined above, they may be used in a quality environment to check for consistency and accuracy in the manufacturing process.

- On the workshop floor to initially set up or calibrate a machine, prior to a production run.

- By toolmakers (moldmakers) in the process of manufacturing precision tooling.

- In metal engineering workshops, where a typical application is the centering of a lathes workpiece in a four jaw chuck. The DTI is used to indicate the run out of the work piece, with the ultimate aim of reducing it to a suitably small range by small chuck jaw adjustments.

- In areas other than manufacturing where accurate measurements need to be recorded, eg:- physics.

2.10.5 Gauges

The gauges are fixed-dimension instruments. These are generally used to check the particular dimensions of a work piece within their tolerance. The gauges should be made so that a minimum time and skill is required in their use. These instruments differ from measuring instruments because they have no graduations to measure various lengths and angles and normally adjustment can be made in their use. The gauges are specially made for some particular job, which is to be produced in large quantities.

Plug gauges: The plug gauges are used to test the accuracy of holes. A standard plug gauge has its diameter finished to the standard size. It is used in general engineering workshop, tool rooms etc., where production quantities are not too great and accuracy is more important.

The limit plug gauges are used where large quantities are to be produced and the variation in the standard size is allowed.

The single-ended limit plug gauge has separate "Go" and "Not go" members, the progressive limit plug gauges has "Go" and "Not go" members on the same side of a handle and the double ended limit plug gauge has "Go" member at one end and "Not go" member on the other end of the handle. It may be noticed that the "Not go" end has less width than the "Go" end. The "Go" and "Not go" ends are detachable so that they may be renewed separately when worn

Snap Gauges: The snap gauges are used for checking the external dimensions; they are made with openings to fit over a part to be checked. The part may be cylindrical or flat. The most commonly used types of snap gauges are:

Solid or non-adjustable snap gauge

Adjustable snap gauge

Double ended snap gauge

Thread Gauges: The thread gauges are used to check the pitch diameter of screw threads. The internal limit thread gauge is a plug gauge having screws on both ends. One end has a "Go" member and the other has "Not go" member. These gauges are used for checking the pitch diameter of internal threads. The external limit thread gauge, may be of ring type or snap type. These gauges are used for checking pitch diameter of external threads.

Screw Pitch Gauge: The screw pitch gauge is used to check the pitch of a screw. It is similar to a feeler gauge except that the leaves are notched on one edge according to the various pitches and contours of the specific threads. The free end of the leaf is generally stamped to indicate the pitch of the screw or number of threads per inch. These gauges are made for the American, British and the International standard or metric form of threads.

In using the screw pitch gauge, place successive leaves over the thread until one of them is found to coincide with the thread. The pitch of the thread (or the number of threads per inch) is read directly from the stamping on the leaf.

Slip Gauges: The slip gauges (also known as precision gauge blocks) are used for checking the accuracy of measuring instruments such as micrometers, calipers, snap gauges, dial indicators etc. These are also used for setting the sine bar for angular measurements, for accurate measurements in die manufacture and in various precisions measuring machines like tool room microscope etc. The slip gauge consists of alloy steel blocks of section about 30 mm by 10 mm. These are hardened before being finished to size. In order to obtain a very high order of accuracy both as regard to flatness or parallelism of opposite face, they are ground and lapped. The slip gauges are made in the following four grades:

- Workshop or production grade
- Inpection grade
- Calibration grade
- Reference grade

Slip gauges are often called Johannsen gauges also, as Johannsen originated them. When two gauges are wrung together and the overall dimension of a pile made of two or more blocks so joined is exactly the sum of the constituent gauges. It is on the property of wringing units together for building up combinations that the success of system depends, since by combining gauges selected from a suitably arranged combination, almost any dimension may be built-up.These may be used as reference standards for transferring the dimensions of the unit of length from the primary standard to gauge blocks of lower accuracy and for the verification and graduation of measuring apparatus, and length measures for the regulation and adjustment of indicating measuring apparatus and for direct measurement of linear dimensions of industrial component.

Basic forms of slip gauges: Slip gauges with three basic forms are commonly found. These are rectangular, square with centre hole, and square without centre hole.

Rectangular form is the more widely used because rectangular blocks are less expensive to manufacture, and adapt themselves better to applications where space is restricted or excess weight is to be avoided.

For certain applications, square slip gauges, though expensive, are preferred. Due to their large surface area, they wear longer and adhere better to each other when wrung to high stacks. Square blocks with centre holes are used to permit the use of tie rods as an added assurance against the wrung stocks falling apart while handling.

The greatest emphasis in the slip gauges is laid on the two gauging surfaces on opposite sides of the blocks. The length between measuring surfaces, flatness and surface conditions of measuring surfaces are the most important features of slip gauges which have to be controlled within desired limits.

2.10.6 Sine Bar

It is used either to measure angles more precisely than a bevel protector. It is generally used in conjuction with slip gauges. A sine bar consists of a hardened and ground steel bar, which is stepped at the ends. A rollar is fastened, in each step, with a screw. These rollers should touch the both faces of the step. The distance between the centers of two-roller (L) specifies a sine bar. The most commonly used sine bar has this distance as 100 mm. Some holes are drilled in the body of the sine bar in order to reduce the weight and to facilitate handling.

Measuring Angles with sine bar: In measuring the angles or locating any work to a given angle, the use of sine bar is made as follows:

- The work having tapered top and flat base is placed on the surface plate.
- One of the rollers is placed on the surface plate, while the other roller is made to rest on slip gauges of height H in such a way that the top of the work piece coincides with the sine bar as shown in Fig. 2.48.
- Let the sine bar is set at an angle $Sin = H/L$
- In order to obtain better results, both the rollers of the sine bar are places on slip gauges of heigth H_1 and H_2. In such cases

$$Sin = H_1 - H_2/L$$

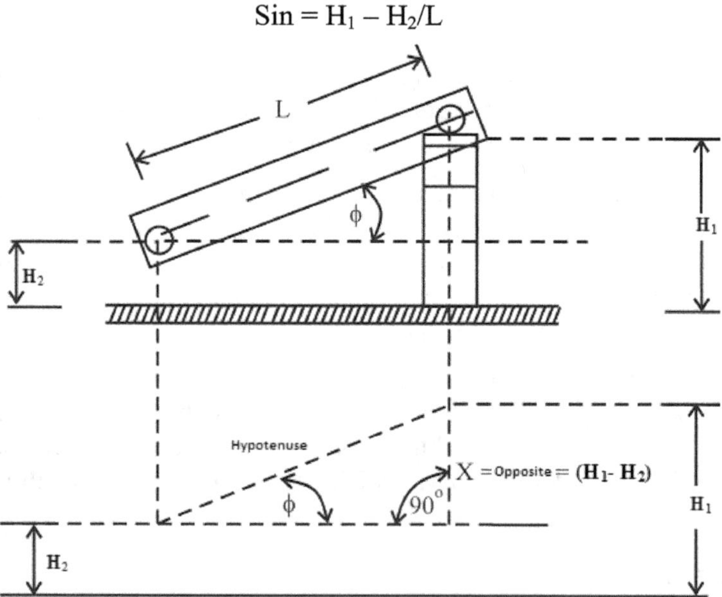

Fig. 2.48 Sine Bar and Gauge Block

2.10.7 Angle gauge blocks

These blocks are used to set up any angle to the nearest 3 minutes. These are commonly used to measure the angle in the die insert. The angle gauge block is made of hardened steel and the measuring faces are lapped and polished to a high degree of accuracy and flatness. These blocks are about 75 mm long and 15 mm wide. There are thirteen separate angle gauge blocks in conjuction with one square block and one parallel straight edge. These are available in the following three series:

- First series: 1, 3, 9, 27 and 41 degrees
- Second series: 1', 3', 9' and 27'
- Third series: 3", 6", 18" and 30"

Note:

- Since the angle gauge blocks can not always be directly applied to work, therefore these are being used as a reference and as an aid to other angle measuring devices.

- The centers of sine bar pins rest on the centers of gage block stacks (Fig. 2.48).

2.10.8 Combination set

Combination set is the most useful instrument in engineering applications. As the name indicates this instrument consists of set of several instruments. It fulfills all the requirements of a square, set square, meter square, height gauge, centre square, bevel protector, marking gauge etc. It also provides the work to be kept horizontal.

It consists of square head, central head, a bevel protector head, spirit level and beam. Fig. 2.49 shows all the parts of a combination set.

The different parts of a combination set are made of cast iron. These parts are manufactured with high accuracy.

Working: The different parts are used separately. A clamping bolt with a nut and a spiral spring is provided in each part so as these could be secured in any desired position on the whole length of the beam.

The working of different parts is explained as:

(a) ***Steel Rule:*** The beam of the instrument which acts as a rule marked either in inches or centimeters or in both for measuring length and height. A centre groove along the either length of the beam is provided to hold different parts at any desired position. Length of the rule varies from 200 to 6000 mm.

(b) ***Square Head:*** The square head has two edges one at 90^0 and the other at 45^0 to the rule. Thus it can be used both as square (90^0) and as a meter (45^0). It is also provided with spirit level. Spirit level enables the work to be kept horizontal.

(c) ***Centre Head:*** The centre head with the rule fastened to it is called a centre head. It has two arms at right angles to one another. It is so arranged that the blade exactly bisects the angle between the machined surfaces. This may be used to find the centre of a round bar or shaft

(d) ***Protractor Head:*** The protractors head can be used with the rule to measure angles as to measure the slopes of the surfaces. The scale on the protractor may be divided into degrees or a vernier attached whereby the angles can be measured in degrees and minutes. A spirit level is also fitted to set the work horizontally or setting at an angle.

Fig. 2.49 Combination set

Hence this set provides the best way of measuring different parameters as length, angle surface leveling etc., with little effort.

This set saves the time spent in picking up different instruments for measuring the length, angle etc. by a worker. So from the time and accuracy point of view it is the best assistance of workman in workshop and industries.

2.11   Lathe

"Lathe" itself is a machine tool which holds the work between two rigid and strong supports called centers, or in a chuck or in a face plate while the latter revolves. The chuck or the face plate is mounted on the projected end of the machine spindle. The cutting tool is rigidly held and supported in a tool post and is fed against the revolving works. While the work revolves about its own axis, the tool is made to move either parallel to or at an inclination with the axis to cut the desirable material. In doing so it produces a cylindrical surface, if it is fed parallel to axis, or will produce a tapered surface if it is fed at an inclination.

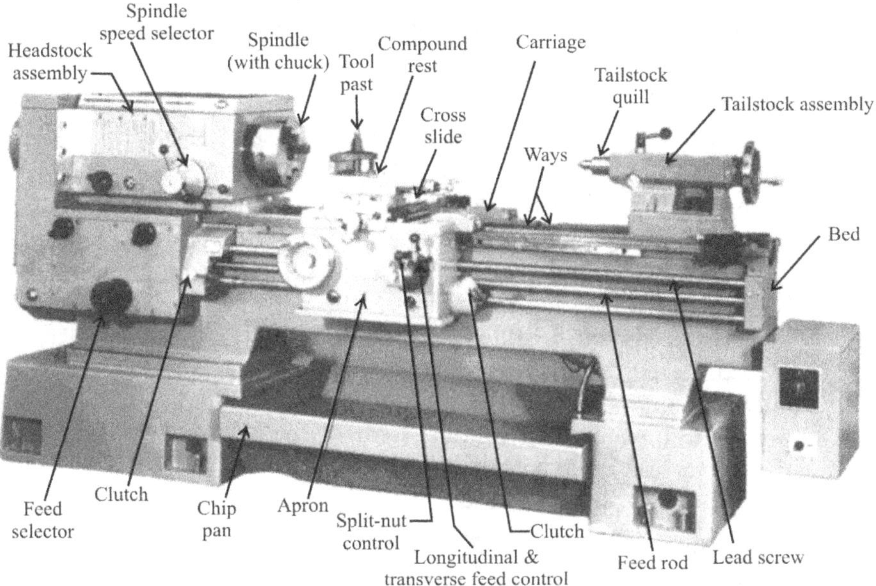

Fig. 2.50 Lathe Machine

2.11.1   Main components of lathe

Engine lathes have the following general functional parts (Fig.2.50):

(i)   *Lathe bed:* lathe bed is the foundation of the working parts of the lathe (Fig. 2.50). It provide the means for holding the tailstock and carriage, which slide along the ways, in alignment with the permanently attached headstock.

(ii)   *Headstock:* The headstock is located on the operator's left end of the lathe bed. It contains the main spindle and oil reservoir and the gearing mechanism for obtaining various spindle speeds and for transmitting power to the feeding and threading mechanism. The headstock mechanism is driven by an electric motor connected either to a belt or pulley system or to a geared system.

(iii) *Spindle:* The main spindle is mounted on bearings in the headstock and is hardened and specially ground to fit different lathe holding devices. The spindle has a hole through its entire length to accommodate long workpiece. The hole in the nose of the spindle usually has a standard Morse taper which varies with the size of the lathe.

(iv) *Tailstock:* The tailstock is located on the opposite end of the lathe from the headstock. It supports one end of the work when machining between centers, supports long pieces held in the chuck, and holds various forms of cutting tools, such as drills, reamers, and taps.

(v) *Carriage:* The carriage includes the apron, saddle, compound rest, cross slide, tool post, and the cutting tool. It sits across the lathe ways and in front of the lathe bed. The function of the carriage is to carry and move the cutting tool. The saddle carries the cross slide and the compound rest. The cross slide is mounted on the dovetail ways on the top of the saddle and the compound rest is mounted on the cross slide. The compound rest is used extensively in cutting steep tapers and angles for lathe centers. The cutting tool and tool holder are secured in the tool post which is mounted directly to the compound rest.

2.11.2 Classification of lathe

- Bench lathe
- Engine lathe
- Automatic lathe
- Tool room lathe
- Capstan lathe
- Rotary lathes
- Turret lathe

2.11.3 Lathe Operation

Different operations such as Turning, Drilling, Boring, Knurling, Milling, and Grinding can be performed on lathe.

2.12   Drilling Machines

Drilling machine is one of the most important machine tool used in a workshop to produce holes in solid objects. The process of making a hole is called drilling. In drilling operation, work is clamped to the table and the rotating cutting tool, called drill is fed onto it (Fig. 2.51).

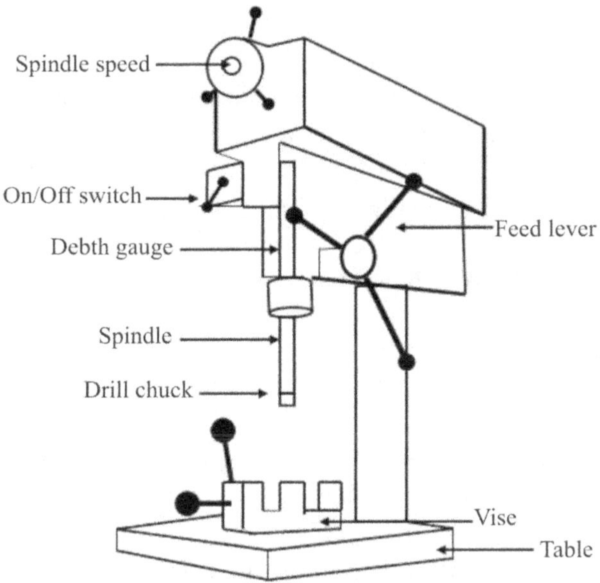

Fig. 2.51 Drilling Machine

2.12.1   Types of drilling machine

Drilling machines are manufactured in types and sizes to suit the different type of work. The different types of drilling machines are:

  (a)   Hand drill
  (b)   Portable drilling machine
  (c)   Sensitive drilling machine
  (d)   Pillar drilling machine
  (e)   Radial drilling machine
  (f)   Gang drilling machine
  (g)   Multi-spindle drilling machine
  (h)   Numerically controlled drilling machine

2.12.2 Components of drilling machine

The principal parts of the radial drilling machine are:

- (i)   Base
- (ii)   Column
- (iii) Radial arm
- (iv)   Drill head
- (v)   Spindle sped and feed mechanism

- **(i)   *Base:*** It is a large rectangular casting and is made rigid to support column and the table. T-slot are provided for clamping the work piece directly so that it will serve the function of the table.

- **(ii) *Column:*** It is a cylindrical casting mounted vertically to the base. It supports the radial arm which can be raised or lowered vertically to accommodate work pieces of different heights. A separate motor is used for raising and lowering the arm.

- **(iii) *Radial arm:*** The radial arm is mounted horizontally on the column. It may be swung around to any position over the work table.

- **(iv) *Drill head:*** It is mounted on the radial arm. It can be slide along the arm to locate the spindle with respect to the work.

- **(v)   *Spindle speed and feed mechanism:*** A motor fitted directly over the drill head drives the drill spindle. Through gear box, different spindle speeds and feeds can be obtained.

2.12.3 Drilling operation

Different operations such as Drilling, Reaming, Boring, Counter boring, Counter sinking, Spot facing and Tapping can be performed on drilling machine. The drilling machine is used for drilling heavy work and especially for the jobs where high degree of accuracy is required.

2.13 Milling Machine

A milling machine is a machine tool used for the shaping of metal and other solid materials. Its basic form is that of a rotating cutter which rotates about the spindle axis (similar to a drill), and a table to which the workpiece is affixed. The cutter and workpiece move relative to each other, generating a tool path along which material is to be removed. The movement is precisely controlled, usually with slides and leadscrews or analogous technology. Often the movement is achieved by moving the table while the cutter rotates in one place, but regardless of how the parts of the machine slide, the result that matters is the relative motion between cutter and workpiece. Milling machines may be operated manually or by CNC (computer numerical control). Milling machines can perform a vast number of operations, some of them with quite complex toolpaths, such as slot cutting, planing, drilling, diesinking, rebating, routing, etc.

The work piece clamped on the table is fed slowly against the milling cutter, rotating at a high speed. The metal is removed by series of cutting edges in the form of discontinuous chips. The work piece can be fed in longitudinal vertical and crosswise direction (Fig. 2.52).

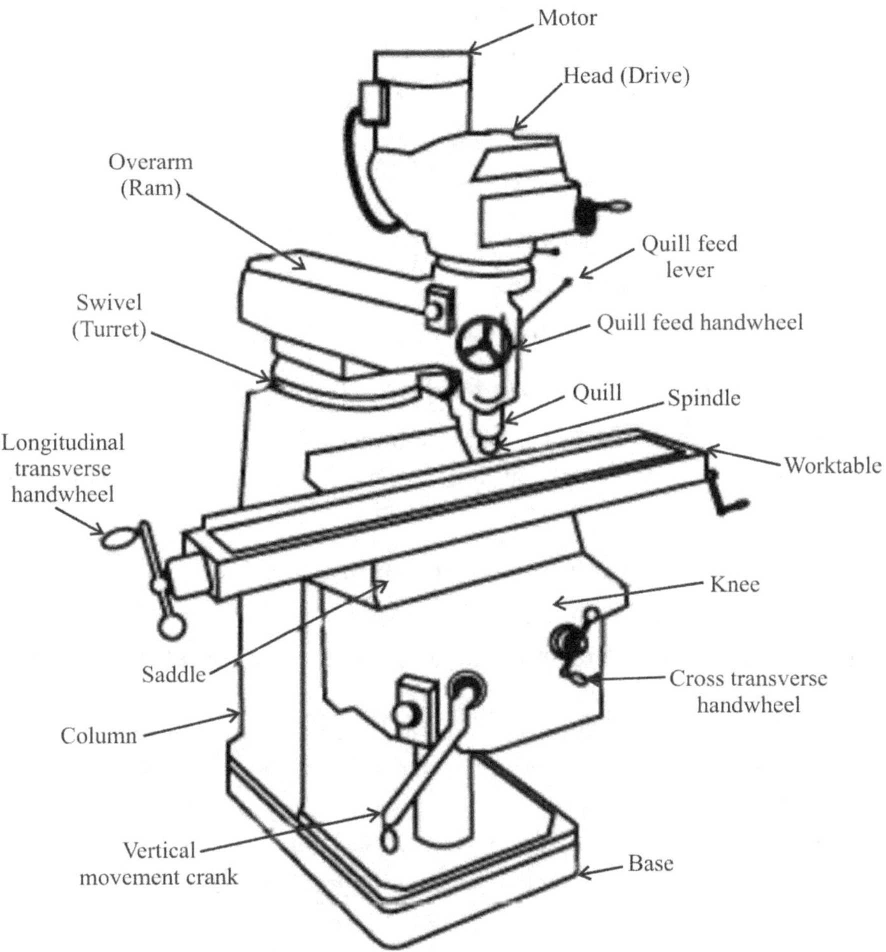

Fig. 2.52 Milling Machine

2.13.1 Types of Milling Machine

The milling machine may be classified according to:

1. Column and knee type – Hand milling

   Plain milling machine

   Universal milling machine

   Vertical milling machine

2. Fixed bed type

   Simplex milling machine

   Duplex milling machine

   Triplex milling machine

3. Planar type

4. Special type

   Hand milling machine

   Plain milling machine

   Universal milling machine

2.13.2 Operation of Milling Machine

Milling machines can perform a vast number of operations, from simple (e.g., slot and keyway cutting, planing, drilling) to complex (e.g., contouring, diesinking).

2.13.3 Advantages of milling machine

(i) The milling machine can perform large variety of operations.

(ii) Due to multiple point cutting tools, it removes metal at a very fast rate.

(iii) The milling machine can hold and operate one or more number of cutters. Therefore, the number of surfaces can be machined simultaneously.

(iv) It gives superior accuracy and surface finish.

(v) Heavy cuts can be taken without appreciable sacrifice in surface finish or accuracy.

(vi) Wide varieties of cutters are available for different types of works.

(vii) Parts which can be broached, can easily be milled.

2.13.4   Limitations

(i)   Milling machine is preferred only for small and medium sized job.

(ii)   Milling operation is too slow for long and heavy jobs.

2.14   Shaping Machine

A shaping machine is used to machine surfaces. It can cut curves, angles and many other shapes. It is a reciprocating type of machine, used to produce small flat surfaces with the help of single point cutting tool. The tool reciprocates over the stationary work piece. The flat surface may be horizontal, inclined or vertical. The reciprocating motion of the tool is obtained either by the crank & slotted lever quick return motion mechanism or whit-worth quick return motion mechanism.

The single point cutting tool reciprocates over the work piece rigidly fixed on a table. The tool held in the tool holder mounted on the ram moves forwards and backwards in a straight line over the work piece rigidly held in a vice clamped over the work table. Each time the tool moves forward it cuts a slice of metal from the work piece. Each time the tool moves backward the tool lifts clear of the workpiece.The work remains stationary during the forward (cutting) stroke of the tool but moves across by one cross traverse increment during the return (non cutting) stroke. The appearance of the machined surface is of a succession of closely straight line cuts (Fig. 2.53).

2.14.1   Principle parts of shaper

Different parts of shaping machine are shown in Fig. 2.53.

Base: It is a heavy structure of cast iron, which supports other parts of shaper.

Column: It is a box like casting. It carries the driving mechanism and is provided with two machined guide ways on which the ram reciprocates.

Ram: It is a reciprocating member. It carries a tool slide on its head and a mechanism for adjusting the stroke length.

Tool Head: It is attached to the front of the ram with the help of a nut and a bolt. It holds the tool rigidly. It also provides vertical and angular movement to the tool for cutting.

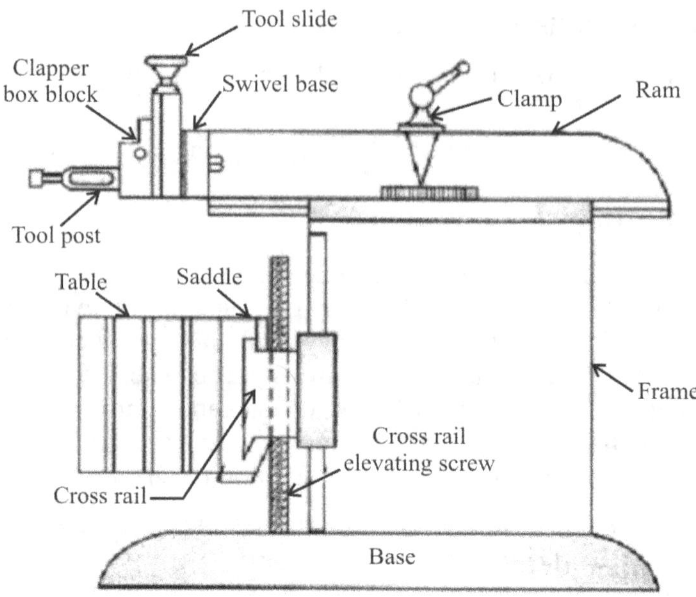

Fig. 2.53 Shaping Machine

Cross Rail: It is mounted on the front vertical portion of column. It is used for lowering or raising the table to accommodate work piece of different sizes.

Table: It holds the work piece. It can be moved horizontally or vertically with the help of cross rail.

2.14.2   Types of shaper

According to type of mechanism used for reciprocating motion of ram

    (a)   Crank and stotter link mechanism.

    (b)   Bull gear sliding block mechanism.

    (c)   Hydraulic shaper mechanism.

According to the position and travel ram

    (d)   Horizontal ram type.

    (e)   Vertical ram type.

    (f)   Traveling ram type.

According to the design of table

    (g)   Standard shaper.

    (h)   Universal shaper.

According to the type of cutting stroke

(i)   Push type shaper.

(j)   Draw cut type shaper.

Review Questions

1. Briefly explain the devices used for measuring the pressure of a fluid.

2. State the basic principle behind the theory of venturimeter.

3. What is a pitot tube? Sketch a standard pitot tube.

4. Explain the principle of venturimeter with a neat sketch. Derive the expression for the rate of flow of fluid through it.

5. Derive an expression to find the discharge through an orifice meter.

6. Co-efficient of discharge of venturimeter is always greater than orificemeter. Why?

7. Describe giving a sketch of micro manometer. Explain how it could be used for measuring small pressure difference.

8. What are the advantages of using venturimeter and orifice meter in fluid flow measurements?

9. Explain the principle of temperature measurement.

10. Describe various methods of temperature measurement.

11. Explain the working of a thermocouple (thermoelectric) and radiation pyrometer with neat sketches.

12. Explain the working of a Bordon pressure gauge with a neat sketch.

13. Explain the construction, working and uses of a micrometer.

14. Explain Sine Bar.

15. Write short notes on the following: (i) Errors of measurement (ii) Uncertainty analysis.

16. Explain the following properties of any measuring instruments: (a) Hysteresis (b) Sensitivity (c) Response time (d) Precision and accuracy.

17. Explain the construction, working and uses of the following: (i) Venturimeter (ii) Rotameter (iii) Prony brake dynamometer (iv) Vernier.

18. Describe Vernier calliper with neat sketch. How small/linear measurements are done with it?

19. Write short notes on the following measuring instruments: (i) vernier calliper (ii) dial gauge (iii) sine bar (iv) slip gauges.

20. Explain the construction and working principle of a milling machine with neat sketch.

21. With the help of a simple diagram explain various parts of a lathe machine and also enumerate various operations which can be performed on it.

22. Draw a neat sketch of simple shaper machine and lable its important parts.

23. Explain the construction and working principle of a drilling machine with neat sketch.

CHAPTER 3

Fluids

Fluids: Fluid properties, pressure, density and viscosity; pressure variation with depth, static and kinetic energy; Bernoulli's equation for incompressible fluids, viscous and turbulent flow, working principle of fluid coupling, pumps, compressors, turbines, positive displacement machines and pneumatic machines. Hydraulic power & pumped storage plants for peak load management as compared to base load plants.

3.1 Fluid

In the view of fluid mechanics, all matter consists of only two states i.e., fluid and solid. The difference between these two is an interesting work out. The technological difference lies with the reaction of the two to an applied shear or tangential stress.

A solid can resist a shear stress by a static deformation; a fluid cannot. Any shear stress applied to a fluid, no matter how small, will result in motion of that fluid. The fluid moves and deforms continuously as long as the shear stress is applied. As a result, we can say that a fluid at rest must be in a state of zero shear stress.

There are two classes of fluids – *liquids* and *gases*; the difference between them is the effect of cohesive forces. A liquid, being composed of relatively close-packed molecules with strong cohesive forces, tends to retain its volume and will form a free surface in a gravitational field if unconfined from above. Free-surface flows are dominated by gravitational effects. Since gas molecules are widely spaced with negligible cohesive forces, a gas is free to expand until it comes across confining walls. A gas has no definite volume, and when left to itself without confinement. Gases cannot form a free surface, and thus gas flows are rarely concerned with gravitational effects other than buoyancy.

Fig. 3.1 shows a solid block resting on a rigid plane and stressed by its own weight. The solid goes down into a static deflection, shown as a highly exaggerated dashed line, resisting shear without flow.

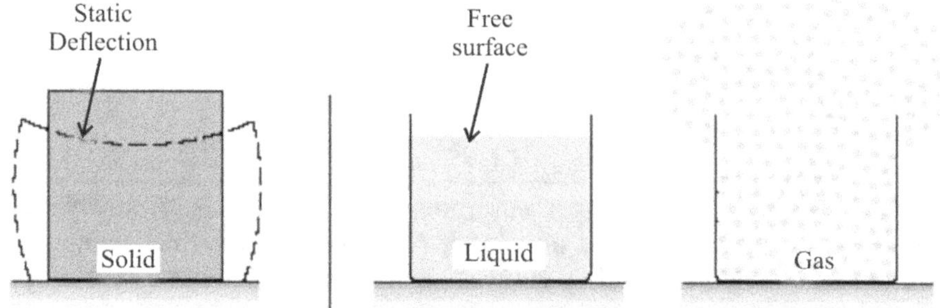

Fig. 3.1 Solid, liquid and gas

By contrast, the liquid and gas at rest (Fig. 3.1) require the supporting walls in order to eliminate shear stress. The walls exert a compression stress of $-p$. The liquid retains its volume and forms a free surface in the container. If the walls are removed, shear develops in the liquid and a big splash results.

3.2   Fluid Mechanics

Fluid mechanics is a branch of science which deals with property and behavior of fluids at rest and in motion (Fig. 3.2).

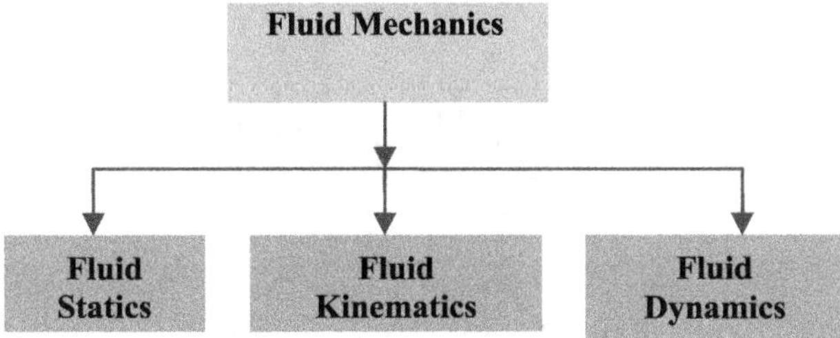

Fig. 3.2 Fluid mechanics

- Fluid Statics deals with action of forces on fluids at rest or in equilibrium.

- Fluid Kinematics deals with geometry of motion of fluids without considering the cause of motion.

- Fluid Dynamics deals with the motion of fluids considering the cause of motion.

3.3 Fluid Properties

3.3.1 Pressure

Pressure is the (compression) stress at a point in a static fluid (Fig. 3.3). Differences or *gradients* in pressure often drive a fluid flow, especially in ducts. Pressure is the force per unit area applied in a direction perpendicular to the surface of fluid.

It is usually more convenient to use pressure rather than force to describe the influences upon fluid behavior. The SI unit for pressure is the Pascal (Pa), equal to one Newton per square meter (N/m^2 or $kg \cdot m^{-1} \cdot s^{-2}$).

Gauge pressure: Gauge pressure is the pressure relative to the local atmospheric or ambient pressure. Pressure is a scalar quantity. The pressure at any given point of a non-moving (static) fluid is called the hydrostatic pressure. For an object sitting on a surface, the force pressing on the surface is the weight of the object, but in different orientations it might have a different area in contact with the surface and therefore exert a different pressure.

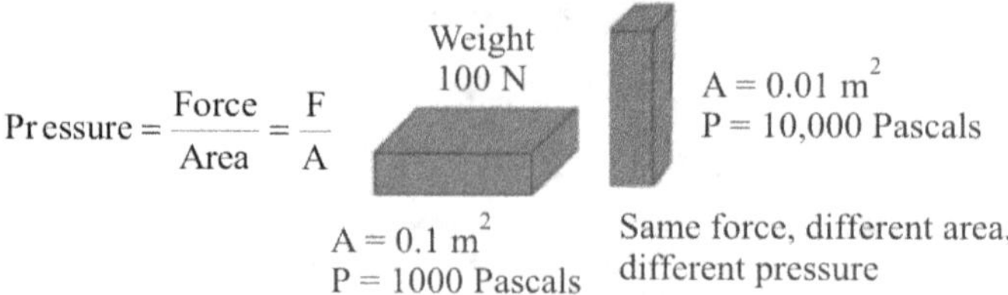

Fig. 3.3 Definition of pressure

Atmospheric pressure: Air above the surface of liquids exerts pressure on the exposed surface of the liquid and normal to the surface. This pressure exerted by the atmosphere is called atmospheric pressure. Atmospheric pressure at a place depends on the elevation of the place and the temperature. Atmospheric pressure is measured using an instrument called 'Barometer' and hence atmospheric pressure is also called Barometric pressure.

Unit: kPa . 'bar' is also a unit of atmospheric pressure 1bar = 100 kPa.

Absolute pressure: Absolute pressure at a point is the intensity of pressure at that point measured with reference to absolute vacuum or absolute zero pressure. Absolute pressure at a point can never be negative since there can be no pressure less than absolute zero pressure as shown in Fig. 3.4.

Absolute pressure at 'A'

Fig. 3.4 Absolute pressure, atmospheric pressure and gauge pressure.

If the intensity of pressure at a point is measured with reference to atmospheric pressure, then it is called gauge pressure at that point. Gauge pressure at a point may be more than the atmospheric pressure or less than the atmospheric pressure. Accordingly gauge pressure at the point may be positive or negative. Negative gauge pressure is also called vacuum pressure. From the figure, it is evident that,

Absolute pressure at a point = Atmospheric pressure ± Gauge pressure.

3.3.2 Temperature

Temperature T is a measure of the internal energy level of a fluid. It may vary considerably during high-speed flow of a gas. Engineers/Scientists often use Celsius or Fahrenheit scales for convenience, many applications in this text require *absolute* (Kelvin or Rankine) temperature scales:

$$°R = °F + 459.69$$

$$K = °C + 273.16$$

3.3.3 Density

The mass per unit volume of material is called the density, which is generally expressed by the symbol ρ (lowercase Greek rho). The density of a gas changes according to the pressure, but that of a liquid may be considered unchangeable in general. The units of density are kg/m^3 (SI). The density of water at 4°C and 1 atm (101.325 kPa, standard atmospheric pressure) is 1000 kg/m^3.

The density of water increases only 1 percent if the pressure is increased by a factor of 220. Thus most liquid flows are treated analytically as nearly "incompressible." The heaviest common liquid is mercury, and the lightest gas is hydrogen.

If we compare their densities at 20°C and 1 atm:

Mercury: $\rho = 13,580$ kg/m^3

Hydrogen: $\rho = 0.0838$ kg/m^3

They differ by a factor of 162,000.

3.3.4 Specific weight

The *specific weight* of a fluid, denoted by γ (lowercase Greek gamma), is its weight per unit volume. Just as a mass has a weight $W = mg$, density and specific weight are simply related by gravity:

$$\gamma = \rho/g$$

The units of γ are weight per unit volume, in N/m^3. In standard earth gravity, $g = 9.807$ m/s^2. Thus, e.g., the specific weights of air and water at 20°C and 1 atm are approximately

$$\gamma_{air} = (1.205 \text{ kg/m}^3) \times (9.807 \text{ m/s}^2) = 11.8 \text{ N/m}^3$$

$$\gamma_{water} = (998 \text{ kg/m}^3) \times (9.807 \text{ m/s}^2) = 9790 \text{ N/m}^3$$

Specific weight is very useful in the hydrostatic-pressure applications.

3.3.5 Specific gravity

The ratio of the density of a fluid ρ to the density of water ρ_w (for liquid) and air (for gases), is called the specific gravity, which is expressed by the symbol s:

$$s = \rho_{liquid}/ \rho_w = \rho_{liquid}/1.205 \text{ kg/m}^3$$

$$s = \rho_{gas}/ \rho_{air} = \rho_{gas}/998 \text{ kg/m}^3$$

For example,

the specific gravity of mercury (Hg) is $s_{Hg} = 13{,}580/998 \approx 13.6$.

3.3.6 Specific volume

The reciprocal of density, i.e. the volume per unit mass, is called the specific volume, which is generally expressed by the symbol v,

$$v = 1/ \rho \text{ m}^3/\text{kg}$$

3.3.7 Viscosity

When a fluid is sheared, it begins to move at a strain rate inversely proportional to a property called its coefficient of viscosity μ. Consider a fluid element sheared in one plane by a single shear stress, as in Fig. 3.5.

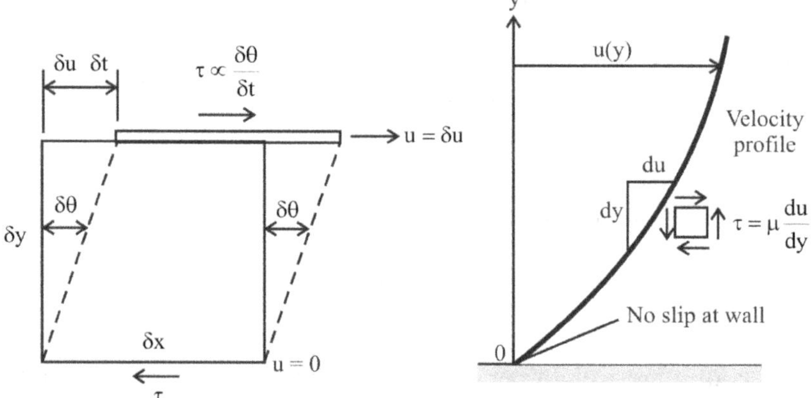

Fig. 3.5. Shear strain in fluid

The shear strain angle $\delta\theta$ will continuously grow with time as long as the stress τ is maintained, the upper surface moving at speed δu larger than the lower. Such common fluids as water, oil, and air show a linear relation between applied shear and resulting strain rate

$$\tau \propto \frac{\delta\theta}{\delta t} \qquad\qquad .....(3.1)$$

From the Fig. 3.5,

$$\tan\delta\theta = \frac{\delta u\,\delta t}{\delta y} \qquad\qquad .....(3.2)$$

Taking limit in infinitesimal changes, the above equation becomes a relation between shear strain rate and velocity gradient

$$\frac{d\theta}{dt} = \frac{du}{dy} \qquad\qquad .....(3.3)$$

From Eq. 3.1 and 3.2, the applied shear is also proportional to the velocity gradient for the common linear fluids.

$$\tau \propto \frac{du}{dy} \qquad\qquad .....(3.4)$$

If the constant of proportionality is the viscosity coefficient (μ).

$$\tau = \mu\frac{du}{dy} \qquad\qquad .....(3.5)$$

From Eq. 3.5,

$$\mu = \frac{\tau}{\left(\dfrac{du}{dy}\right)} \qquad\qquad .....(3.6)$$

Thus viscosity is defined as the shear stress required to produce unit rate of shear strain.

3.3.8 Surface tension

Surface tension is the property of the surface of a liquid that allows it to resisit an external force. It is defined as the force along a line of unit length. It is measured in force per unit length (N/m).

3.4 Classifications of Fluids

- Newtonian fluid
- Non Newtonian fluid

- Plastic Fluid
- Thixotropic fluids
- Ideal fluid
- Real fluid
- Ideal plastic fluid

3.4.1 Newtonian fluids

A fluid which obeys Newton's law of viscosity ($\tau = \mu \dfrac{du}{dy}$) is called Newtonian fluid. In such fluids shear stress varies directly as shear strain. In this case the stress strain curve is a straight line passing through origin. The slope of the line gives dynamic viscosity of the fluid (Fig. 3.6).

Eg: Water, Kerosene.

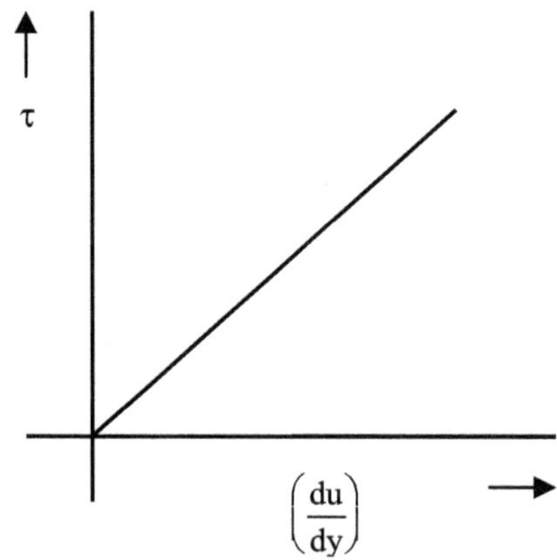

Fig. 3.6 Newtonian fluids

3.4.2 Non-Newtonian fluid

Fluids which do not follow the linear law of Eq. (3.6) are called *nonnewtonian fluid.* or

A fluid which does not obey Newton's law of viscosity is called non-Newton fluid as shown in Fig. 3.7. For such fluids,

$$\tau = \mu \left(\frac{du}{dy}\right)^{n}$$

where n is a index that may vary from zero to greater than unity.

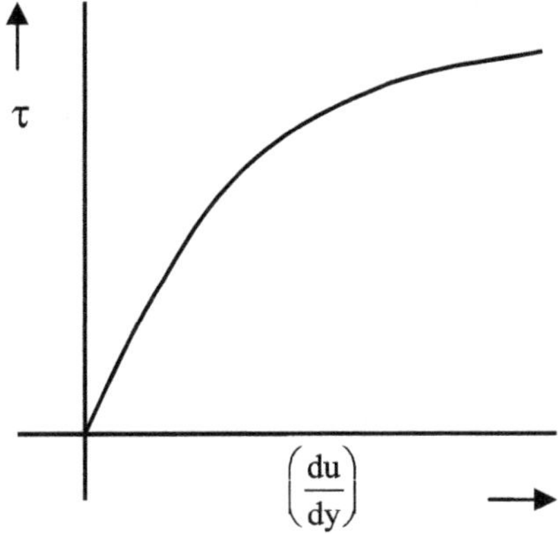

Fig. 3.7 Non-newtonian fluid

3.4.3 Plastic fluids

In this case the strain starts after certain initial stress (τ_0) and then the stress-strain relationship will be linear (Fig. 3.8). τ_0 is called initial yield stress. Sometimes they are also called Bingham's Plastics.

Eg: Industrial sludge.

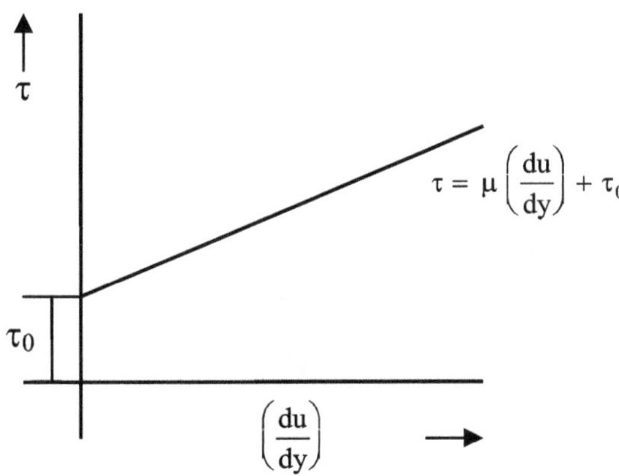

Fig. 3.8 Plastic fluids

3.4.4 Thixotropic fluids

These require certain amount of yield stress to initiate shear strain. After wards stress-strain relationship will be non-linear (Fig. 3.9).

Eg: Printers ink.

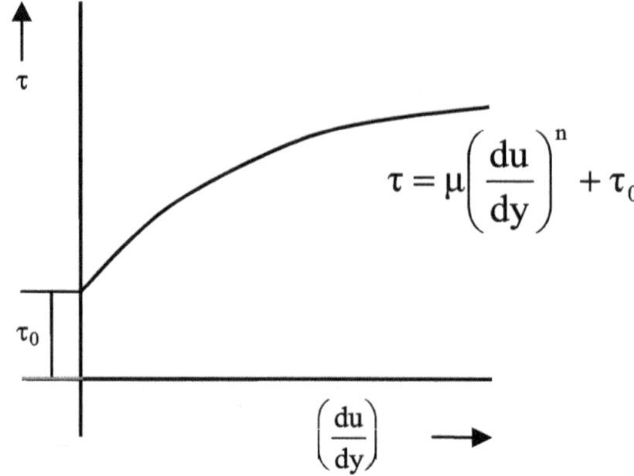

Fig. 3.9 Thixotropic fluids

3.4.5 Ideal fluid

Any fluid for which viscosity is assumed to be zero is called Ideal fluid (Fig. 3.10). For ideal fluid $\tau = 0$ for all values of $\dfrac{du}{dy}$

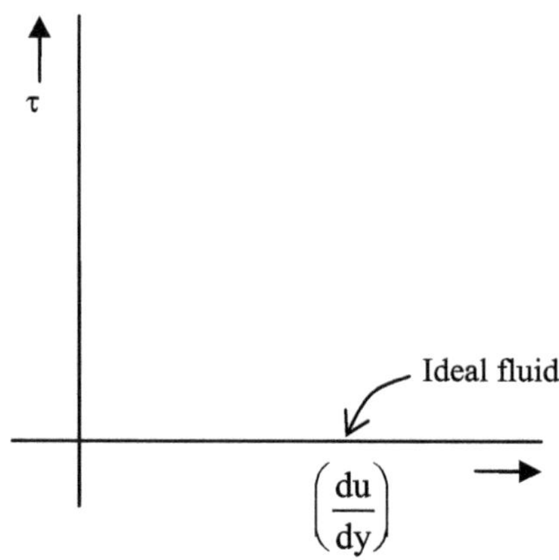

Fig. 3.10 Ideal fluid

3.4.6 Real fluid

Any fluid which posses certain viscosity is called real fluid. It can be Newtonian or non-Newtonian, Thixotropic or Ideal plastic (Fig. 3.11).

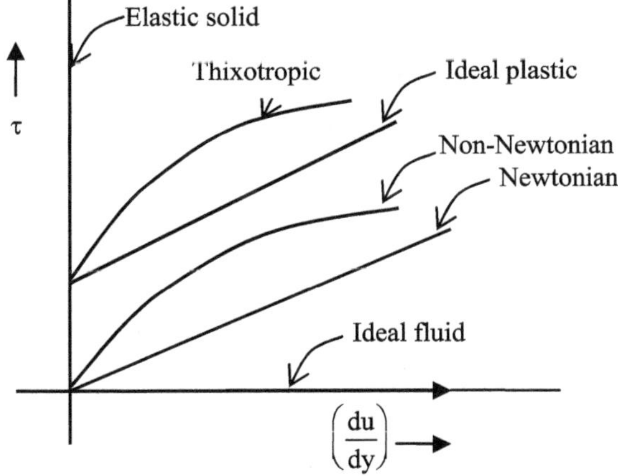

$$\left(\frac{du}{dy}\right) \longrightarrow$$

Fig. 3.11 Real fluid

3.4.7 Real versus ideal fluids

A fluid which has viscosity is a real fluid. All fluids in practice is *real fluids*. A fluid which is assumed to have no viscosity – which is not real, does not exist – known as the *ideal fluid*. This is a useful concept when theoretical solutions are considered and it does help to achieve some practically useful solutions.

3.4.8 Newtonian versus non-newtonian fluids

Even among fluids which are accepted as fluids there can be wide differences in behaviour under stress. Fluids obeying Newton's law where the value of μ is constant, known as *Newtonian fluids*. If μ is constant the shear stress is linearly dependent on velocity gradient. This is true for most common fluids. Fluids in which the value of μ is not constant are known as *Non-Newtonian fluids*. There are several categories of these, and they are outlined briefly below. These categories are based on the relationship between shear stress and the velocity gradient (rate of shear strain) in the fluid. These relationships can be seen in the Fig. 3.11.

Shear stress vs. Rate of shear strain du/dy

Each of these lines can be represented by the equation

$$\tau = A + B\left(\frac{\delta u}{\delta y}\right)^n$$

where,

A, B and n are constants. For Newtonian fluids $A = 0$, $B = \mu$ and $n = 1$.

Below is brief description of the physical properties of the several categories:

Plastic: Shear stress must reach a certain minimum before flow commences.

Bingham plastic: As with the plastic above a minimum shear stress must be achieved. With this classification $n = 1$. An example is sewage sludge.

Pseudo-plastic: No minimum shear stress necessary and the viscosity decreases with rate of shear, e.g. colloidial substances like clay, milk and cement.

Dilatant substances; Viscosity increases with rate of shear e.g. quicksand.

Thixotropic substances: Viscosity decreases with length of time shear force is applied e.g. thixotropic jelly paints.

Rheopectic substances: Viscosity increases with length of time shear force is applied

Viscoelastic materials: Similar to Newtonian but if there is a sudden large change in shear they behave like plastic.

3.5   Pressure Variation with Depth

One might guess that the deeper we go into a liquid or gas, the greater the pressure on us from the surrounding fluid will be. The reason for the increased pressure is that the deeper into a fluid we go, the more fluid, and thus the more weight, we have over top of us. We can calculate the variation of pressure with depth by considering a volume of fluid of height h and cross-sectional area A as shown in Fig. 3.12.

h = height of volume of fluid

w = weight of volume of fluid

A = cross-sectional area

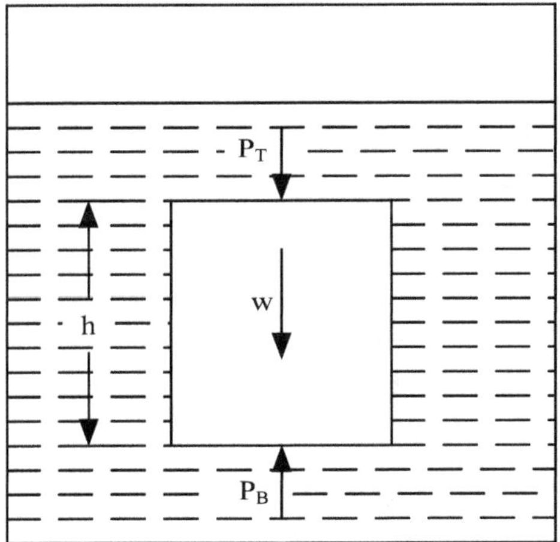

Fig. 3.12 Variation of pressure with depth

If this volume of fluid is to be in equilibrium, the net force acting on the volume must be zero. There are three external forces acting on this volume of fluid. These forces are:

1. The force $P_T A$ due to the pressure on top of the volume of fluid. If the fluid is open to the air, $P_T = P_O = 1.01 \times 10^5$ Pa, which is atmospheric pressure.

2. The weight of the volume of fluid, $w = M\,g$. Remembering the definition of density, $\rho = M/V$, and that the volume of the fluid may be calculated as $V = A\,h$, we can write the weight of the fluid as $w = \rho\,g\,h\,A$.

3. The force pushing up on the bottom of the volume of fluid, $P_B A$, due to the fluid below the volume under consideration.

If we take the up direction to be positive and add the forces.

we get,

$$P_B\,A - \rho\,g\,h\,A - P_T\,A = 0,$$

which gives,

$$P_B = P_T + \rho\,g\,h.$$

This provides the general formula relating the pressures at two different points in a fluid separated by a depth h.

hydrostatic law of pressure: To study the variation of intensity of pressure in a static mass of fluid

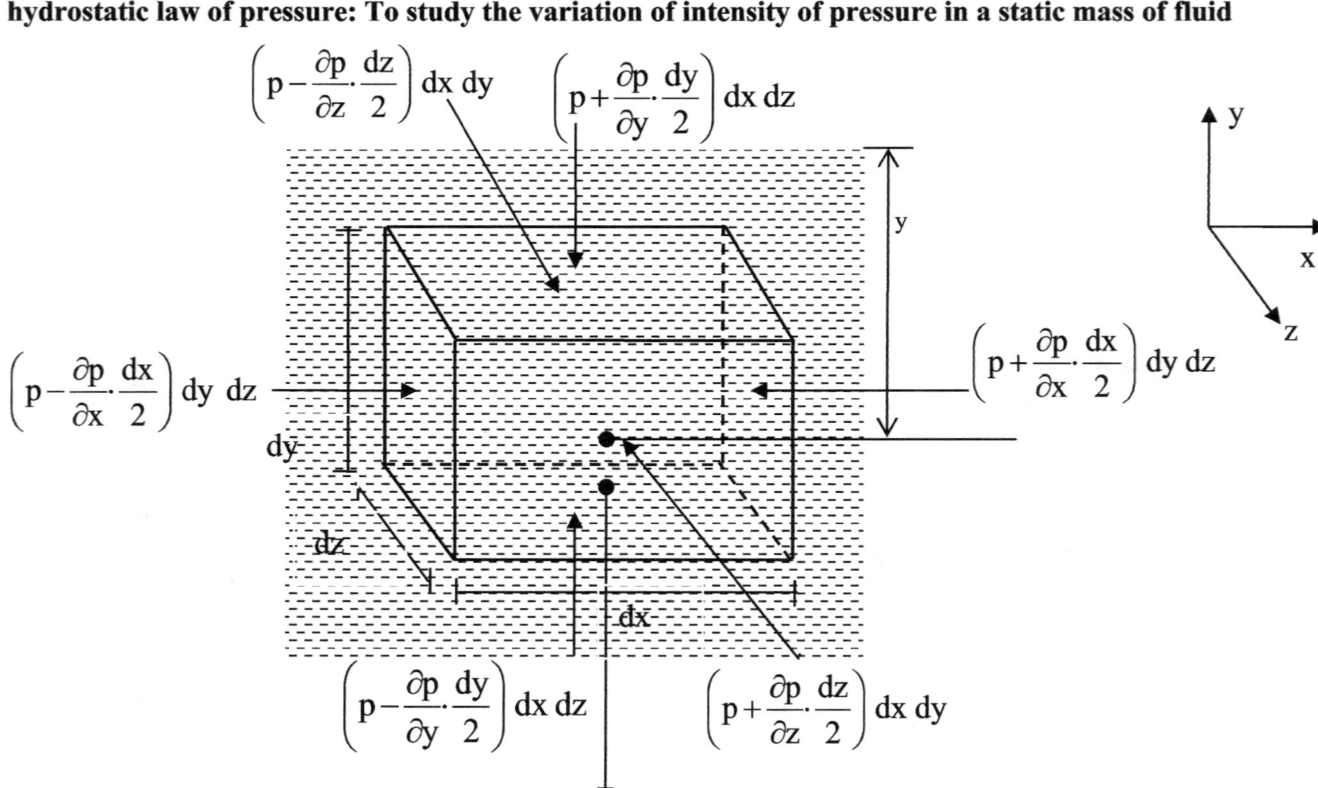

Fig. 3.13 Variation of fluid pressure

Let us consider a point 'M' at a depth 'y' below the free surface of the liquid of specific weight 'γ'. (dx dy dz) is the elemental volume of the fluid considered around the point 'M'.

Fig. 3.13 shows forces acting on the element including self weight. The element of fluid is in equilibrium and hence we can apply conditions of equilibrium.

$$\Sigma\,Fx = 0$$

$$+\left[p - \frac{\partial p}{\partial x}\cdot\frac{dx}{2}\right]dy\ dz - \left[P + \frac{\partial p}{\partial x}\cdot\frac{dx}{2}\right]dy\ dz = 0$$

i.e $\qquad p - \dfrac{\partial p}{\partial x}\cdot\dfrac{dx}{2} - p - \dfrac{\partial p}{\partial x}\cdot\dfrac{dx}{2} = 0$.

$$-2\cdot\frac{\partial P}{\partial x}\cdot\frac{dx}{2} = 0$$

$$\therefore\ \frac{\partial P}{\partial x} = 0$$

$\therefore$ Rate of change of intensity of pressure along x-direction is zero. In other words there is no change in intensity of pressure along x-direction inside the fluid.

$$\Sigma\,Fz = 0$$

$$+\left[p - \frac{\partial p}{\partial z}\cdot\frac{dz}{2}\right]dxdy - \left[p + \frac{\partial p}{\partial z}\cdot\frac{dz}{2} = 0\right]dxdy = 0$$

$$\therefore\ \frac{\partial p}{\partial z} = 0$$

$\therefore$ The Rate of change of intensity of pressure along z direction is zero. In other words there is no change in intensity of pressure along z-direction.

$$\Sigma\,Fy = 0$$

$$+\left[p - \frac{\partial p}{\partial y}\cdot\frac{dy}{2}\right]dx\ dz - \left[p + \frac{\partial p}{\partial y}\cdot\frac{dy}{2}\right]dx\ dz - \gamma\,dx\,dy\,dz = 0$$

i.e
$$p - \frac{\partial p}{\partial y} \cdot \frac{dy}{2} - p - \frac{\partial P}{\partial y} \cdot \frac{dy}{2} = \gamma \, dy$$

i.e
$$-\frac{\partial p}{\partial y} dy = \gamma \, dy$$

$$\therefore \frac{\partial p}{\partial y} = -\gamma$$

Negative sign indicates that the pressure increases in the downward direction i.e., as the depth below the surface increases intensity of pressure increases.

$$\therefore \frac{\partial p}{\partial y} = \gamma$$

$$\therefore \partial p = \gamma . \partial y$$

Integrating

$$p = \gamma \, y + C$$

at
$$y = 0; \, p = p_{atm}$$

$$p_{atm} = \gamma \times 0 + C$$

$$\therefore p_{atm} = C$$

$$p = \gamma \times y + p_{atm}$$

The above equation is called hydrostatic law of pressure.

3.6   Static and Kinetic Energy in the Fluid

The energy in the flow is composed of several energies. The *kinetic* energy arises because of the directed motion of the fluid; the *pressure* energy is due to the random motion within the fluid; and the kinetic energy of a system arises as the sum of the motions of all the system's particles, whether it is the motion of atoms, molecules, atomic nuclei, electrons, or other particles. The potential energy includes all energies given by the mass of particles, by the chemical composition, i.e., the chemical energy stored in chemical bonds having the potential to undergo chemical reactions.

3.7   Bernoulli's Equation for Incompressible Fluids

In fluid dynamics, **Bernoulli's principle** states that for an inviscid flow, an increase in the speed of the fluid occurs simultaneously with a decrease in pressure or a decrease in the fluid's potential energy.

Bernoulli's principle can be applied to various types of fluid flow; in fact, there are different forms of the Bernoulli equation for different types of flow. The simple form of Bernoulli's principle is valid for incompressible flows (e.g., most liquid flows) and also for compressible flows (e.g., gases). Bernoulli's principle can be derived from the principle of conservation of energy. This states that in a steady flow the sum of all forms of mechanical energy in a fluid along a streamline is the same at all points on that streamline provided that the sum of kinetic energy and potential energy remain constant. If the fluid is flowing out of a reservoir the sum of all forms of energy is the same on all streamlines because in a reservoir the energy per unit mass (the sum of pressure and gravitational potential $\rho\, g\, h$) is the same everywhere.

The Bernoulli equation for incompressible fluids can be derived by integrating the Euler equations, or applying the law of conservation of energy in two sections along a streamline, ignoring viscosity, compressibility, and thermal effects (Fig. 3.14).

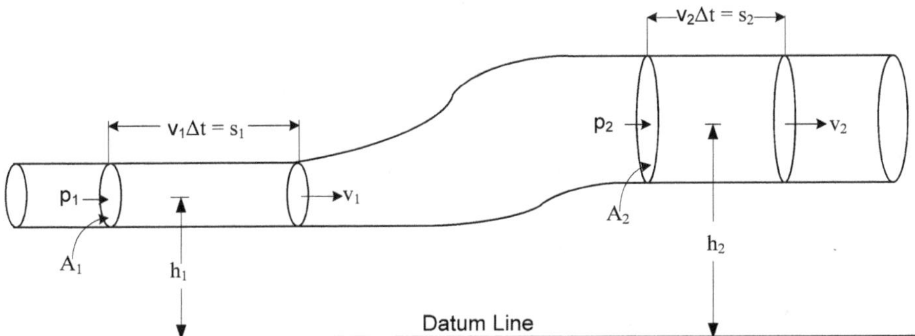

Fig. 3.14 Bernoulli's principle

The simplest derivation is to first ignore gravity and consider constrictions and expansions in pipes.

Let the x axis be directed down the axis of the pipe. The equation of motion for a parcel of fluid on the axis of the pipe is

$$m\frac{dv}{dt} = -F$$

$$\rho\, A\, dx\,\frac{dv}{dt} = -A\, dp$$

$$\rho\frac{dv}{dt} = -\frac{dp}{dx}$$

In steady flow, $v = v(x)$ so

$$\frac{dv}{dt} = \frac{dv}{dx}\frac{dx}{dt} = \frac{dv}{dx}\,v = \frac{d}{dx}\frac{v^2}{2}$$

With ρ constant, the equation of motion can be written as

$$\frac{d}{dx}\left(\rho\frac{v^2}{2} + p\right) = 0$$

OR

$$\frac{v^2}{2} + \frac{p}{\rho} = C$$

where C is a constant, sometimes referred to as the Bernoulli constant. We deduce that where the speed is large, pressure is low. In the above derivation, no external work-energy principle is invoked. Rather, the work-energy principle was inherently derived by a simple manipulation of the momentum equation. The derivation that follows includes gravity and applies to a curved trajectory, but a work-energy principle must be assumed.

Applying conservation of energy in form of the work-kinetic energy theorem we find that:

The change in KE of the system equals the net work done on the system;

$$W = \Delta KE$$

Therefore,

The work done by the forces in the fluid + decrease in potential energy = increase in kinetic energy.

The work done by the forces is

$$F_1 s_1 - F_2 s_2 = p_1 A_1 v_1 \Delta t - p_2 A_2 v_2 \Delta t$$

The decrease of potential energy is

$$m g h_1 - m g h_2 = \rho g A_1 v_1 \Delta t\, h_1 - \rho g A_2 v_2 \Delta t\, h_2$$

The increase in kinetic energy is

$$\frac{1}{2} m v_2^2 - \frac{1}{2} m v_1^2 = \frac{1}{2} \rho A_2 v_2 \Delta t\, v_2^2 - \frac{1}{2} \rho A_1 v_1 \Delta t\, v_1^2$$

Putting these together,

$$p_1 A_1 v_1 \Delta t - p_2 A_2 v_2 \Delta t + \rho g A_1 v_1 \Delta t\, h_1 - \rho g A_2 v_2 \Delta t\, h_2$$
$$= \frac{1}{2} \rho A_2 v_2 \Delta t\, v_2^2 - \frac{1}{2} \rho A_1 v_1 \Delta t\, v_1^2$$

OR

$$\frac{\rho A_1 v_1 \Delta t\, v_1^2}{2} + \rho g A_1 v_1 \Delta t\, h_1 + p_1 A_1 v_1 \Delta t$$
$$= \frac{\rho A_2 v_2 \Delta t\, v_2^2}{2} + \rho g A_2 v_2 \Delta t\, h_2 + p_2 A_2 v_2 \Delta t$$

After dividing by Δt, ρ and $A_1 v_1$ ($=$ rate of fluid flow $= A_2 v_2$ as the fluid is incompressible):

$$\frac{v_1^2}{2} + g h_1 + \frac{p_1}{\rho} = \frac{v_2^2}{2} + g h_2 + \frac{p_2}{\rho}$$

or, as stated in the first paragraph:

$$\frac{v^2}{2} + g h + \frac{p}{\rho} = C$$

Further division by g implies

$$\frac{v^2}{2g} + h + \frac{p}{\rho g} = C$$

A free falling mass from a height h (in vacuum), will reach a velocity

$$v = \sqrt{2 g h} \quad \text{OR} \quad h = \frac{v^2}{2g}$$

The term $v^2/2g$ is called the *velocity head.*

The hydrostatic pressure or *static head* is defined as

$$p = \rho g h, \text{ or } h = \frac{p}{\rho g}$$

The term $p/\rho g$ is also called the *pressure head.*

A way to see how this relates to conservation of energy directly is to multiply by density and by unit volume (which is allowed since both are constant) yielding:

$$v^2 \rho + P = \text{Constant} \qquad \text{and}$$

$$mV^2 + P.\text{volume} = \text{Constant}$$

3.7.1   Incompressible flow equation

In most flows of liquids, and of gases at low Mach number, the mass density of a fluid parcel can be considered to be constant, regardless of pressure variations in the flow. For this reason the fluid in such flows can be considered to be incompressible and these flows can be described as incompressible flow. Bernoulli performed his experiments on liquids and his equation in its original form is valid only for incompressible flow. A common form of Bernoulli's equation, valid at any arbitrary point along a streamline where gravity is constant, is:

$$\frac{v^2}{2} + g h + \frac{p}{\rho} = \text{Constant}$$

where:

v is the fluid flow speed at a point on a streamline,

g is the acceleration due to gravity,

h is the elevation of the point above a reference plane, with the positive z-direction pointing upward so in the direction opposite to the gravitational acceleration,

P is the pressure at the point, and

ρ is the density of the fluid at all points in the fluid.

3.8   Classification of Fluid Flow

3.8.1 Compressible vs Incompressible flow

All fluids are compressible to some extent, changes in pressure or temperature will result in changes in density. However, in many situations the changes in pressure and temperature are sufficiently small that the changes in density are negligible. In this case the flow can be modeled as an incompressible flow. Otherwise the more general compressible flow equations must be used.

3.8.2   Viscous vs inviscid flow

Viscous flow is those in which fluid friction has significant effects on the fluid motion.

The Reynolds number, which is a ratio between inertial and viscous forces, can be used to evaluate whether viscous or inviscid equations are appropriate to the problem.

On the contrary, high Reynolds numbers indicate that the inertial forces are more significant than the viscous (friction) forces. Therefore, we may assume the flow to be an inviscid flow (no viscosity). An approximation in which we neglect viscosity completely, compared to inertial terms.

3.8.3   Steady vs unsteady flow

When all the time derivatives of a flow field vanish, the flow is considered to be a steady flow. Steady-state flow refers to the condition where the fluid properties at a point in the system do not change over time. Otherwise, flow is called unsteady.

3.8.4   Laminar Vs turbulent flow

Laminar flow or viscous flow, sometimes known as streamline flow, occurs when a fluid flows in parallel layers, with no disruption between the layers. In such a flow the viscous forces are more predominant compared to inertia Forces. Stream lines are practically parallel to each other or flow takes place in the form of telescopic tubes. This type of flow occurs when Reynolds's number Re < 2000. In laminar flow velocity increases gradually from zero at the boundary to Maximum at the center. In nonscientific terms laminar flow is "smooth," while turbulent flow is "rough." Laminar flow is regular and smooth and velocity at any point practically remains constant in magnitude & direction. Therefore, the flow is also known as stream line flow. There will be no exchange of fluid particles from one layer to another. Thus there will be no momentum transmission from one layer to another.

Ex: Flow of thick oil in narrow tubes, flow of Ground Water, Flow of Blood in blood vessels (Fig. 3.15).

Turbulent flow is characterized by recirculation, eddies, and apparent randomness. It should be noted, however, that the presence of eddies or recirculation alone does not necessarily indicate turbulent flow, these phenomena may be present in laminar flow as well. Mathematically, turbulent flow is often represented via a Reynolds decomposition, in which the flow is broken down into the sum of an average component and a perturbation component (Fig. 3.15).

3.8.5 Transition flow

In such a type of flow the stream lines get disturbed a little. This type of flow occurs when 2000 < Re < 4000 and as can be shown in Fig. 3.15.

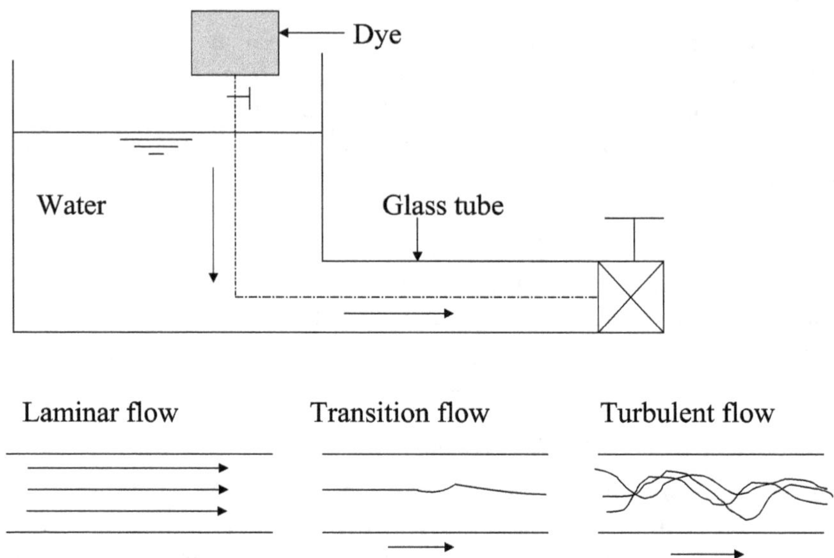

Fig. 3.15 Laminar, Transition and Turbulent Flow

3.9 Fluid Couplings

Fluid couplings work on the hydrodynamic principle. As we can seen in Fig. 3.16, it consists of a pump generally known as impeller and a turbine generally known as rotor, both enclosed suitably in a casing. The impeller and rotor is bowl shaped and have large number of radial vanes. They face each other with an air gap. The impeller is suitably connected to the prime mover while the rotor has a shaft bolted to it. This shaft is further connected to the driven machine through the suitable arrangement.

Oil is filled in the fluid coupling from the filling plug provided on its body. A fusible plug is provided on the fluid coupling which blows off and drains out oil from the coupling in case of sustained overloading.

There is no mechanical internal connection between the impeller and the rotor (i.e. driving and driven units) and the power is transmitted by virtue of the fluid filled in the fluid coupling. The impeller when rotated by the prime mover imparts the velocity and energy to the fluid which is converted into mechanical energy in the rotor thus rotating it. The fluid follows a closed circuit of flow from the impeller to rotor through the air gap at the outer periphery and from the rotor to impeller again through the air gap at the inner periphery. To enable the fluid to flow from the impeller to rotor it is essential that there is a difference in "Head" between the two and thus it is essential that there is a difference in RPM known as Slip, between the two.

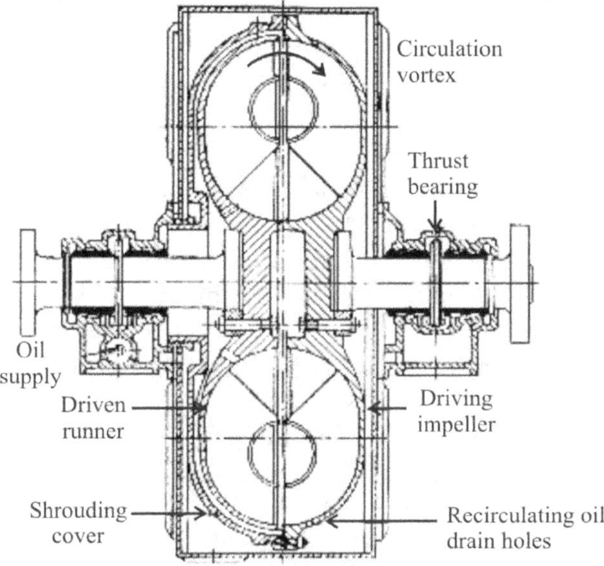

Fig. 3.16 Fluid couplings.

Slip is an important and inherent characteristic of a fluid coupling resulting in several desired advantage. As the slip increases, more and more fluid can be transferred from the impeller to the rotor and more torque can be transmitted. However when the rotor is at standstill, maximum fluid is transmitted from the impeller to rotor and maximum torque is transmitted from the coupling. This maximum torque is limiting torque. The fluid coupling also acts a torque limiter.

Fluid couplings have centrifugal characteristics during starting, thus enabling no-load start-up of prime mover which is of great importance.

The slipping characteristic of fluid coupling provides a wide range if choice of power transmission characteristics which also results in speed variation, smooth and controlled acceleration, clutching and declutching operations and other characteristics of load limiting, shock loads and peak load absorption and dampening etc.

3.9.1   Advantages of using fluid couplings

Following are the advantages of using fluid coupling:

   (i)   Wear-free power transmission because of absence of mechanical connection between the input and output elements.

   (ii)   No-load start-up of motor, irrespective of machine load and motor peak torque utilized for machine acceleration.

   (iii)   Motor selection to operating HP rather than starting loads. Saving on energy cost and capital cost.

   (iv)   Simple control of maximum or limiting torque by easy variation of oil filling.

   (v)   Added protection for motor and driven machine by limiting torque to a predetermined safe value

   (vi)   Automatic unloading of prime mover in case of any sustained overload, by blowing-off of fusible plug on coupling, thus draining out oil.

   (vii)   Automatic synchronization and load sharing in case of multi-motor drives.

   (viii)   Effective dampening of shocks, load fluctuations and torsional vibrations. Helpful in improving the drive duty conditions in machines subjected to shock loads, frequent load changes, startups and reversals. Reduced fatigue on transmission components/elements.

   (ix)   Smooth and controlled acceleration of driven machines. Important for machines such as conveyors, wire drawing and textile machines.

   (x)   Steeples speed variation in wide range obtained in scoop type fluid couplings.

   (xi)   Economy on machine and drive designs.

   (xii)   Energy saving.

3.10   Pump

A pump is a machine which converts mechanical energy to fluid energy, the fluid being incompressible. This action is opposite to that in hydraulic turbines. In pumps, the movement of blade system moves the fluid, which is always in contact with blade thereby converting mechanical energy of blade system to kinetic energy. For perfect conversion, the moving blade should be in contact with the fluid at all places. In other words, the moving blade system should be completely immersed in fluid.

3.10.1 Classification of Pumps

Pumps are classified in different ways. One classification is according to the type as positive displacement pumps and rotodynamic pumps. Fig. 3.17 gives the clear understanding regarding the classification of pumps.

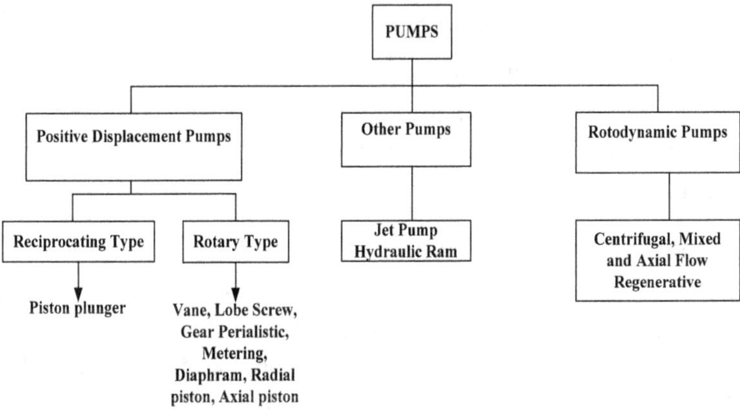

Fig. 3.17 Classification of pumps

3.10.2 Centrifugal pump

Centrifugal pump is a rotodynamic pump that uses a rotating impeller to increase the pressure of a fluid. Centrifugal pumps are commonly used to move liquids through a piping system. The fluid enters the pump impeller along or near to the rotating axis and is accelerated by the impeller, flowing radially outward into a diffuser or casing, from where it exits into the downstream piping system (Fig. 3.18). Centrifugal pumps are used for large discharge through smaller heads.

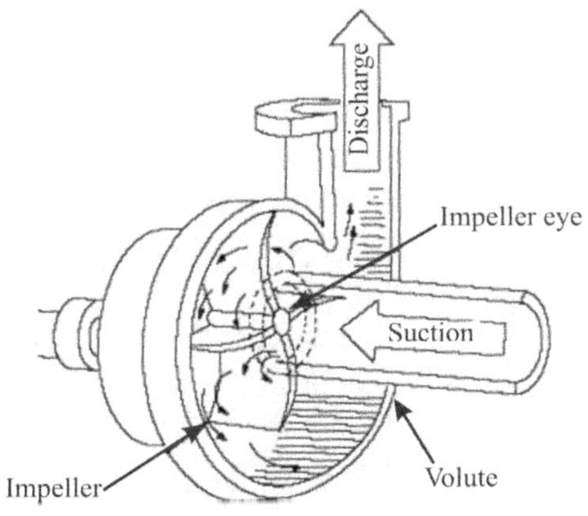

Fig. 3.18 Centrifugal pump

There are floowing problems occurs during using of centrifugal pumps.

(i)   Cavitations (low pressure/high vacuum condition) the NPSH (net positive suction head) of the system is too low for the selected pump.

(ii)   Wear of the Impeller can be worsened by suspended solids.

(iii) Corrosion inside the pump caused by the fluid properties.

(iv)   Overheating due to low flow.

(v)   Leakage along rotating shaft.

(vi)   Lack of prime – centrifugal pumps must be filled (with the fluid to be pumped) in order to operate.

(vii) Surge (Jerk).

3.10.3   Reciprocating Pump

The liquid enters a pumping chamber via a suction valve and is pushed out via a delivery valve by the action of the piston or diaphragm. Reciprocating pumps are generally very efficient and are suitable for very high heads at low flows. This type of pump is self priming as it can draw liquid from a level below the suction flange even if the suction pipe is not evacuated. The pump delivers reliable discharge flows and is often used for metering duties delivering accurate quantities of fluid. Fig.3.19 shows the schematic view of different parts of reciprocating pump.

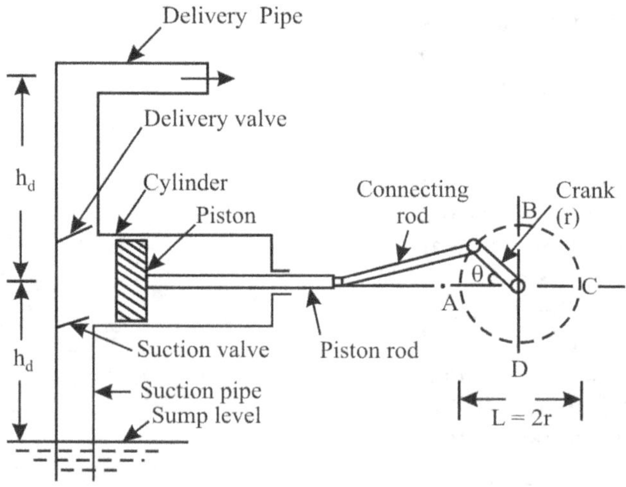

Fig. 3.19 Single acting reciprocating pump

Comparison between Centrifugal pumps and Reciprocating Pumps

Centrifugal pumps	Reciprocating Pumps
The discharge is continuous and smooth.	The discharging is fluctuating and pulsating.
It can handle large quantity of liquid.	Handles small quantity of liquid.
It is used for large discharge through small heads	It is meant for small discharge at high heads.
Cost of centrifugal pump is less as compared to reciprocating pump.	Cost of reciprocating pump is approximately four times the centrifugal pump.
Runs at high speeds.	Runs at low speed.
Efficiency is high.	Efficiency is low.
Needs smaller area and cost of installation is less.	Needs large floor area and installation cost is high.
Low maintenance cost.	High maintenance cost.
It can be used for lifting highly viscous liquids.	Used only for lifting pure water or less viscous fluids.

3.11   Air Compressor

Air Compressor is one of the major energy consuming utility in any industrial operation and is mainly classified in two types. The classification is based on their construction and operation features.

The two main classification of Air Compressor are.

(i)   Positive Displacement Type

(ii)   Dynamic Type

3.11.1   Positive displacement type air compressor

Positive displacement type compressors are those which mechanically displace a fixed volume of air into a reduced volume. The constant volume can be delivered by compressor when operated at fixed speed. The discharge pressure is determined by the system load conditions. If the consumption of air is more, then discharge pressure falls below the cut-off pressure and compressor works on load condition.

These types of Air Compressor further classified as:

(a)   Reciprocating Air Compressor

(b)   Rotary Air Compressor

(a) *Reciprocating Air Compressor:* In this type, compressed air is generated by the to & fro movement of piston inside the cylinder (compression chamber). Each movement compresses a fixed quantity of free air at a specific pressure. Single stage or double stage or multiple stages and single acting or double acting is the further classification of Reciprocating Air Compressor. A Vertical type Air Compressor is used for the application of compressed air ranging in between 50 to 150 cfm and the Horizontal balance type is most suitable for application ranging from 200 to 5000 cfm.

The reciprocating air compressor, illustrated in Fig. 3.20, is the most common design employed today.

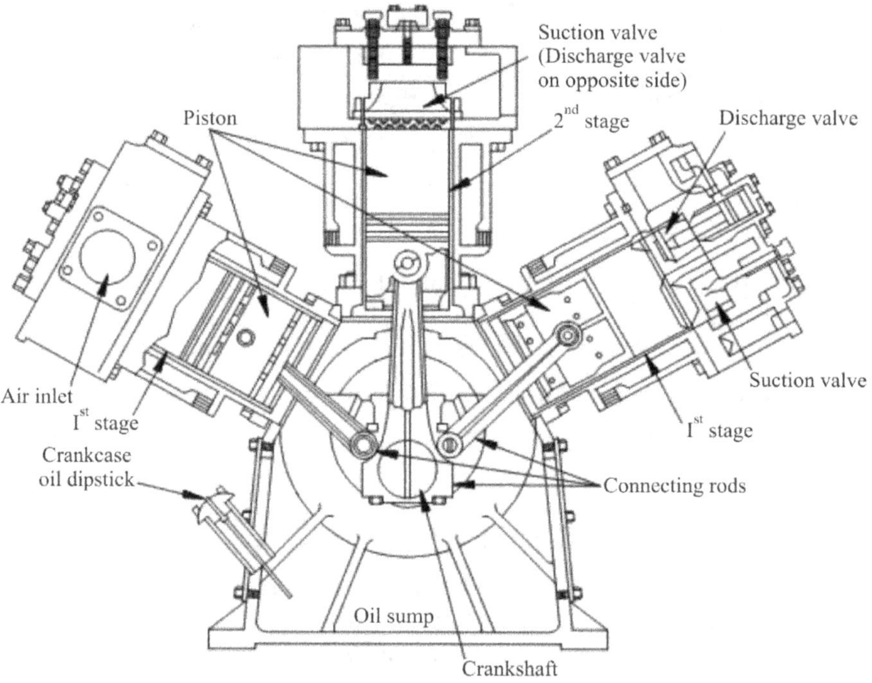

Fig. 3.20 Reciprocating air compressor

The reciprocating compressor normally consists of the following elements.

The compressing element, consisting of air cylinders, heads and pistons, and air inlet and discharge valves.

A system of connecting rods, piston rods, crossheads, and a crankshaft and flywheel for transmitting the power developed by the driving unit to the air cylinder piston.

A self-contained lubricating system for bearings, gears, and cylinder walls, including a reservoir or sump for the lubricating oil, and a pump, or other means of delivering oil to the various parts. On some compressors a separate force-fed lubricator is installed to supply oil to the compressor cylinders.

A regulation or control system designed to maintain the pressure in the discharge line and air receiver (storage tank) within a predetermined range of pressure.

An unloading system, which operates in conjunction with the regulator, to reduce or eliminate the load put on the prime mover when starting the unit.

(b) *Rotary Air Compressor:* Air is compressed between two rotating screw (male & female screw) and casing. The compression took place in gradually reducing space between these two parts.

Air is compressed between casings & rotating rotor by the help of moveable rotor blade. This type of Air Compressor can be further classified as Single Stage or Double Stage Rotary Air Compressor.

The rotary compressor is adaptable to direct drive by induction motors or multicylinder gasoline or diesel engines. The units are compact, relatively inexpensive, and require a minimum of operating attention and maintenance. They occupy a fraction of the space and weight of a reciprocating machine of equivalent capacity. Rotary compressor units are classified into three general groups, slide vane-type, lobe-type, and liquid seal ring-type.

The rotary slide vane-type, as illustrated in Fig.3.21, has longitudinal vanes, sliding radially in a slotted rotor mounted eccentrically in a cylinder. The centrifugal force carries the sliding vanes against the cylindrical case with the vanes forming a number of individual longitudinal cells in the eccentric annulus between the case and rotor. The suction port is located where the longitudinal cells are largest. The size of each cell is reduced by the eccentricity of the rotor as the vanes approach the discharge port, thus compressing the air.

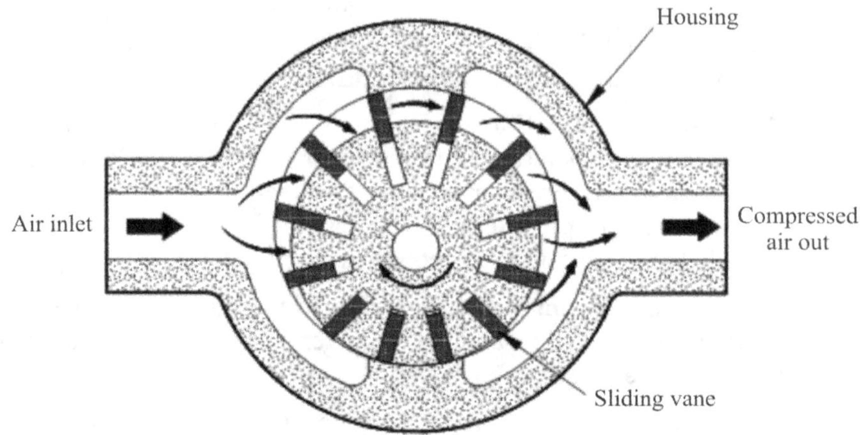

Fig. 3.21 Rotary slide vane air compressor

3.11.2 Dynamic type air compressor

The Dynamic Type Air Compressors mechanically impart velocity to the air. This action is produced by the Impellers rotating at a high speed, in enclosed casing. The air is forced into a progressively reduced volume. The volumetric flow will vary inversely with the differential pressure across the compressor.

This type of compressor is further classified as

   (a) Centrifugal Air Compressor

   (b) Axial Flow Air Compressor

   (c) Roots Blower Compressor

   (a) *Centrifugal Air Compressor:* The centrifugal compressor, originally built to handle only large volumes of low pressure gas and air (maximum of 40 psig), has been developed to enable it to move large volumes of gas with discharge pressures up to 3,500 psig. However, centrifugal compressors are now most frequently used for medium volume and medium pressure air delivery. One advantage of a centrifugal pump is the smooth discharge of the compressed air.

The centrifugal force utilized by the centrifugal compressor is the same force utilized by the centrifugal pump. The air particles enter the eye of the impeller, designated D in Fig. 3.22. As the impeller rotates, air is thrown against the casing of the compressor. The air becomes compressed as more and more air is thrown out to the casing by the impeller blades. The air is pushed along the path designated A, B, and C in Fig. 3.22. The pressure of the air is increased as it is pushed along this path.

(b) *Axial Flow Air Compressor:* The Axial Flow type Air Compressor is essentially a large capacity, high speed machine with characteristics quite different from the centrifugal air compressor. Each stage consist of two rows of blades, one row rotating & the next row of blade is stationery. The rotor blades impart velocity and pressure to the air as the motor turns, the velocity being converted to pressure in stationery blades.

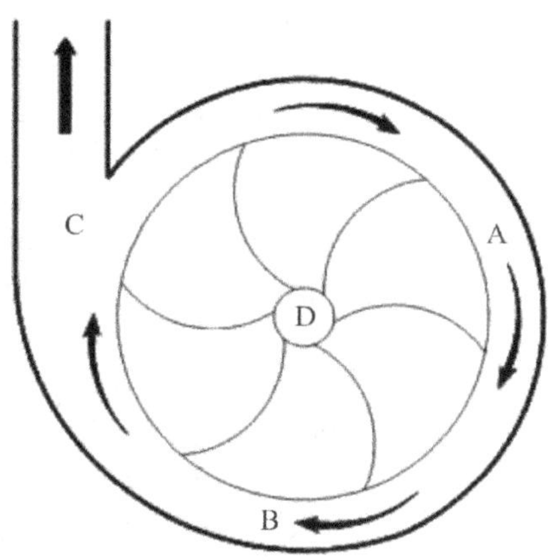

Fig. 3.22 Simplified centrifugal compressor

The major components of an axial flow compressor are: rotor with rows of blades, and a stator with rows of stationary blades (Fig. 3.23). The moving blades accelerate the air, and the stationary blades direct the flow of air into the next row of moving blades

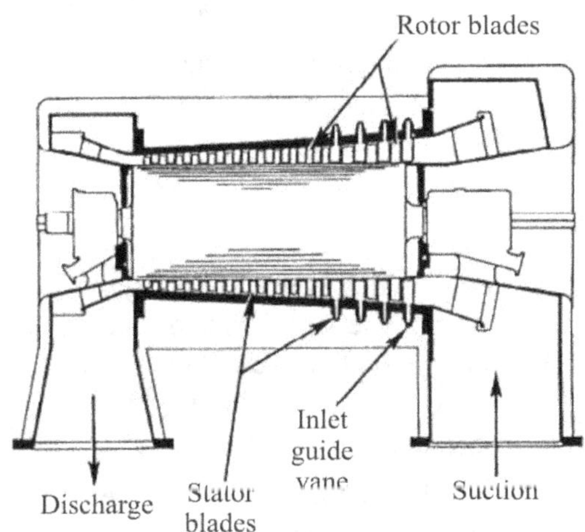

Fig. 3.23 Axial flow air compressor

(d) *Roots Blower Compressor:* This type is generally called as blower. The discharge air pressure obtained from this type of machine is very low. The discharge pressure of 1 bar can be obtained in Single Stage and pressure of 2.2 bar is obtained from Double Stage. The discharge pressure achieved by two rotors which have separate parallel axis and rotate in opposite directions. This is the example of Positive Displacement Compressor in Rotary Type Air Compressor.

The arrangement of the Root-Blower compressor is shown in Fig.3.24. It consists of two rotors with lobes rotating in opposite directions. The lobes of the rotors are of enewable s, hypocycloid or involute profiles, because this ensures correct mating. The high pressure side of the compressor is sealed from low pressure side for all angular positions of the rotors. A small clearance is provided between the rotors and the cylinder surface to reduce wear due to friction. The leakage through this clearance increases with increasing pressure ratio and reduces the efficiency of the compressors.

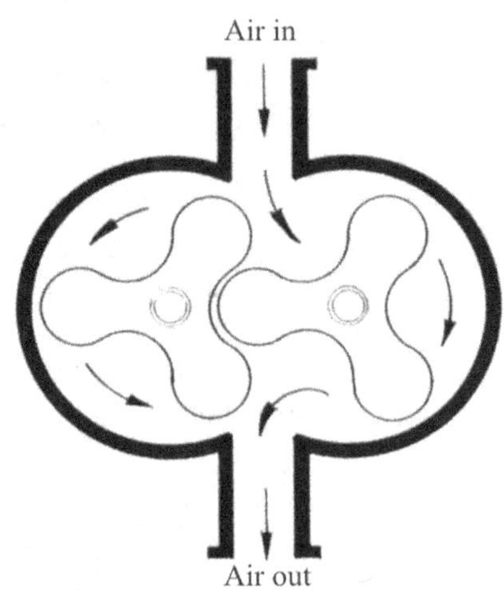

Fig. 3.24 Roots blower compressor

3.12 Turbine

A turbine is a rotary engine that extracts energy from a fluid flow and converts it into useful work. The simplest turbines have one moving part, a rotor assembly, which is a shaft or drum with blades attached. Moving fluid acts on the blades, or the blades react to the flow, so that they move and impart rotational energy to the rotor.

Water Turbines: Water turbines are used to convert the energy of falling water into mechanical energy. A working fluid contains potential energy (pressure head) and kinetic energy (velocity head). The fluid may be compressible or incompressible. Several physical principles are employed by turbines to collect this energy:

The principal types of water turbines are:

(i) Impulse Turbines   (ii) Reaction Turbines

3.12.1 Impulse turbines

Impulse turbines change the direction of flow of a high velocity fluid jet. The resulting impulse spins the turbine and leaves the fluid flow with diminished kinetic energy. In an impulse turbine entire pressure of water is converted into kinetic energy in a nozzle and the velocity of the jet drives the wheel. There is no pressure change of the fluid in the turbine rotor blades. Before reaching the turbine the fluid's pressure head is changed to velocity head by accelerating the fluid with a nozzle. Pelton wheels and de Laval turbines use this process exclusively. Impulse turbines do not require a pressure casement around the runner since the fluid jet is prepared by a nozzle prior to reaching turbine. Such turbines are used for high heads (Fig. 3.25).

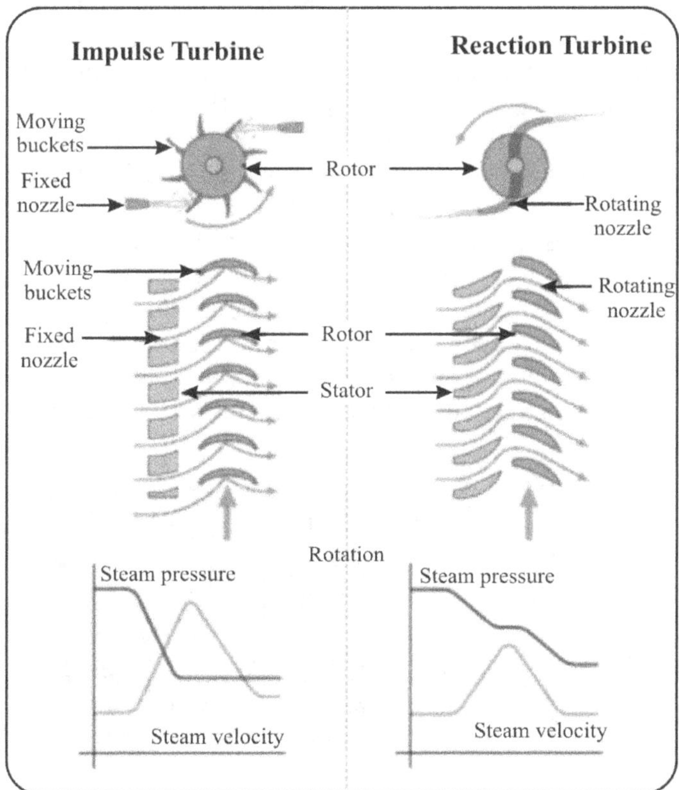

Fig. 3.25 Working principle of impulse and reaction turbine

3.12.2 Reaction turbines

Reaction turbines are used for low and medium heads. In a reaction turbine water enters the runner partly with pressure energy and partly with velocity head. These turbines develop torque by reacting to the fluid's pressure or weight. The pressure of the fluid changes as it passes through the turbine rotor blades. A pressure casement is needed to contain the working fluid as it acts on the turbine stage(s) or the turbine must be fully immersed in the fluid flow (wind turbines). The casing contains and directs the working fluid and, for water turbines, maintains the suction imparted by the draft tube (Fig. 3.25).

The **Pelton wheel** is among the most efficient types of water turbines. It was invented by Lester Allan Pelton in the 1870s. The Pelton wheel extracts energy from the impulse (momentum) of moving water, as opposed to its weight like traditional overshot water wheel. Although many variations of impulse turbines existed prior to Pelton's design, they were less efficient than Pelton's design; the water leaving these wheels typically still had high speed, and carried away much of the energy. Pelton paddle geometry was designed so that when the rim runs at ½ the speed of the water jet, the water leaves the wheel with very little speed, extracting almost all of its energy, and allowing for a very efficient turbine. It consists of a wheel fitted with elliptical bucket along its periphery as shown in Fig.3.26. The force of water jet striking the buckets on the wheel drives the turbine. The quantity of water jet falling on the turbine is controlled by means of a needle or spear valve placed in the tip of the nozzle. The movement of the spear is controlled by the governor. If the load on the turbine decreases, the governor pushes the needle into the nozzle, there by reducing the quantity of water striking the buckets. Reverse action takes place if the load on the turbine increases.

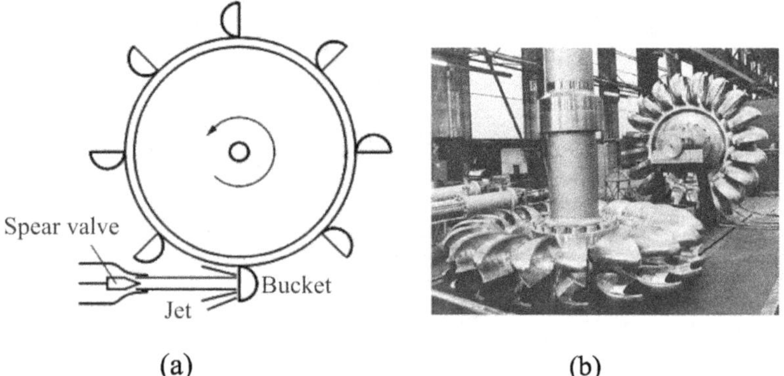

(a)   (b)

Fig. 3.26 Pelton wheel turbine

A Francis turbine is used for low to medium heads. It consists of an outer ring of stationary guide blades fixed to the turbine casing and an inner ring of rotating blades forming the runner. The guide blades control the flow of water to the turbine. Water flows radially inwards and changes to a downward direction while passing through the runner. As the water passes over the rotating blades of the runner, both pressure and velocity of water is reduced. This causes a reaction force which drives the turbine (Fig. 3.27).

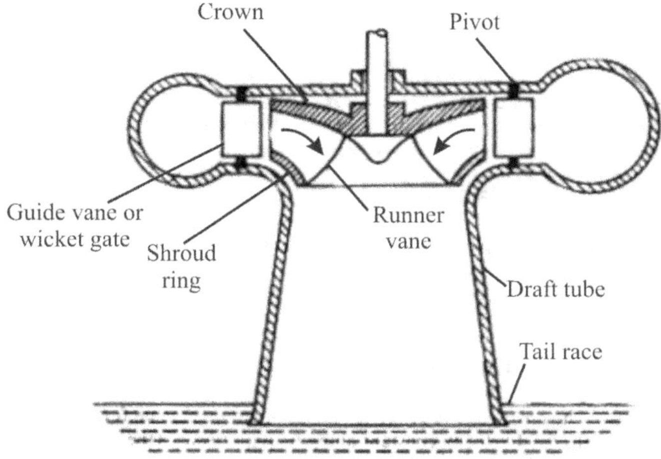

Fig. 3.27 Francis turbine

Kaplan turbine is a type of reaction turbine which gives us axial flow. It is a flow in which the water enters parallel to the shaft. It is a turbine which can operate on a very low head with a large flow rate.

It is very familiar to the Francis turbine and its overall path of water flow is from radial entrance to axial exit. Firstly there is a scroll casing in which water enters to the runner or rotor of the turbine. At the periphery there is a way between the rotor of the Kaplan turbine and the guide vanes or wicket gate as shown in Fig. 3.28. It is in the shape that water enters radial flow but is forced to get axial direction. Its rotor is very similar to propeller of the ship. These rotor blades are mounted on the hub and there are 3 to 8 blades and the hub are also mounted on the shaft. These blades can be adjusted according the direction of flow and available water head. After that it has a draft tube which is used to reduce the velocity of turbine.

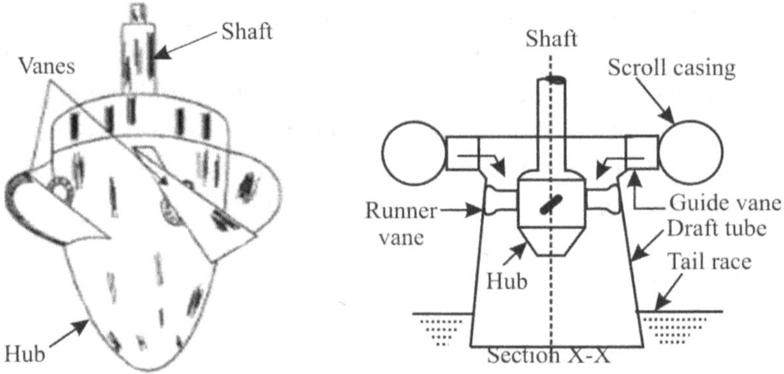

Fig. 3.28 Kaplan turbine

3.13   Pneumatic Machine

All systems have three basic parts:

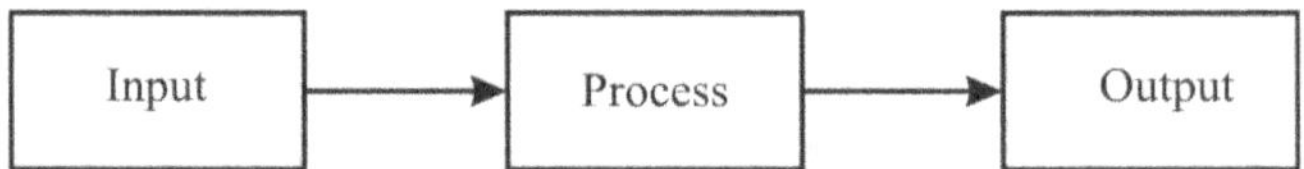

Input: The compressed air from the hand pump or compressor

Process: The valve directing the air to the base or rod end of the cylinder

Output: The piston rod extending or retracting. Tubing connects the hand pump or   compressor to the valves and cylinders to complete the system.

A pneumatic machine uses compressed air to run. Usually the compressed air will create motion by spinning a turbine. Examples of a pneumatic machine are: Dentists drill, Wheel nut remover (as used in race track pit crews) etc.

- Air is compressed by a pump. When we push on a handle of a bicycle pump, for example, we are compressing the air and forcing it into a smaller space. Compressed air uses a pneumatic cylinder to create movement. When compressed air enters the base of a pneumatic cylinder, it pushes on the piston and makes the piston rod extend (move out). Air on the other side of the piston is vented into the atmosphere. (See Fig.3.29 (a)).

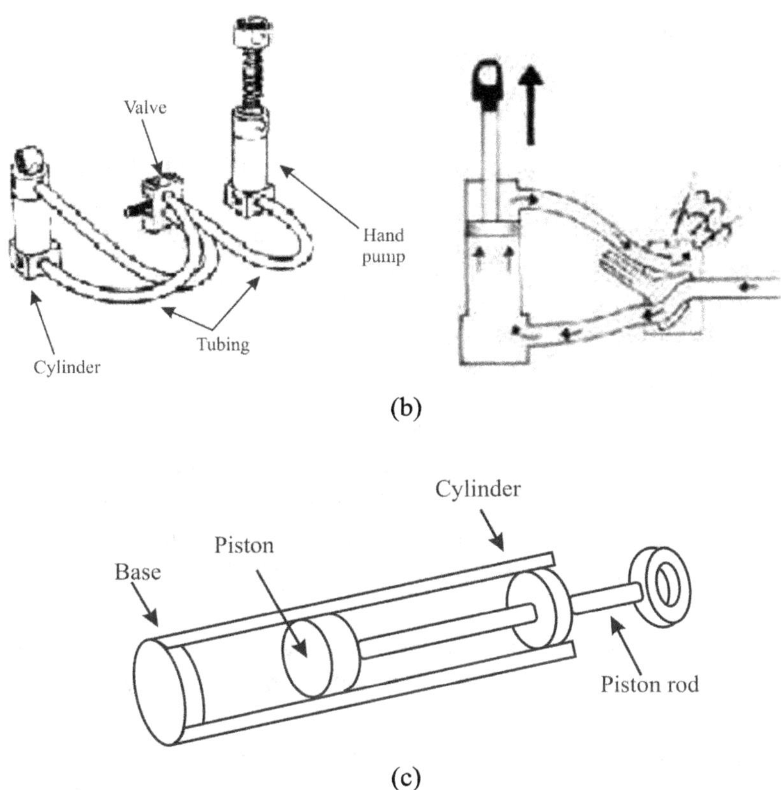

Fig. 3.29 Pneumatic machine

When compressed air enters the rod end of a pneumatic cylinder, it makes the piston rod retract (move in). (Fig.3.29 (c))

We use a valve to control movement. It works like a switch to direct air through tubing to the base or the rod end of the cylinder.

Creating movement and transferring forces are two important tasks given to machines. In industry, these tasks are often performed by either pneumatic (compressed air) or hydraulic (pressurized oil) devices. The term fluid power includes both pneumatic and hydraulic devices.

3.13.1 Pneumatic Logic

Pneumatic logic systems are often used to control industrial processes, consisting of primary logic units such as:

(a)   And Units.

(b)   Or Units.

(c)   'Relay or Booster' Units.

(d)   Latching Units.

(e)   'Timer' Units.

Pneumatic logic is a reliable and functional control method for industrial processes. In recent years, these systems have largely been replaced by electrical control systems, due to the smaller size and lower cost of electrical components. Pneumatic devices are still used in processes where compressed air is the only energy source available or upgrade cost, safety, and other considerations outweigh the advantage of modern digital control.

3.13.2 Applications of pneumatic machine

(i)   In dentistry applications, pneumatic drills are lighter, faster, and simpler than an electric drill of the same power rating (because the prime mover, the compressor, is separate from the drill and pumped air is capable of rotating the drill bit at extremely high rpm).

(ii)   Pneumatic transfer systems are employed in many industries to move powders and devices. Pneumatic tubes can carry objects over distances.

(iii) Pneumatic devices are very useful in material handling and material processing. They can push, pull, lift, and close and open doors. They can hold and remove materials, and position pieces for manufacturing.

(iv)   Pneumatic devices are also used where electric motors cannot be used for safety reasons, such as deep in a mine where explosive dust or gases may be present.

Examples of pneumatic tools:

(a)   Pneumatic drill (jackhammer) used by road workers.

(b)   Pneumatic nailgun.

(c)   Pneumatic switches.

(d)   Pneumatic actuators.

(e)   Air compressors.

(f)   Vacuum pumps.

(g)   Barostat systems used in Neuro gastroenterology and for researching electricity.

(h)   Cable Jetting – a way to install cables in ducts.

(i)   Pneumatic mail systems.

(j)   Air brakes on buses, trains, and trucks.

3.13.3   Advantages of pneumatic machine

- Air is cheap and plentiful. Hence, there is usually no need for a return line for the working fluid and leaks of the working fluid tend not to be messy.

- Air is easily compressed and can be stored in tanks.

- The working fluid is very light in weight so supply hoses are not heavy and Pneumatic systems can be interconnected easily with hoses, pipe, or tubing.

- Pneumatic systems can produce large linear movement.

- Compressed air is relatively environmentally friendly.

- Air is compressible; the equipment is less likely to be damaged by shock. The air in pneumatics absorbs excessive force, whereas the fluid of hydraulics directly transfers force.

3.13.4   Disadvantages of pneumatic machine

Industrial compressed air systems:

- Must keep the compressed air clean and dry. This requires special filtering.

- Using air at high pressure requires safety precautions.

- Sometimes spray a fine mist of lubricating oil into the atmosphere. This has an impact on air quality.

- May not be able to produce very large forces. In such cases, a higher pressure hydraulic system might be necessary

- Controlling is difficult because of the non-linearity of the system.

- These systems have larger size and are much costly as compared to the electrical components.

3.14   Hydraulic Power

Hydro power works by converting the movement or falling of water into useful forms of energy such as electricity. Many forms of hydro power exist and have been used by humans throughout history. Some examples of this are watermills which powered machinery such as sawmills, irrigation which used gravity, and dams which use falling water to power turbines.

3.14.1 Hydraulic Press

Hydraulic press works on the basis of Pascal's Law. A hydraulic press consists of basic components used in a hydraulic system that includes the cylinder, pistons, the hydraulic pipes, etc as shown in Fig. 3.30. The system comprises of two cylinders, the fluid (usually oil) is poured in the cylinder having a small diameter. This cylinder is known as the slave cylinder.

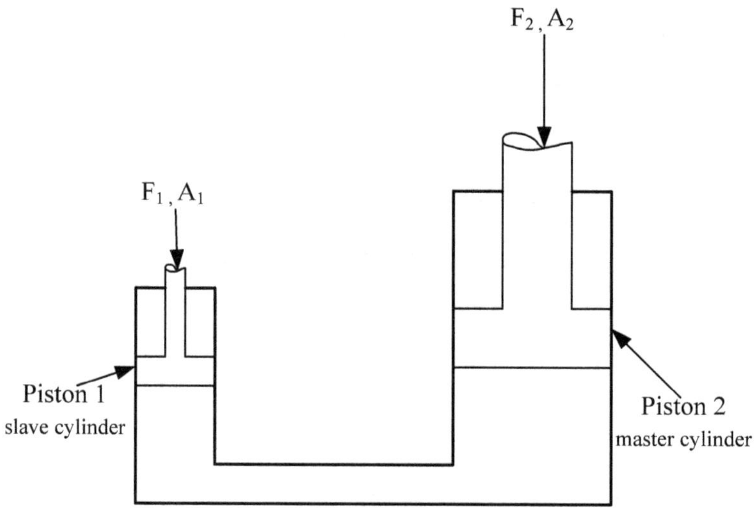

Fig. 3.30 Hydraulic Press

The piston in this cylinder is pushed so that it compresses the fluid in it that flows through a pipe into the larger cylinder. The larger cylinder is known as the master cylinder. The pressure is exerted on the larger cylinder and the piston in the master cylinder pushes the fluid back to the original cylinder.

Force increase with hydraulics:

$$F_2 = F_1 (A_2/A_1)$$

where, F_1, F_2, A_1, A_2 are forces and areas of piston 1 and 2 respectively.

The force applied on the fluids by the smaller cylinder results in a larger force when pushed in the master cylinder. The hydraulic press is mostly used for industrial purposes where a large pressure is required for compressing metals into thin sheets. An industrial hydraulic press uses the material to be worked upon along with the help of the press plates to crush or punch the material into a thin sheet.

3.14.2 Hydraulic Crane

Hydraulic crane is a device used for rising or transferring heaving loads. It is widely used in workshop, warehouses and dock sidings.

A hydraulic crane consists of mast, tie, jib, guide pulley and a jigger as shown in schematic Fig. 3.31. The jib and tie are attached to the mast. The jib can be raised or lowered in order to decreases or increase the radius of action of the crane. The mast along with the jib can revolve about a vertical axis and thus the load attached to the rope can be transferred to any place within the area of the crane's action. The jigger

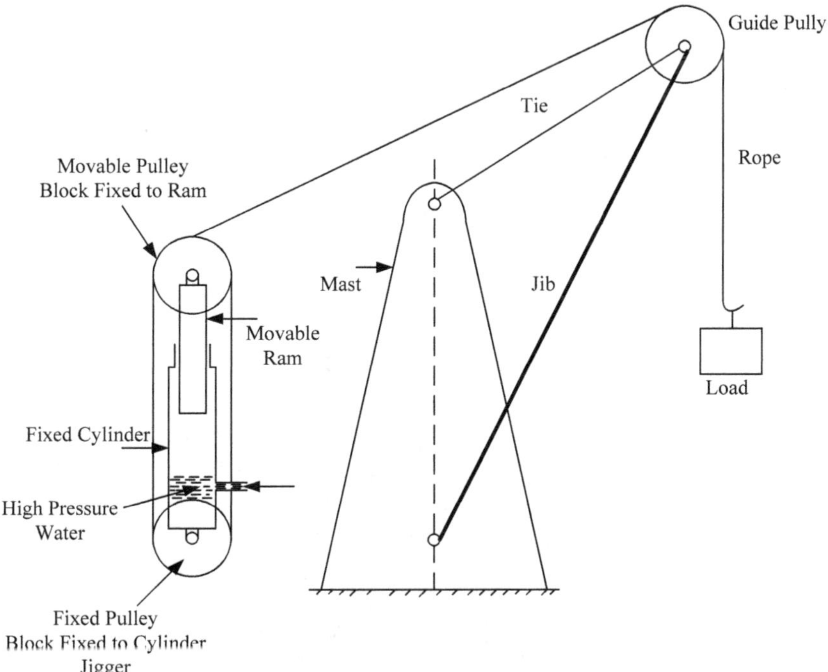

Fig. 3.31 Hydraulic crane

consists of a movable ram sliding in a fixed cylinder and is used for lifting or lowering the heavy loads. One end of the ram is in contact with water and the outer end is connected to set of movable pulley block. Another pulley block, called the fixed pulley block is attached to the fixed cylinder. The pulley block, attached to the ram, moves up and down while the pulley block, attached to the fixed cylinder, and is not having any movement.

3.14.3   Hydraulic Intensifier

In most of the hydraulic machinery used, the usual pressure of 80 to 100 psi may not be sufficient to operate certain spool valves and other mechanisms. To cater to the need for a high pressure requirement for a comparatively short period of time, pumps and accessories are definitely not the solution. But the substitute can be hydraulic intensifiers which can increase the pressure from 100 psi to 40,000 psi, using small volumes of fluid.

A hydraulic intensifier is a device which is used to increase the intensity of pressure of any hydraulic fluid or water, with the help of the hydraulic energy available from a huge quantity of water or hydraulic fluid at a low pressure. These devices are very important in the case of hydraulic machines, mainly hydraulic presses, which require water or hydraulic fluid at very high pressure which cannot be obtained from the main supply directly.

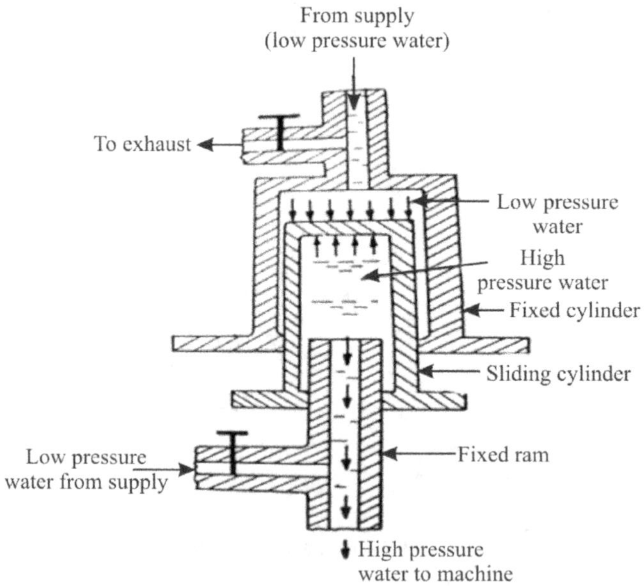

Fig. 3.32 Hydraulic intensifiers

There are different types based on the medium of hydraulic fluids used and the number of strokes used to intensify to the desired pressure. They are single-stroke, differential cylinder intensifiers, oil-oil intensifiers, air-air intensifiers, and oil-air intensifiers. Recent developments are so vast that huge pressures are achieved by using combinations of the above types.

There are three main parts in the hydraulic intensifiers to be noted. They are: fixed ram, hollow inverted sliding cylinder, fixed inverted cylinder.

A hydraulic intensifier consists of fixed ram through which the water, under a high pressure, flows to the hydraulic machine (Fig. 3.32). A hollow inverted sliding cylinder, containing water under high pressure, is mounted over the fixed ram. The inverted sliding cylinder is surrounded by another inverted fixed cylinder which contains water from the main supply at a lower pressure.

A large quantity of water at a low pressure from the supply enters the inverted fixed cylinder. The weight of this water presses the sliding cylinder in the downward direction. The water inside the inverted sliding cylinder gets compressed due to the downward movement of the sliding cylinder and its pressure thus increases. This high pressure water is forced out of the sliding cylinder through the fixed ram, to the hydraulic machine.

3.15   Hydro Power

Hydropower is the force or energy of moving water. Water continuously moves all the way through a huge worldwide cycle, in which it evaporates (due to the movement of the Sun) from oceans, seas and other water reservoirs, forms clouds, precipitates as rain or snow, then flows reverse to the ocean as shown in Fig.3.33. The energy of this water cycle, which is driven by the sun, is tapped most resourcefully with hydropower.

Runoff in rivers is a part of the hydrologic cycle in which water evaporates from the sea and moves through the atmosphere to land where it precipitates, and then returns back to the sea by overland and subterranean routes.

The use of water to produce mechanical power is a very old practice. A flowing stream can make a paddle turn, but a waterfall can turn a blade fast adequate to produce electricity. The real key in the size of water power is the physical height difference achieved between source and sink – the distance through which the water falls. There are many different ways to harness the energy in water. The most common way of capturing this energy is hydroelectric power, electricity created by falling water.

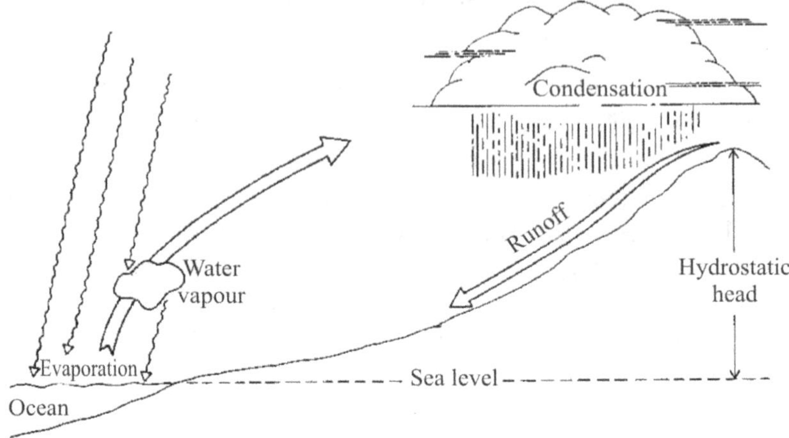

Fig. 3.33 Layout of Hydro Power

Water flows from a high potential energy (high ground) to lower potential energy (lower ground); the potential energy difference is partially converted into electric energy through the use of a generator.

3.15.1 Hydro power Potential

There are two major factors that determine the generating potential at any specific site: the amount of water flow per time unit and the vertical height that water can be made to fall (head). Head may be natural due to the topographical condition or may be formed artificially by means of dams.

Water flow on the other hand is a direct result of the strength, distribution and duration of rainfall, but is also a function of direct evaporation, transpiration, infiltration into the ground, the area of the particular drainage basin, and the field-moisture capacity of the soil.

The principal advantages of using hydro power are:

1.   Its huge renewable domestic resource base.

2.   The absence of polluting emissions during operation. Hydro power produces essentially no carbon dioxide or other harmful emissions, in contrast to burning fossil fuels, and is not a significant contributor to global warming through CO_2.

3. Its capability in some cases to respond quickly to utility load demands

4. Its very low operating costs.

5. Hydroelectric projects also include beneficial effects such as recreation in reservoirs or in tail water below dams.

6. Best source to support the development of other   enewable such as wind, solar, etc.

3.15.2 Problems of Hydro Power

The main reasons that hydro power plants are not build everywhere are that they are costly and require large bodies of water relatively close to inhabitants. Other arising problems are the effects of dams on river ecosystems and social problems related to relocation of inhabitants.

Environmental Aspects of Hydro Power Plants: A watercourse is an ecological system where changes within one component may create a series of spread-effects. For instance, changes in the water flow may affect the quality of the water and the production of fish downstream. In the tropics there may be great seasonal variations as to the amount of precipitation, and in dry periods evaporation from lakes and reservoirs may be considerable. This may affect the water level of the reservoirs more dramatically than in temperate areas. During the construction phase the transport of mud and sediments will be especially large downstream from the construction area. Excavation and tunneling may lead to greatly reduced water quality and problems for those dependent on the water.

Excessive Fertilization: Whenever nutrients are trapped in a reservoir, the result may be excessive fertilization – eutrophication – in the reservoir. It may lead to an increased growth of algae or large amounts of higher-order aquatic plants. A substantial production of organic matter in the reservoir, or the supply of external organic matter, may cause anaerobic conditions – lack of oxygen – in the deep-water layers.

Transport of Nutrients: A reservoir serves as a trap for nutritious elements and mud flowing in, possibly leading to a considerable reduction of the total transport of nutrients downstream. In addition, the annual variations in supply downstream may undergo changes. This may reduce the biological production all the way to the sea.

Fish: The composition of fish species may be altered, since reproduction for some species may be hindered if the operation involves changes in the water level during the spawning period.

Flora and Fauna: Submerging and water-flow changes, moreover, will lead to changes in the fauna and vegetation beyond the watercourse as such.

Population Movements: Large hydro power plants with dams require large reservoir and discharge areas. Many people have to be evacuated to make room for these areas. This could lead to a completely new situation for people who have lived in a relatively small, protected environment. Housing, land distribution, working conditions and way of life may change radically.

Health: Large hydro power plants can increase the extent of water-related diseases. The reservoir may improve the living and breeding conditions of disease-causing organisms (pathogens) and their intermediate hosts. Among water-related diseases one could mention typhus, cholera, dysentery and several tapeworm and roundworm diseases.

Dam Breach – Uncontrolled Flooding: A dam breach seldom occurs, but owing to the enormous consequences, which it may involve, the impacts of a breach should be assessed. The risk of casualties and damaged property or technical installations must be considered the most serious consequences, but the impacts on the natural environment can also be considerable.

Technology: In hydro power plants the kinetic energy of falling water is captured to generate electricity. A turbine and a generator convert the energy from the water to mechanical and then electrical energy. The turbines and generators are installed either in or adjacent to dams, or use pipelines (penstocks) to carry the pressured water below the dam or diversion structure to the powerhouse. The power capacity of a hydro power plant is primarily the function of two variables: (1) flow rate expressed in cubic meters per second (m^3/s), and (2) the hydraulic head, which is the elevation difference the water falls in passing through the plant. Plant design may concentrate on either of these variables or both.

3.15.3 Components of Hydro Power Plant

Most conventional hydro power plants include following major components as shown in Fig. 3.34:

Dam: Controls the flow of water and increases the elevation to create the head. The reservoir that is formed is, in effect, stored energy.

Turbine: Turned by the force of water pushing against its blades.

Generator: Connects to the turbine and rotates to produce the electrical energy.

Transformer: Converts electricity from the generator to usable voltage levels.

Transmission lines: Conduct electricity from the hydro power plant to the electric distribution system.

In some hydro power plants also another component is present – penstock, which carries water from the water source or reservoir to the turbine in a power plant.

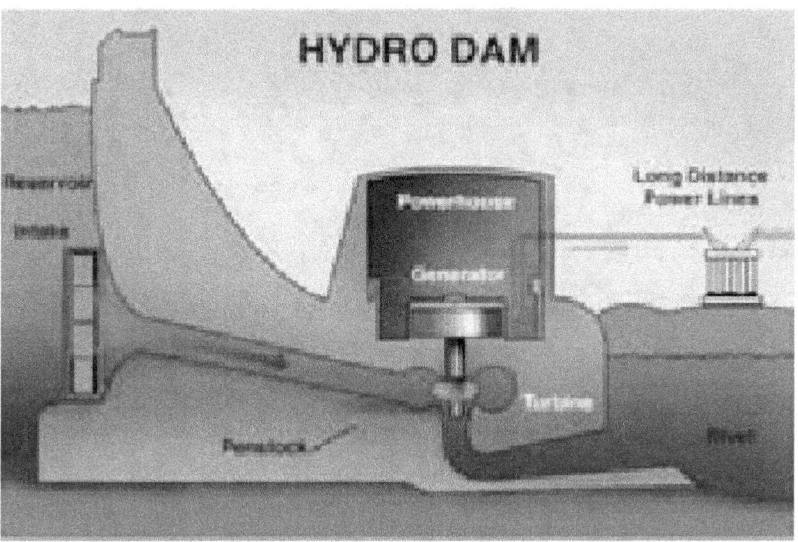

Fig. 3.34 Components of Hydro Power Plant

3.15.4 Types of Hydro power Facilities

Hydro power technology can be categorized into two types: conventional and pumped storage. Another way of classification of hydro power plants is according to:

(i)   Rated power capacity (big or small)

(ii) Head of water (low, medium and high heads)

(iii) The type of turbine used (Kaplan, Francis, Pelton etc.)

(iv) The location and type of dam, reservoir.

3.15.5 Conventional Hydro power Plants

Conventional hydro power plants use water energy from a river, stream, canal system, or reservoir to produce electrical energy as shown in Fig. 3.35.

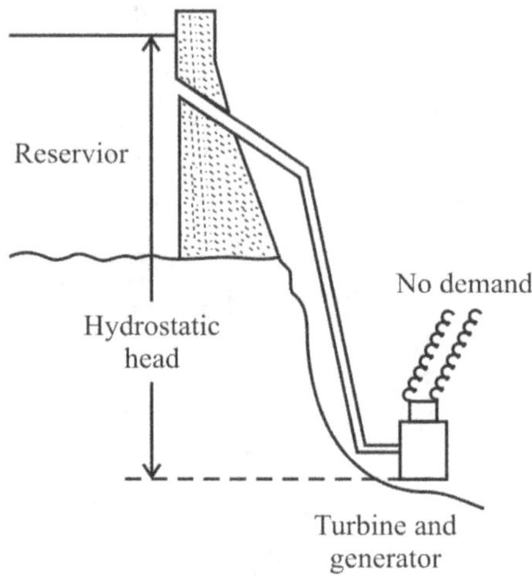

Fig. 3.35 Conventional Hydroelectric Power Generation

Conventional hydro power can be further divided between impoundment and diversion hydro power.

Impoundment hydro power – uses dam to store water. Water may be released either to meet changing electricity needs or to maintain a constant water level.

Diversion hydro power – channels a portion of the river through a canal or penstock, but may require a dam.

3.15.6 Pumped Storage Plants

Pumped storage hydro-electricity is an extremely simple principle. To start with, two reservoirs at different altitudes are required. Water stored at height offers valuable potential energy. During periods of high electrical demand, the water is released to the lower reservoir to generate electricity (Fig.3.36 a). When the water is released, kinetic energy is created by the discharge through high-pressure shafts, which direct the water through turbines connected to generator/motors. The turbines power the generators to create electricity. After the generation process is complete, water is pumped back to the upper reservoir for storage and readiness for the next cycle (Fig. 3.36 b). The process frequently takes place overnight when electricity demand is at its lowest. The Schematic view of pump storage plant is also shown in Fig. 3.37.

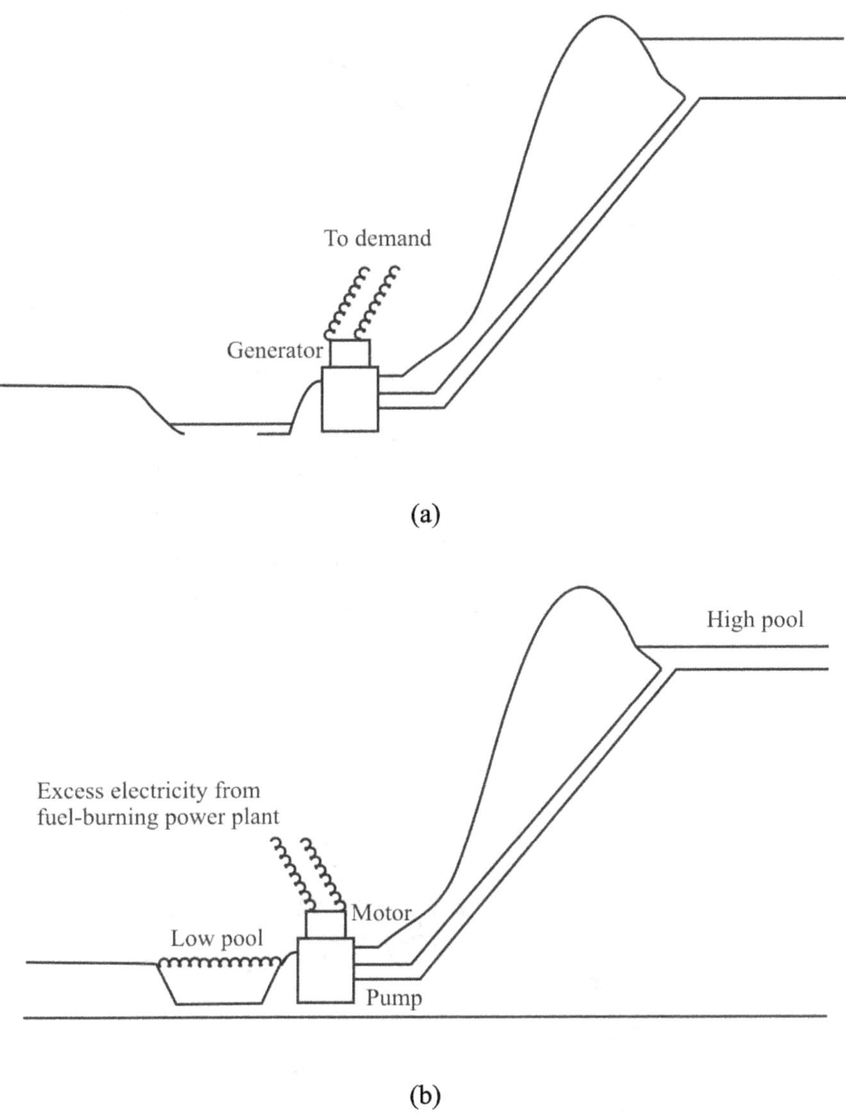

Fig. 3.36 (a) Generating electricity at the time of peak demand (b) Pumping during the time of low demand

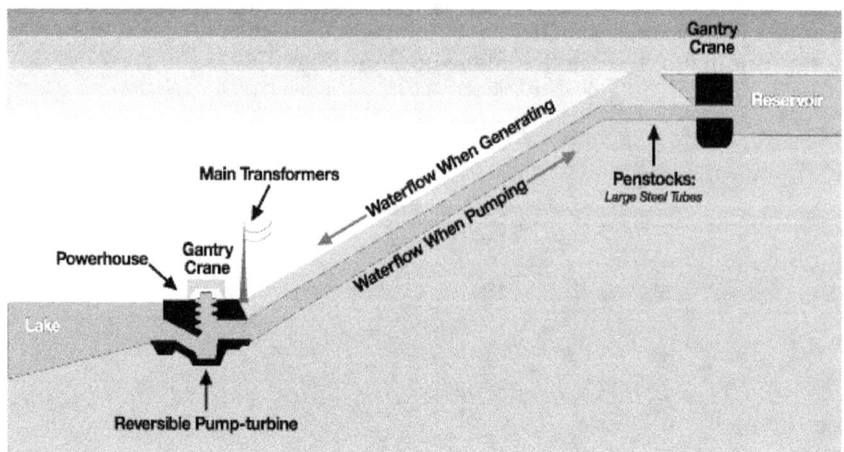

Fig. 3.37 Schematic View of Pump Storage Plant

While pumped storage facilities are net energy consumers, they are cherished by a utility because they can be rapidly brought on-line to operate in a peak power production mode. This process benefits the utility by increasing the load factor and reducing the cycling of its base load units. In most cases, pumped storage plants run a full cycle every 24 hours. Power spectrum of pump storage power plant is given in Fig. 3.38.

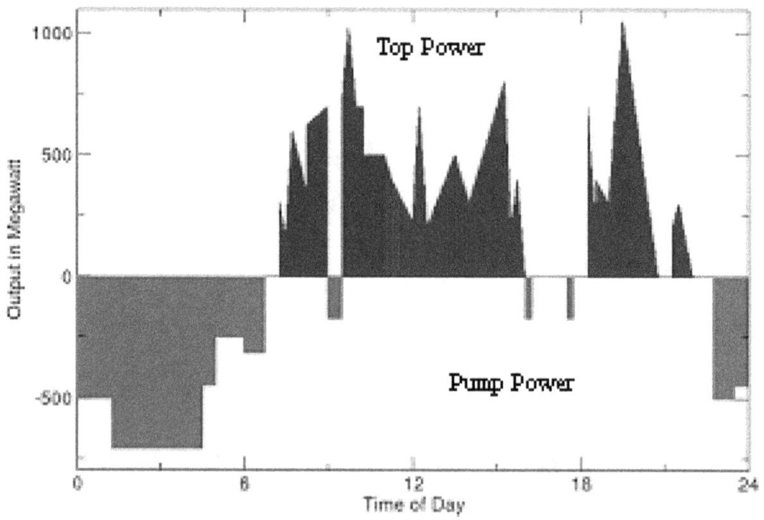

Fig. 3.38 Power Spectrum of Pump Storage Power Plant

3.15.7 Base load power plant

A base load power plant (or base load power station) is one that provides a steady flow of power regardless of total power demand by the grid. These plants run at all times through the year except in the case of repairs or scheduled maintenance.

Power plants are designated base load based on their low cost generation, efficiency and safety at set outputs. Base load power plants do not change production to match power consumption demands since it is always cheaper to run them rather than running higher cost combined cycle plants or combustion turbines. Typically these plants are large enough to provide a majority of the power used by a grid, making them slow to fire up and cool down. Thus, they are more effective when used continuously to cover the power base load required by the grid.

Each base load power plant on a grid is allotted a specific amount of the base load power demand to handle. The base load power is determined by the load duration curve of the system. For a typical power system, the rule of thumb is that the base load power is usually 35-40% of the maximum load during the year.

Peaks or spikes in customer power demand are handled by smaller and more responsive types of power plants.

Base load power plant usage: Nuclear and coal power plants may take many hours, if not days, to achieve a steady state power output. On the other hand, they have low fuel costs. Since they require a long period of time to heat up to operating temperature, these plants typically handle large amounts of base load demand. Different plants and technologies may have differing capacities to increase or decrease output on demand: nuclear plants are generally run at close to peak output continuously (apart from maintenance, refueling and periodic refurbishment), while coal-fired plants may be cycled over the course of a day to meet demand. Plants with multiple generating units may be used as a group to improve the "fit" with demand, by operating each unit as close to peak efficiency as possible.

3.15.8 Peak Load Power Plant

Natural gas and oil power plants are much faster to start, but have much higher fuel costs. These plants are typically scheduled to handle peak power demands since they can be ready to supply power in 30 minutes or less. They are more expensive to operate than coal power plants, primarily due to higher fuel costs.

Load balancing: By moving peak loads to non-peak times, the output of base load power plants can be used more efficiently. Power companies use several techniques to shift demand and balance the load:

Demand pricing: A power company installs special electric meters, particularly for its heaviest users, such as industry. These meters record not only how much electricity was consumed in all, but at what time the consumption took place. A higher price is assessed for power consumed during peak times than off-peak times, which encourages users to shift their operating patterns to demand the most electricity at off-peak times, if possible.

Load shedding: In this case, it is remote-control switchgear that power companies use to remove large loads from the system. In ordinary usage, it is large industrial users whose power is cut during periods of exceptional demand. In emergencies, the power demand of large numbers of consumers can be shed from the system, either by planned rolling blackouts or by unplanned power cuts.

Brownouts: Power companies can, in times of exceptional demand, deliberately reduce the line voltage on the power grid. Also, sudden spikes in demand can cause unplanned reductions in line voltage. Because of Ohm's law, reducing the voltage reduces the demand for power, lessening peak load. This is not an ideal method, however. Lowered voltage reduces the efficiency of electrical equipment and can, if lowered far enough, damage electric motors and other devices.

Solved Numerical Examples

Example 1: A sea level, low-speed wind tunnel of circular cross section with a diameter upstream of the contraction of 20 metre and a test section diameter of 10 metre. The test section is vented to the atmosphere (sea level pressure is 2116 N/m^2). If the working section velocity is 180 mph, calculate the following: Assume $\rho = 1.2$ kg/m^3.

(a) Upstream section velocity.

(b) Upstream pressure.

Solution:

Let's draw a wind tunnel first.

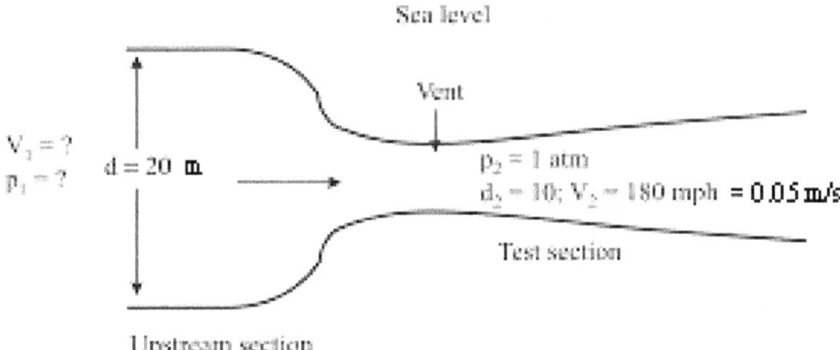

(a) From continuity equation,

$$\frac{V_2}{V_1} = \frac{S_1}{S_2} = \frac{(\pi/4)d_1^2}{(\pi/4)d_2^2} = \frac{20^2}{10^2} = 4$$

Then $\qquad V_1 = \frac{V_2}{4} = \frac{180}{4} = 45 \text{ mph} = 0.0125 \text{ m/s}$

From the Bernoulli equation,

$$p_1 + \frac{\rho}{2}V_1^2 = p_2 + \frac{\rho}{2}V_2^2$$

$$p_1 = p_2 + \frac{\rho}{2}\left(V_2^2 - V_1^2\right)$$

$$= 2116 + \frac{1.2}{2}\left[(0.05)^2 - (0.0125)^2\right]$$

$$= 2116 \text{ N/m}^2$$

***Example* 2:** An airfoil (red shape) is moving through the air at 355 km/h at sea level. The atmospheric pressure is 102,325 N/m^2 and temperature is 27 °C. At a point on the airfoil upper surface the local velocity is 420 km/h.

(a) Determine the pressure at the upper surface point.

(b) If the pressure in part (a) is the average upper surface pressure, how much lift per square meter (referred to ambient pressure) is being provided by the upper surface ?

(c) If the average speed on the lower surface is 320 km/h, what is the pressure and the average lift per square meter on the lower surface ?

(d) What is the total lift per square meter of wing ?

Solution:

Let's draw first the airfoil,

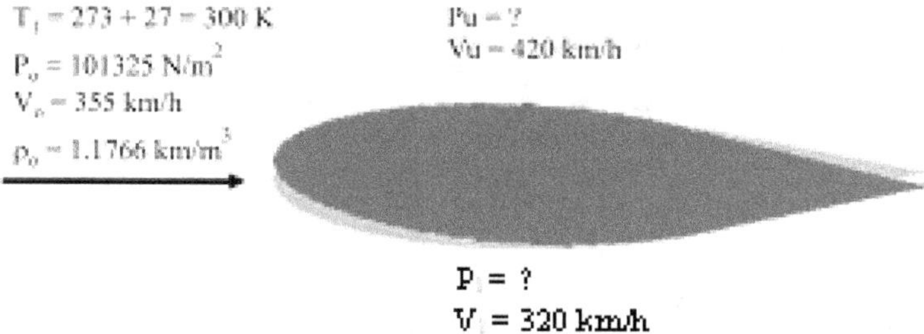

In the SI system, temperature must be expressed in Kelvin and velocities in m/s.

$$355 \text{ km / h} = 355 \times \frac{1000}{3600} \text{ m / s} = 98.61 \text{ m / s}$$

$$420 \text{ km / h} = 116.67 \text{ m / s}$$

$$320 \text{ km / h} = 88.89 \text{ m / s}$$

(a) From the Bernoulli equation,

$$p_o = p_o + \frac{\rho}{2}\left(V_o^2 - V_0^2\right) = 101325 + \frac{1.1766}{2}\left[(98.61)^2 - (116.67)^2\right]$$

$$= 101325 - 2287 = 99038 \text{ N/m}^2 \text{ upper surface pressure}$$

(b) Lift per square meter, upper surface $= 101325 - 99038 = 2287$ N/m^2.

(c) On the lower surface,

$$p_1 = p_o + \frac{\rho}{2}\left(V_o^2 - V_1^2\right) = 101325 + \frac{1.1766}{2}\left[(98.61)^2 - (88.89)^2\right]$$

$$= 101325 + 1072 = 102397 \text{ N/m}^2 \text{ lower surface pressure}$$

Lower surface lift $= 102397 - 101325 = 1072$ N/m^2.

(d)   Total lift per square meter $= 2287 + 1072 = 3359$ N/m^2.

***Example* 3:** A rectangular tank 10 m × 5 m and 3.25 m deep is divided by a partition wall parallel to the shorter wall of the tank. One of the compartments contains water to a depth of 3.25 m and the other oil of specific gravity 0.85 to a depth of 2 m. Find the resultant pressure on the partition.

***Solution*:** The problem can be solved by considering hydrostatic pressure distribution diagram for both water and oil as shown in Fig.

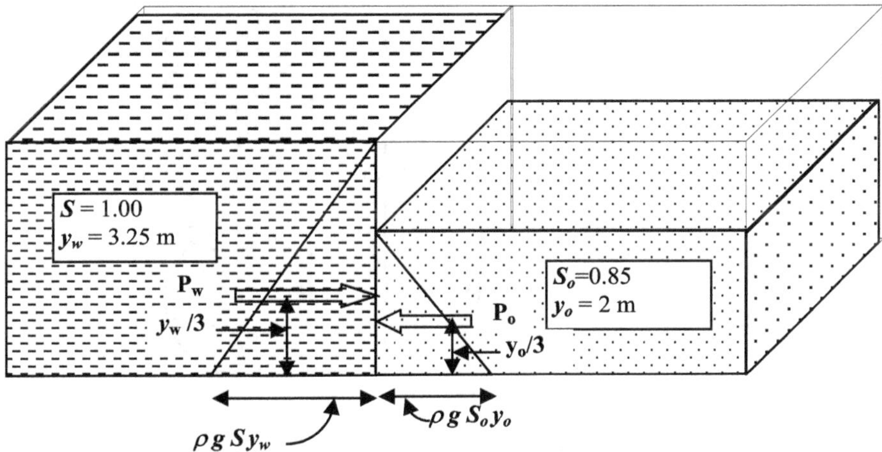

From hydrostatic law, the pressure intensity **p** at any depth **y$_w$** is given by

$$p = S_o \, \rho \, g \, y_w$$

Where ρ is the mass density of the liquid

Pressure force $P = p \times$ Area

$P_w = 1000 \times 10 \times 3.25 \times 5 \times 3.25 = 528.125$ kN ($\rightarrow$)

Acting at 3.25/3 m from the base

$P_o = 0.85 \times 1000 \times 10 \times 2.0 \times 5 \times 2.0 = 170$ kN ($\leftarrow$)

Acting at 2/3 m from the base.

Net Force $P = P_w - P_o = 528.125 - 170 = 358.125$ kN ($\rightarrow$)

Location: Let **P** act at a distance **y** from the base. Taking moments of P_w, P_o and P about the base, we get

$$P \times y = P_w \times y_w/3 - P_o \times y_o/3$$

$$358.125\,y = 528.125 \times (3.25/3) - 170 \times (2/3)$$

or $y = 1.28$ m

***Example* 4:** A vertical gate closes a circular tunnel of 5 m diameter running full of water, the pressure at the bottom of the gate is 0.5 MPa. Determine the hydrostatic force.

***Solution*:**

Assume $\rho = 1000$ kg/m^3 and $g = 10$ m/s^2

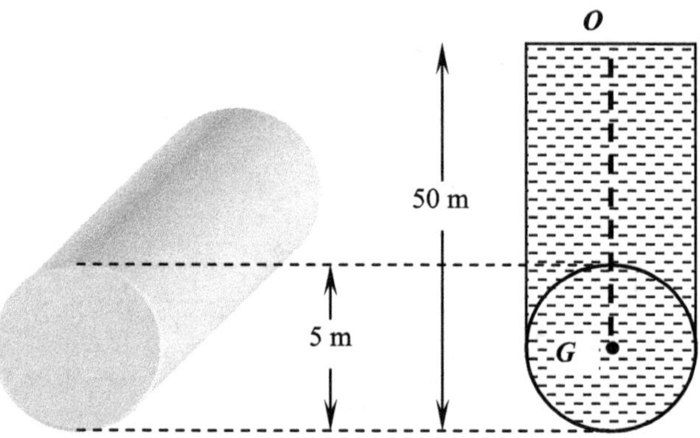

Pressure intensity at the bottom of the gate is,

$$p = \rho g y$$

Where y is the depth of point from the free surface.

$$0.5 \times 10^6 = 1000 \times 10 \times y$$

$$y = 50 \text{ m}$$

Hence the free surface of water is at 50 m from the bottom of the gate

$$A = \frac{\pi D^2}{4} = \frac{\pi \times 5^2}{4} = 19.635 \text{ m}^2$$

$$\bar{y} = OG = 50 - 2.5 = 47.5 \text{ m}$$

Radius of cylinder = 2 m We know that the total pressure force is given by

$$P = \rho g A \bar{y} = 1000 \times 10 \times 19.635 \times 47.5 = 9326.625 \text{ kN}$$

Example 5: A cylinder holds water in a channel as shown in Fig. Determine the weight of 1 m length of the cylinder.

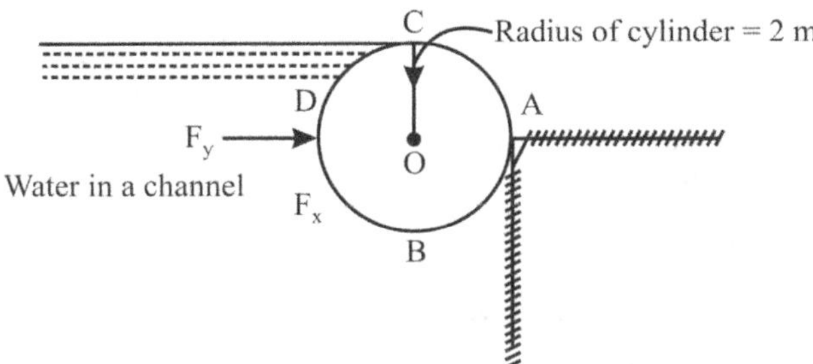

Solution:

Radius of Cylinder = R = 2 m

Length of cylinder = 1 m

Weight of Cylinder = W

Horizontal force exerted by water = F_x

F_x = Force on vertical area BOC

$\qquad$ = ρ g A y = 1000 × 10 × (4 × 1) × (2/2) = 40 kN ($\rightarrow$)

The vertical force exerted by water = F_y = Weight of water enclosed in BDCOB

$$F_y = \rho \ g \left(\frac{\pi \times 2^2}{4} \right) \times L$$

$$= 1000 \times 10 \times 3.142 \times 1$$

$$= 31.416 \ kN \ (\uparrow)$$

Example 6: Calculate specific weight, mass density, specific volume and specific gravity of a liquid having a volume of 4 m³ and weighing 29.43 kN. Assume missing data suitably.

Solution:

$$\gamma = \frac{W}{V} \qquad\qquad \gamma = ?$$

$$= \frac{29.43 \times 10^3}{4} \qquad\qquad \rho = ?$$

$$\gamma = 7357.58 \ N/m^3 \qquad\qquad \forall = ?$$

$$S = ?$$

$$V = 4 \ m^3$$

$$W = 29.43 \ kN$$

$$= 29.43 \times 10^3 N$$

To find ρ - Method 1:

$$W = mg$$

$$29.43 \times 10^3 = m \times 9.81 \qquad\qquad \text{Method 2:}$$

$$m = 3000 \ kg \qquad\qquad \gamma = \rho \ g$$

$$\therefore \rho = \frac{m}{v} = \frac{3000}{4} \qquad\qquad 7357.5 = \rho \ 9.81$$

$$\rho = 750 \ kg/m^3 \qquad\qquad \rho = 750 \ kg/m^3$$

(i)
$$\forall = \frac{V}{M} \qquad\qquad\qquad \rho = \frac{M}{V}$$

$$= \frac{4}{3000} \qquad\qquad\qquad \forall = \frac{V}{M}$$

$$\forall = 1.33 \times 10^{-3} \ m^3/kg \qquad \forall = \frac{1}{\rho} = \frac{1}{750}$$

$$\forall = 1.33 \times 10^{-3} \ m^3/kg$$

$$S = \frac{\gamma}{\gamma_{Standard}} \qquad\qquad\qquad S = \frac{\rho}{\rho_{Standard}}$$

$$= \frac{7357.5}{9810} \qquad\qquad\qquad S = \frac{750}{1000}$$

$$S = 0.75 \qquad\qquad\qquad\qquad S = 0.75$$

***Example* 7:** Calculate specific weight, density, specific volume and specific gravity and if one liter of Petrol weighs 6.867 N.

$$\gamma = \frac{W}{V} \qquad\qquad\qquad V = 1 \ Litre$$

$$= \frac{6.867}{10^{-3}} \qquad\qquad\qquad V = 10^{-3} \ m^3$$

$$\gamma = 6867 \ N/m^3 \qquad\qquad W = 6.867 \ N$$

$$S = \frac{\gamma}{\gamma_{Standard}} \qquad\qquad \rho = s\,g$$

$$= \frac{6867}{9810} \qquad\qquad 6867 = \rho \times 9.81$$

$$S = 0.7 \qquad\qquad \rho = 700 \text{ kg/m}^3$$

$$\forall = \frac{V}{M}$$

$$M = 6.867 \div 9.81$$

$$= \frac{10^{-3}}{0.7}$$

$$\forall = 1.4 \text{x} 10^{-3} \text{ m}^3 / \text{kg} \qquad\qquad M = 0.7 \text{ kg}$$

***Example* 8:** Specific gravity of a liquid is 0.7. Find i) Mass density ii) specific weight. Also find the mass and weight of 10 liters of liquid.

$$S = \frac{\gamma}{\gamma_{Standard}} \qquad \gamma = \rho g \qquad S = 0.7$$

$$V = ?$$

$$\rho = ?$$

$$0.7 = \frac{\gamma}{9810} \qquad 6867 = \rho \times 9.81 \qquad M = ?$$

$$W = ?$$

$$\gamma = 6867 \text{ N} / \text{m}^3 \qquad \rho = 700 \text{ kg} / \text{m}^3 \qquad V = 10 \text{ litre}$$

$$= 10 \times 10^{-3} \text{ m}^3$$

$$S = \frac{\rho}{\rho_{Standard}} \qquad \gamma = \frac{W}{V}$$

$$0.7 = \frac{\rho}{1000} \qquad 6867 = \frac{W}{10^{-2}}$$

$$\rho = 700 \text{ kg} / \text{m}^3 \qquad W = 68.67 \text{ N}$$

$$\text{or}$$

$$\rho = \frac{M}{V} \qquad W = m\,g$$

$$700 = \frac{M}{10 \times 10^{-3}}$$

$$= 7 \times 9.81$$

$$W = 68.67 \text{ N}$$

$$M = 7 \text{ Kg}$$

***Example* 9:** A vertical cylinder 300 mm in diameter is fitted at the top with a tight but frictionless piston and filled with water at 70^0 C. The outer portion of the piston is exposed to atmospheric pressure of 101.3 kPa. Calculate the minimum force applied on the piston that will cause water to boil at 70^0 C. Take Vapour Pressure of water at 70^0 C as 32 k Pa.

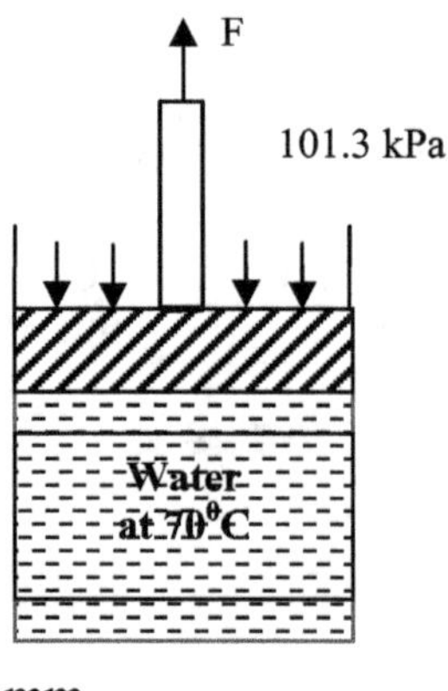

$$D = 300 \text{ mm}$$

$$= 0.3 \text{ m}$$

F Should be applied such that the pressure is reduced from 101.3 kPa to 32 kPa. Therefore reduction in pressure required

$$= 101.3 - 32$$

$$= 69.3 \text{ kPa}$$

$$= 69.3 \times 10^3 \text{ N/m}^2$$

$$\therefore \qquad F \, / \, \text{Area} = 69.3 \times 10^3$$

$$F \, / \, \frac{\pi}{4} \times (0.3)^2 = 69.3 \times 10^3$$

$$F = 4.9 \times 10^3 \text{ N}$$

$$F = 4.9 \text{ kN}$$

***Example* 10:** Viscosity of water is 0.01poise. Find its kinematics viscosity if specific gravity is 0.998.

Kinematics viscosity = ? $\mu = 0.01P$

$S = 0.998$ $= 0.01 \times 0.1$

$$S = \frac{\rho}{\rho_{standrad}}$$ $\mu = 0.001 \dfrac{NS}{m^2}$

$$\therefore KV = \frac{\mu}{\rho}$$

$$0.998 = \frac{\rho}{1000}$$ $= \dfrac{0.001}{998}$

$$\rho = 998 \ kg/m^3$$ $KV = 1 \times 10^{-6} \ m^2/s$

***Example* 11:** A Plate at a distance 0.0254 mm from a fixed plate moves at 0.61 m/s and requires a force of 1.962 N/m^2 area of plate. Determine dynamic viscosity of liquid between the plates.

U = 0.61 m/s

Y = 0.0254 mm
= 0.0254 × 10⁻³ m

$$\tau = 1.962 \ N/m^2$$

$$\mu = ?$$

Assuming linear velocity distribution

$$\tau = \mu \ \frac{U}{Y}$$

$$1.962 = \mu \times \frac{0.61}{0.0254 \times 10^{-3}}$$

$$\mu = 8.17 \times 10^{-5} \frac{NS}{m^2}$$

***Example* 12:** A plate having an area of 1 m^2 is dragged down an inclined plane at 45^0 to horizontal with a velocity of 0.5 m/s due to its own weight.

There is a cushion of liquid 1 mm thick between the inclined plane and the plate. If viscosity of oil is 0.1 PaS, find the weight of the plate.

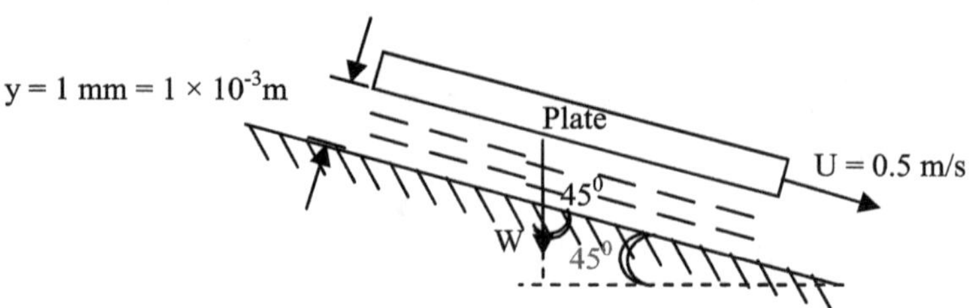

$$A = 1 \text{ m}^2$$

$$U = 0.5 \text{ m/s}$$

$$Y = 1 \times 10^{-3} \text{ m}$$

$$\mu = 0.1 \text{ NS/m}^2$$

$$W = ?$$

$$F = W \times \cos 45^0$$

$$= W \times 0.707$$

$$F = 0.707W$$

$$\tau = \frac{F}{A}$$

$$\tau = \frac{0.707W}{1}$$

$$\tau = 0.707W \text{ N / m}^2$$

Assuming linear velocity distribution,

$$\tau = \mu \frac{U}{Y}$$

$$0.707W = 0.1 \times \frac{0.5}{1 \times 10^{-3}}$$

$$W = 70.72 \text{ N}$$

***Example* 13:** A shaft of ϕ 20 mm and mass 15 kg slides vertically in a sleeve with a velocity of 5 m/s. The gap between the shaft and the sleeve is 0.1 mm and is filled with oil. Calculate the viscosity of oil if the length of the shaft is 500 mm.

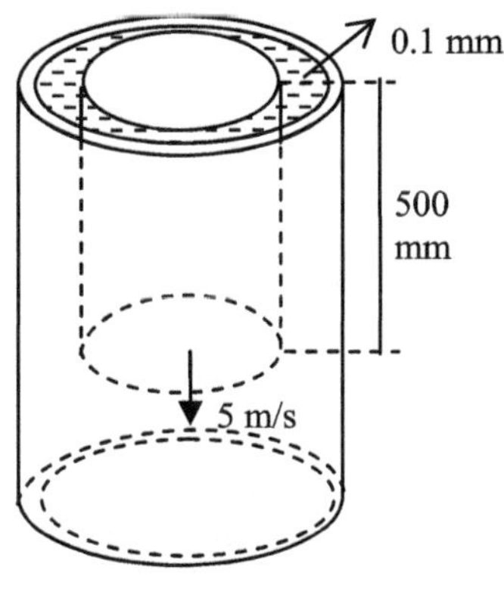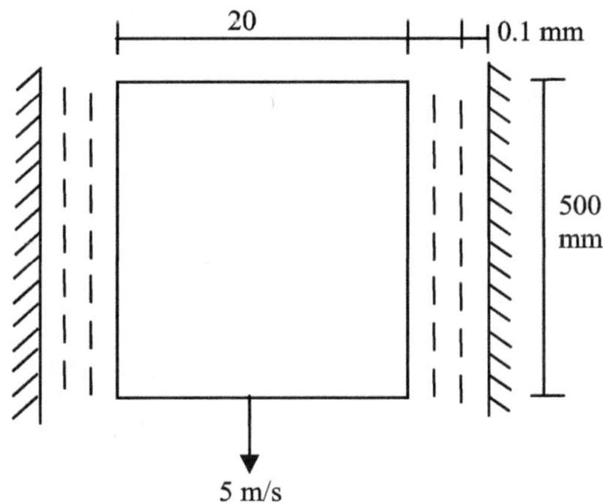

$D = 20 \text{ mm} = 20 \times 10^{-3} \text{ m}$

$M = 15 \text{ kg}$

$W = 15 \times 9.81$

$W = 147.15 \text{ N}$

$y = 0.1 \text{ mm}$

$y = 0.1 \times 10^{-3} \text{ mm}$

$U = 5 \text{ m/s}$

$F = W$

$F = 147.15 \text{ N}$

$\mu = ?$

$A = \pi \, d \, l$

$A = \pi \times 20 \times 10^{-3} \times 0.5$

$A = 0.031 \text{ m}^2$

$$\tau = \mu \frac{U}{Y}$$

$$4746.7 = \mu \times \frac{5}{0.1 \times 10^{-3}}$$

$$\mu = 0.095 \ \frac{NS}{m^2}$$

$$\tau = \frac{F}{A}$$

$$= \frac{147.15}{0.031}$$

$$\tau = 4746.7 \ N/m^2$$

Review Questions

1. Differentiate between (i) Liquid and gases (ii) Real fluids and ideal fluiqs.

2. What do you mean by vacuum pressure?

3. Define the terms: gauge pressure and absolute pressure.

4. Distinguish between Newtonian and Nonnewtonian fluids.

5. State Pascal Law.

6. Show graphically the variation of shear stress with velocity gradient for ideal plastic, ideal, Newtonian, Pseudo plastic fluids.

7. State the Newton's law of viscosity and give examples of its application.

8. Define : (a) Path line and (b) Stream line.

9. Distinguish between laminar and turbulent flow through a pipe.

10. Define stream line and path line.

11. State Bernoulli's theorem and mention the assumptions involved in it.

12. State Bernoulli's theorem. Drive Bernoulli's equation for the flow of an incompressible fluid. Mention the assumptions made for deriving it.

13. State the basic principle of centrifugal pumps.

14. What factors decide whether Kaplan, Francis or a Pelton type turbine would be used in a hydroelectric project?

15. What is an axial piston pump? Sketch and explain. How does this differ from radial piston pump?

16. Describe with a neat sketch the constructional details of an Axial flow pump.

17. Draw and discuss the operating principle of a centrifugal pump.

18. What is a Fluid Coupling and its construction and operating principle?

CHAPTER 4

Thermodynamics

Thermodynamics: First and second law of thermodynamics; steam properties; steam processes at constant pressure, volume, enthalpy & entropy; classification and working of boilers, efficiency & performance analysis; natural and induced draught, calculation of chimney height. Refrigeration, vapor absorption & compression cycles, coefficient of performance (COP), refrigerant properties & Eco-friendly refrigerants.

4.1 Thermodynamics

Thermodynamics is the study of the relationships between *HEAT* (thermos) and *WORK* (dynamics). It deals thus with the energy interactions in physical systems. Thermodynamics is science that deals with transfer of heat and work. Engineering thermodynamics develops the theory and techniques required to use empirical thermodynamic data effectively. Classical thermodynamics can be stated in four laws called the zeroth, first, second, and third laws respectively. The laws of thermodynamics are empirical, i.e., they are deduced from experience, and supported by a large body of experimental evidence.

Engineers use thermodynamics to calculate the fuel efficiency of engines, and to find ways to make more efficient systems like rockets, refineries, or nuclear reactors.

4.1.1 Thermodynamics system

In general, a system is a collection of objects, and there is a lot of subtlety in the way it is defined, as in set theory. However, in thermodynamics, it is a much more straight forward concept.

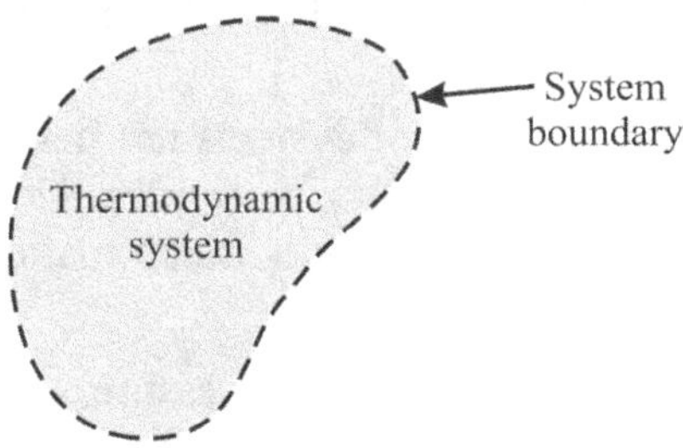

Fig. 4.1 Thermodynamic system

A thermodynamic system is defined as a volume in space or a well defined set of materials (matter) and other than this know as surrounding. The imaginary outer edge of the system is called its boundary. As can be seen from the definition, the boundary can be fixed or moving. Fig. 4.1 shows how thermodynamic system separate from surrounding from system boundary.

A system in which matter crosses the boundary is called an *open system.*

A system in which no matter enters or leaves (i.e., crosses the boundary) is called a *closed system.*

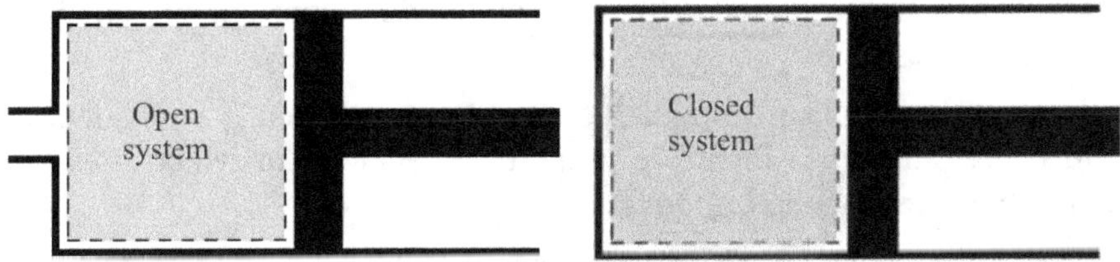

Fig. 4.2 Open & closed system

Fig. 4.2 shows a piston cylinder arrangement, where a gas is compressed by the piston. The dotted lines represent the system boundary. As can be seen, due to an opening in the cylinder, gas can escape outside as the piston moves inwards, and gas enters the system when the piston moves outwards. Thus, it is an open system.

4.1.2   Thermodynamic Properties

Properties of substances are things such as mass, temperature, volume, and pressure. Properties are used to define the current state of a substance. Properties can be *intensive,* if they are point properties (properties that make sense for a point) or *extensive,* if they depend on the amount of matter in the system.

Examples of extensive properties of systems are mass of system, number of moles of a substance in a system, and overall or total volume of a system. These properties depend on how much matter of the system you measure.

Examples of intensive properties are pressure, temperature, density, volume per mass, molar volume (which is volume per mole), and average molecular weight (or molecular mass). These properties are the same regardless of how you vary the amount of mass of the substance.

4.1.3   Properties of Pure Substances

Properties are like the variables for substances in that their values are all related by an equation. The relationship between properties is expressed in the form of an equation which is called an equation of state. Perhaps the most famous state equation is the Ideal Gas Law. The ideal gas law relates the pressure, volume, and temperature of an ideal gas to one another.

Volume: The SI unit for volume is m^3. Volume is an extensive property, but both volume per mass and molar volume are intensive properties since they do not depend on the measured mass of the system. A process during which the volume of the system remains constant is called an *isochoric* process.

Pressure: The SI unit for pressure is Pa (Pascal), which is equivalent to $N/(m^2)$. Pressure is an intensive property. A process in which pressure remains constant is called *isobaric.*

Temperature: The concept of temperature is fundamental to thermodynamics. We know that a body at high temperature will transfer energy to one at lower temperature. Consider two bodies with different temperatures in contact with each other. Net energy transfer will be from the hotter body to the colder body. At some point, the net energy transfer will be zero, and the bodies are said to be in thermal equilibrium. Bodies in thermal equilibrium are defined to have the same temperature.

4.2   Thermodynamic Processes

Process: When a system changes its state from one equilibrium state to another equilibrium state, then the path of successive states through which system has passed, is known as *thermodynamic process.*

Cycle: When a system changes its state from one equilibrium state to another equilibrium state, and finally reaches its initial state, then the system is said to have undergone a *thermodynamic cycle.*

Paths through the space of thermodynamic variables are often specified by holding certain thermodynamic variables constant. It is useful to group these processes into pairs, in which each variable held constant is one member of a conjugate pair.

The pressure-volume conjugate pair is concerned with the transfer of mechanical or dynamic energy as the result of work.

An **isobaric** process occurs at constant pressure. An example would be to have a movable piston in a cylinder, so that the pressure inside the cylinder is always at atmospheric pressure, although it is isolated from the atmosphere.

An **isochoric** process is one in which the volume is held constant, meaning that the work done by the system will be zero. It follows that, for the simple system of two dimensions, any heat energy transferred to the system externally will be absorbed as internal energy. An isochoric process is also known as an isometric process. An example would be to place a closed tin can containing only air into a fire.

An **isothermal process** occurs at a constant temperature. An example would be to have a system immersed in a large constant-temperature bath. Any work energy performed by the system will be lost to the bath, but its temperature will remain constant.

An **isentropic process** occurs at constant entropy. For a reversible process this is identical to an adiabatic process (see below). If a system has entropy which has not yet reached its maximum equilibrium value, a process of cooling may be required to maintain that value of entropy. If a system has an entropy which has not yet reached its maximum equilibrium value, the entropy will increase even though the system is thermally insulated.

An **adiabatic process** is a process in which there is no energy added or subtracted from the system by heating or cooling. For a reversible process, this is identical to an isentropic process. We may say that the system is **thermally insulated** from its environment and that its boundary is a thermal insulator.

4.2.1   Work and heat in thermodynamics processes

A gas in a cylinder can be compressed by pushing a piston. The work done *on* the gas during an infinitesimal volume change is given by

$$dW = -Fdx = -PAdx = -PdV$$

If the gas is compressed, then ΔV is negative and the work done is positive. (Note the negative sign in the above equation.) If the gas expands, then the work done on the gas is negative. (Work is done *by* the gas).

The net work done during a finite volume change is

$$W = -\int_{V_i}^{V_f} PdV$$

Isobaric process: In an *isobaric* process the pressure is constant. Thus, the work done is

$$W = -P\Delta V = P(V_i - V_f)$$

Isovolumeric process: In an isovolumeric (isochoric) process the volume is constant. Thus,

$$W = -P\Delta V = 0$$

If the pressure changes during the compression or expansion, then the work done on the substance is the area under the P versus V curve.

Isothermal process: The work done on a gas during an isothermal (constant temperature) process is

$$W = -\int_{V_i}^{V_f} PdV = -\int_{V_i}^{V_f} \frac{nRT}{V}dV = nRT\ln\left(\frac{V_i}{V_f}\right)$$

PV diagram of Isobaric, isovolumetric and isothermal process are shown in Fig. 4.3.

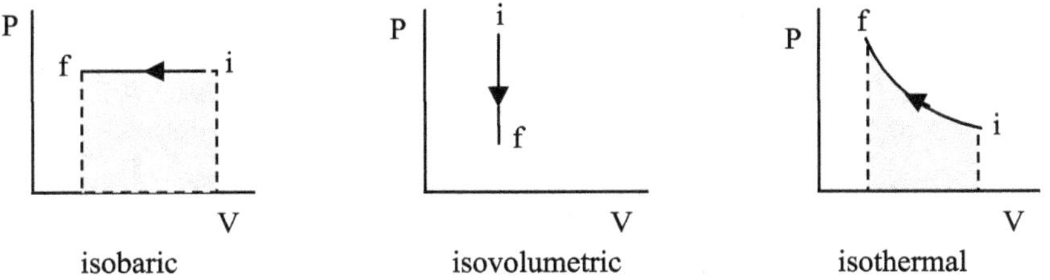

Fig. 4.3 P-V Diagram Thermodynamic Process

4.3 Law of Thermodynamics

4.3.1 Zeroth law of thermodynamics

If two bodies at different temperatures are brought together, the hot body will warm up the cold one. At the same time, the cold body will cool down the hot one. This process will end when the two bodies have the same temperatures. At that point, the two bodies are said to have reached thermal equilibrium. It states that, If two thermodynamic systems are in thermal equilibrium with a third, they are also in thermal equilibrium with each other. If

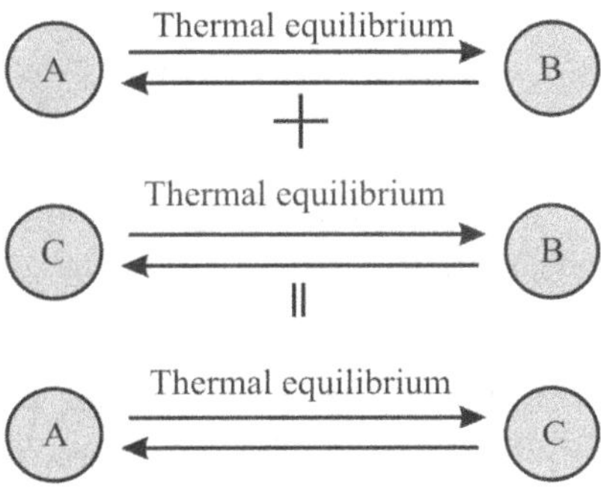

Fig. 4.4 Zeroth Law of Thermodynamics

bodies A and B are each in thermal equilibrium with a third body T, then they are in thermal equilibrium with each other, and have the same temperature (Fig. 4.4). The Zeroth Law of thermodynamics is a basis for the validity of temperature measurement.

4.3.2 First law of thermodynamics

The first law of thermodynamics relates the change in internal energy to the heat absorbed and the work done on a substance. It is essentially a statement of conservation of energy. In any process, the total energy of the universe remains constant. More simply, the First Law states that energy cannot be created or destroyed; rather, the amount of energy lost in a process cannot be greater than the amount of energy gained. The first law states that when heat and work interactions take place between a closed system and the environment, the algebraic sum of the heat and work interactions for a cycle is zero.

Since the first law is another way of stating the conservation of energy, the energy of the system is the sum of the heat and work input, i.e.,

$$\Delta E = Q + W.$$

$$(E = U + K.E. + P.E.)$$

where E represents the heat energy of the system along with the kinetic energy and the potential energy and is called the *total internal energy* of the system

Q is the heat transferred and W is the work done on or by the system

Since these are the only ways energy can be transferred, this implies that the total energy of the system in the cycle is a constant. One consequence of the statement is that the total energy of the system is a property of the system. This leads us to the concept of internal energy.

$$\Delta U = U_f - U_i = Q + W$$

Q is positive if heat is absorbed and is negative if heat is lost by the system.

In an *isolated* system, $Q = 0$ and $W = 0$. Thus, $\Delta U = 0$ and the internal energy remains constant. The internal energy, U, depends on the *state* of the system. Thus, if a system goes through a cycle and returns to its original state, then $\Delta U = 0$ and $Q = -W$.

An *adiabatic process* is one for which no heat flows into or out of a system. For an adiabatic process, $\Delta U = W$. Thus, in an adiabatic expansion the internal energy decreases and in an adiabatic compression the internal energy increases.

4.3.3 Second law of thermodynamics

The first law is a statement of energy conservation. The rise in temperature of a substance when work is done is well known. Thus work can be completely converted to heat. However, we observe that in nature, we don't see the conversion in the other direction spontaneously. The statement of the second law is facilitated by using the concept of heat engines.

Heat engines work in a cycle and convert heat into work. A thermal reservoir is defined as a system which is in equilibrium and large enough so that heat transferred to and from it does not change its temperature appreciably.

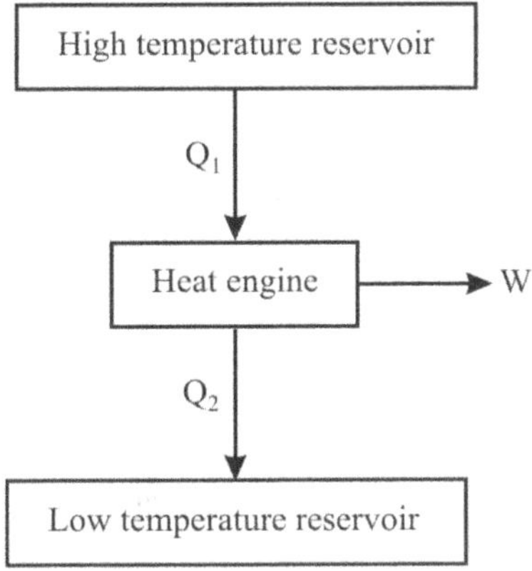

Fig. 4.5 Heat engine

Heat engines usually work between two thermal reservoirs, the low temperature reservoir (sink) and the high temperature reservoir (source) (Fig.4.5). The performance of a heat engine is measured by its thermal efficiency.

Thermal efficiency of heat engine is defined as the ratio of work output to heat input,

i.e., $\quad \eta = W/Q_1 = Q_1 - Q_2 /Q_1 = 1 - Q_2/Q_1 = 1 - T_2/T_1$

Where W is the net work done, and Q_1 is heat transferred from the high temperature reservoir. Q_2 is heat rejected to the low temperature reservoir.

Heat pumps (or) Reversible Heat Engines transfer heat from a low temperature reservoir to a high temperature reservoir using external work, and can be considered as reversed heat engines (Fig. 4.6). The performance of heat pump is measured in terms of coefficient of performance which is defined as the ratio of maximum heat transferred to the amount of work required to produce the desired effect.

i.e., $\quad COP = Q_2/W = Q_2/ Q_1 - Q_2 = T_2 / T_1 - T_2$

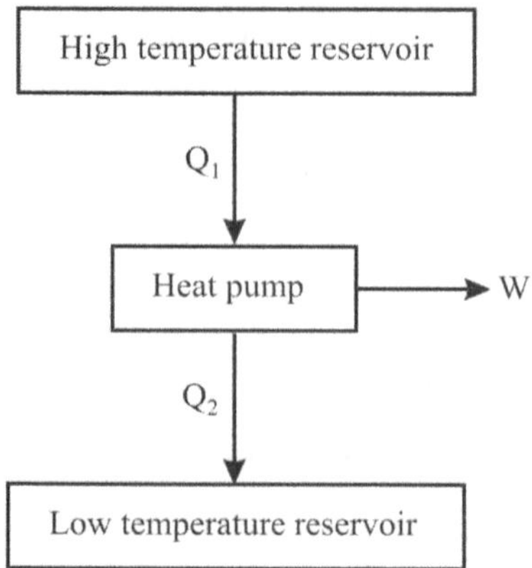

Fig. 4.6 Heat pump

Statement of the Second Law of Thermodynamics:

Kelvin-Planck Statement: It is impossible to construct a heat engine which will operate continuously and convert all the heat it draws from a reservoir into work.

Clausius Statement: It is impossible to construct a heat pump which will transfer heat from a low temperature reservoir to a high temperature reservoir without using external work.

PMM2: A perpetual motion machine of the second kind or PMM2 is one which converts all the heat input into work while working in a cycle. A PMM2 has a η_{th} of 1. But in practice PMM2 is not possible.

Equivalence of Clausius and Kelvin-Planck Statements: Suppose we can construct a heat pump which transfers heat from a low temperature reservoir to a high temperature one without using external work. Then, we can couple it with a heat engine in such a way that the heat removed by the heat pump from the low temperature reservoir is the same as the heat rejected by the heat engine, so that the combined system is now a heat engine which converts heat to work without any external effect (Fig. 4.7). This is thus in violation of the Kelvin-Planck statement of the second law.

Now suppose we have a heat engine which can convert heat into work without rejecting heat anywhere else. We can combine it with a heat pump so that the work produced by the engine is used by the pump. Now the combined system is a heat pump which uses no external work, violating the Clausius statement of the second law.

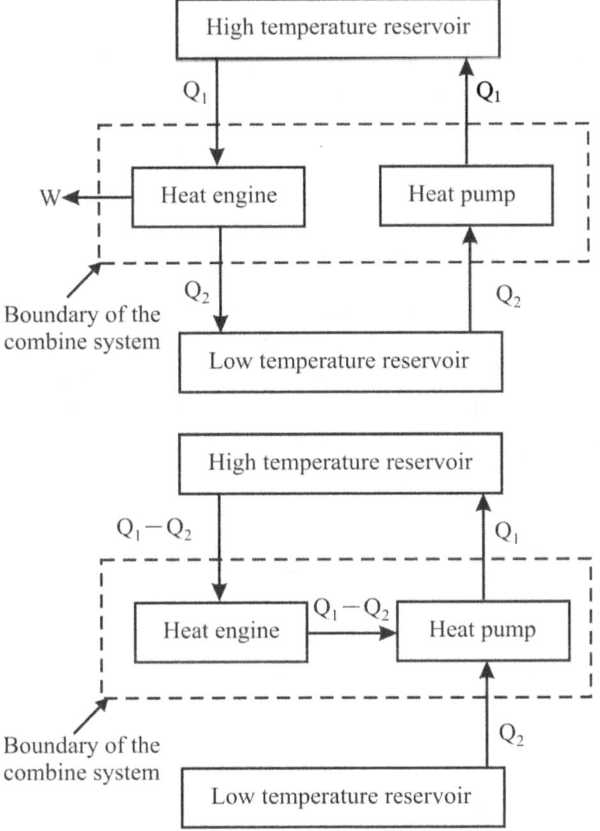

Fig. 4.7 Equivalence of Kelvin-Planck and Clausius Statements

Thus, we see that the Clausius and Kelvin-Planck statements are equivalent, and complimentary to each other. The truth of Kelvin statement necessarily implies the truth of Clausius statement.

4.3.4   Third law of Thermodynamics

As temperature approaches absolute zero, the entropy of a system approaches a constant. The Third Law deals with the fact that there is an absolute constant in the universe known as absolute zero. The third law of thermodynamics is a statistical law of nature regarding entropy and the impossibility of reaching absolute zero of temperature. As a system approaches absolute zero, all processes cease and the entropy of the system approaches a minimum value.

4.4   Steam Properties

Steam does not obey the ideal gas laws. However, the equations derived from the first and second law of thermodynamics hold good for steam. Steam follows different processes as constant pressure process, constant volume process, constant temperature process and constant enthalpy process etc.

4.4.1   Constant Pressure Process

Let unit mass of steam is heated at constant pressure p to form the superheated steam. The temperature of steam remains constant till it is completely vaporized, after this temperature increases. The constant pressure heating is used in boilers for steam generation. The constant pressure heating process can be represented on Pv and Ts diagram as shown in Fig.4.8.

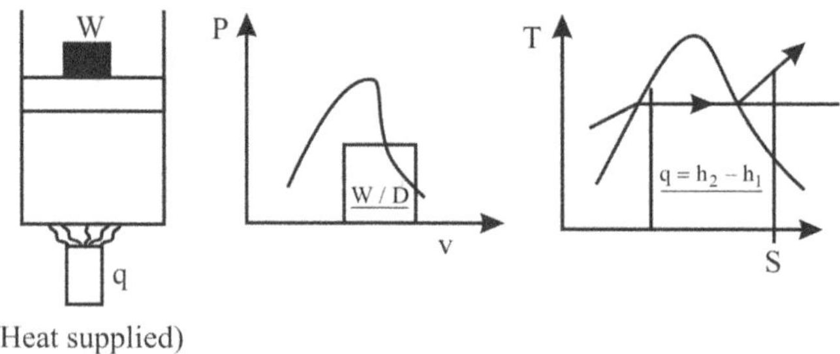

Fig. 4.8 Constant pressure process

Work done at constant pressure:

$$\text{Work done} = \text{sum of p dv under the limit of 1 to 2}$$
$$= P\,(v_2 - v_1)$$
$$= N\,m^3/\,m^2\,kg$$
$$= N.m\,/kg$$
$$= J/kg$$

If p is in bar then unit of work done $\{1\ bar = 100000\ N/m^2\}$

$$W/d = 10^5\ N/m^2\,.\,m^3/\,kg$$
$$= 100\ KJ/kg$$
$$W/D_{1-2} = P\,(v_2 - v_1)$$

where,

$$v_1 = x\,v_1 + (1 - x)\,v_f \qquad \text{(for wet steam)}$$

$$v_1 = x \, v_g \qquad\qquad \text{(v_f is very small so neglected)}$$
$$v_2 = v_{sup} \qquad\qquad \text{(as the steam is superheated)}$$

At given pressure P, v_{sup} and v_g can be found out from the steam table.

Heat transfer

$$Q_{1-2} = du + p \, dv$$
$$= (u_2 - u_1) + p \, (v_2 - v_1)$$
$$= (u + p \, v_2) - (u_1 + p \, v_1)$$
$$Q_{1-2} = h_2 - h_1 \qquad\qquad \{h = u + p \, v\}$$

where, $\qquad h_2 = h_{sup}$

$$h_1 = h + x \, h_{fg}$$

From the above relation it is observed that heat transfer at constant pressure is equal to change in enthalpy.

Change in Internal Energy

$$u_2 - u_1 = h_2 - h_1 - p \, (v_2 - v_1)$$

4.4.2 Constant Volume Process

Let unit mass of wet steam is heated at Constant volume from pressure p_1 to pressure p_2 so that it becomes superheated steam (Fig. 4.9).

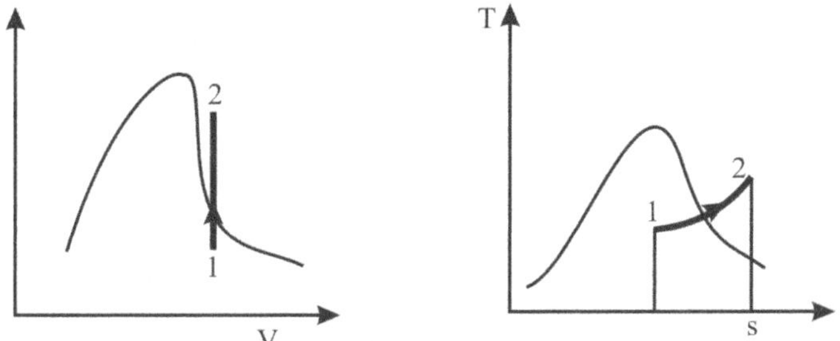

Fig. 4.9 Constant volume process

Constant volume process

$$v_1 = v_2$$
$$x \, v_{g1} = v_{sup\,2} \quad \{v_{f1} \text{ is neglected as very small quantity}\}$$
$$x = v_{sup\,2} \, / \, v_{g1}$$

Work done,

$$W_{1-2} = \text{sum of all } p \, dv$$
$$= 0 \qquad \text{(as constant volume dv=0)}$$

Heat transferred

$$Q_{1\text{-}2} = du + p\,dv$$
$$= du \qquad (as\ dv = 0)$$
$$= u_2 - u_1$$
$$= (h_2 - p_2\,v_{sup\,2}) - (h_1 - p_1\,x\,v_{g1})$$

where, $h_1 = h_{f1} + xh_{fg1}$

and $h_2 = h_{g2} + c_{ps}\,(t_{sup} - t_{sat2})$

Change in internal energy

$$du_{1-2} = u_2 - u_1 = (h_2 - p_2\,V_{sup}) - (h_1 - x\,p_1\,v_{gt})$$

4.4.3 Constant temperature process

When wet steam is heated at constant temperature pressure till it becomes dry saturated steam. In wet region, the constant temperature process is also constant pressure evaporation (Fig. 4.10). After saturation steam becomes saturated region, constant temperature process becomes hyperbolic process (PV = constant i.e., as volume increases pressure decreases.)

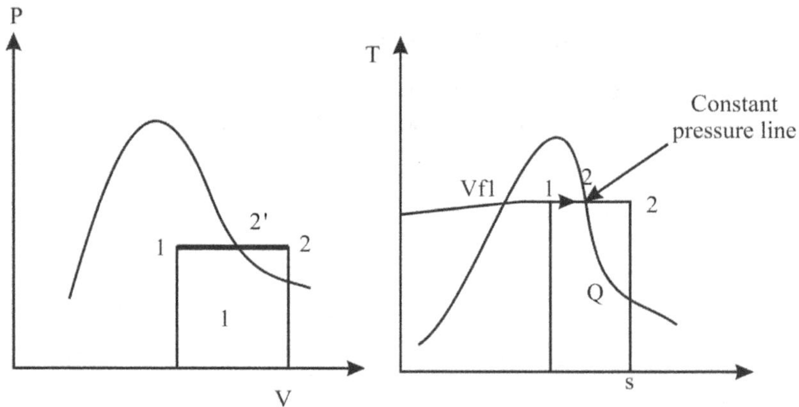

Fig. 4.10 Constant temperature process

Work done

Work done in wet region,

$$W/D_{1-2} = p\,(v_{g2} - x\,v_{g1})$$

In wet region,

$$v_{g2} = v_1 \text{ and } p_1 = p_2$$
$$W/D_{1-2} = p\,v_{g1}\,(1 - x)$$

Work done in hyperbolic process

$$W_{1-2} = \text{sum of p dv under the limits of 1 to 2}$$

$$= p_1\, v_1 \log v_2 / v1 \quad (pv = \text{constant})$$

Heat transferred

$$Q_{1-2} = (u_2 - u_1) + W_{1-2}$$

$$= (u_2 - u_1) + p_1 v_1 \log v_2/v_1$$

$$Q_{1-2} = (h_2 - p_2 v_2) - (h_1 - p_1 v_1) + p_1 v_1 \log v_2/v_1$$

For hyperbolic process, pv= constant

$$p_2 v_2 = p_1 v_1$$

$$Q_{1-2} = (h_2 - h_1) + p_1 v_1 \log v_2/v_1$$

4.4.4 Constant Enthalpy Process

When a fluid passes through a narrow passage it is said to be throttled. During throttling process enthalpy of fluid remained constant, pressure drops and specific volume increases (Fig. 4.11). This process is used to determine the dryness fraction (x) of steam.

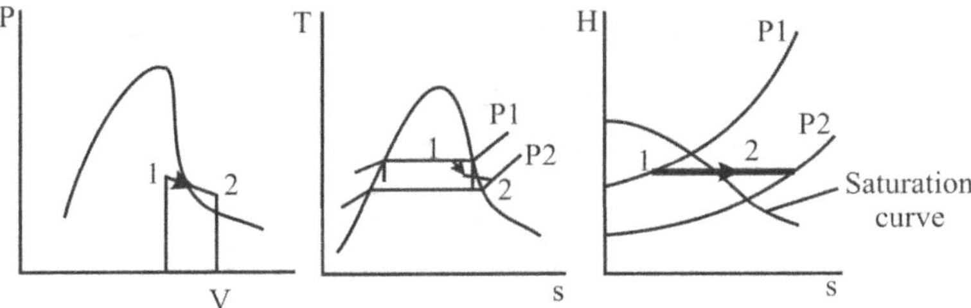

Fig. 4.11 Constant Enthalpy Process

Enthalpy before throttling = Enthalpy after throttling

or,   Enthalpy of wet steam = Enthalpy of superheated steam

$$h_{f1} + x\, h_{fg} = h_{g2} + c_{ps}\,(T_{sup\,2} - T_{sat\,2})$$

$$x = \frac{h_{g2} - h_{f1} + c_{ps}(T_{sup2} - T_{sat2})}{h_{fg1}}$$

4.4.5   Constant Entropy Process

Constant entropy is also known as reversible adiabatic process or isentropic process (Fig. 4.12).

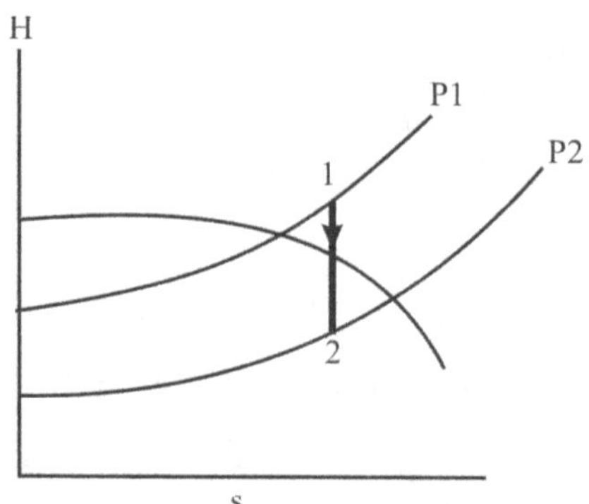

Fig. 4.12 Constant Entropy Process

Now consider 1 kg of wet steam being heated by reversible adiabatic process,

Let,

p_1 = initial pressure of the steam in bar

v_{g1} = specific volume of the dry saturated steam in m^3 / kg

x_1 = initial dryness fraction of the steam

p_2, v_{g2}, x_2 = corresponding values for the final condition of the steam.

Initial entropy of steam before expansion

$$S_1 = S_{f1} + x_1 \; h_{fg1} \; / T_1 \; = S_{f1} + x_1 \; S_{fg1}$$

And Final entropy of steam after expansion

$$S_2 = S_{f2} + x_2 \; h_{fg2} \; / T_2 \; = S_{f2} + x_2 \; S_{fg2}$$

The entropy before expansion is equal to the entropy after expansion,

Therefore, $S_{f1} + x_1 \; h_{fg1} / T_1 \; = S_{f2} + x_2 \; h_{fg2} / T_2$

or, $S_{f1} + x_1 \; S_{fg1} = S_{f2} + x_2 \; S_{fg2}$

Change in Internal Energy

Initial energy of the steam

$$u_1 = h_1 - 100 \; p_1 \; v_1$$

where, V_1 = initial volume of steam = $x_1 \; v_{g1}$

and, Final energy of the steam

$$u_2 = h_2 - 100 \; p_2 \; v_2$$

where, v_2 = final energy of the steam

$$= x_2 \; v_{g2}$$

$$= v_{g2}$$

$$= v_{sup}$$

Change in internal energy

$$du = u_2 - u_1$$

Heat Absorbed or Heat Transferred

Since no heat is added or subtracted during an adiabatic process, therefore

$$Q_{1-2} = 0$$

Work Done During the Process

According the first law of thermodynamics,

the heat absorbed = heat transferred

$$Q_{1-2} = du + W_{1-2}$$

Since, $Q_{1-2} = 0$, during a non flow reversible adiabatic or isentropic process, therefore work done during the process

$$W_{1-2} = - \, du = u_1 - u_2$$

This shows that during a reversible adiabatic or isentropic process done is equal to the change in internal energy.

4.5   Boilers

In simple, a boiler may be defined as a closed vessel in which steam is produced from water by combustion of fuel. According to American Society of Mechanical Engineers (A.S.M.E.) a 'steam generating unit' is defined as:

"A combination of apparatus for producing, furnishing or recovering heat together with the apparatus for transferring the heat so made available to the fluid being heated and vapourised".

Indian boiler regulation: Indian boiler regulation states that a vessel exceeding 22.75 liters in capacity, which is used for generating steam under pressure, will be treated as steam boiler.

4.5.1 Classifications of boilers

(i) *According to contents in the tube:* The steam boilers, according to the contents in the tube may be classified as:

Fire tube or smoke tube boiler: In fire tube steam boilers, the flames and hot gases, produced by the combustion of fuel, pass through the tubes (called multi-tubes) which are surrounded by water. The heat is conducted through the walls of the tubes from the hot gases to the surrounding water.

Examples of fire tube boilers are: Simple vertical boiler, Cochran boiler, Lancashire boiler, Cornish boiler, Scotch marine boiler, Locomotive boiler and Velcon boiler.

Water tube boilers: In water tube steam boilers, the water is contained inside the tubes (called water tubes) which are surrounded by flames and hot gases from outside.

Examples of water tube boilers are: Babcock and Wilcox boiler, Stirling boiler, La-mont boiler, Benson boiler, Yarrow boiler and Loeffler boiler.

(ii) *According to position of the furnace:* The steam boilers, according to the position of the furnace are classified as:

Internally fired boilers: In internally fired boilers, the furnace is located inside the boiler shell. Most of the fire tube boilers are internally fired.

Externally fired boilers: In externally fired boilers, the furnace is arranged underneath in a brick-work setting. Water tube steam boilers are always externally fired.

(iii) *According to the axis of the shell:* The steam boilers, according to the axis of the shell may be classified as:

Vertical boilers: In vertical steam boilers, the axis of the shell is vertical. Simple vertical boiler and Cochran boilers are vertical boilers.

Horizontal boilers: In horizontal steam boilers, the axis of the shell is horizontal. Lancashire boiler, Locomotive boiler and Babcock and Wilcox boilers are horizontal boilers.

(iv) *According to the method of circulation of water and steam*: The steam boilers, according to the method of circulation of water and steam, may be classified as:

Natural circulation boilers: In natural circulation steam boilers, the circulation of water is by natural convection currents, which are set up during the heating of water. In most of the steam boilers, there is a natural circulation of water.

Forced circulation boilers: In forced circulation steam boilers, there is a forced circulation of water by a centrifugal pump driven by some external power. Use of forced circulation is made in high pressure boilers such as La-mont boiler, Benson boiler, Loeffler boiler and Velcon boiler.

(v) *According to the use*: The steam boilers, according to the use are classified as:

Stationary boilers: These are used for power plant steam, for central station utility power plants, for plant process steam etc.

Mobile boilers: These include Locomotive type, and other small units for temporary use at sites (just as in small coal field pits).

(vi) *According to the nature of draught, boilers can be classified as*: Natural draught boilers, Forced draught boilers and Balanced draught boilers.

Natural drought: It is the drought produced by a chimney due to difference of densities between the hot gases inside the chimney and cold atmospheric air outside.

Forced drought: It is also known as mechanical drought and it is of two types namely forced fan drought and forced steam jet drought according to their method of working.

Balanced drought: It is an improved type of drought, and is a combination of induced and forced drought. It is produced by running both induced and forced fans simultaneously.

4.5.2 Selection of a boiler

While selecting a boiler the following factors should be considered:

   (i)   The working pressure and quality of steam required (i.e., whether wet or dry or super heated).

  (ii)   Steam generating rate.

 (iii)   Floor area available.

 (iv)   Accessibility for repair and inspection.

   (v) Comparative initial cost.

  (vi) Erection facilities.

 (vii) The portable load factor.

(viii) The fuel and water available.

  (ix) Operating and maintenance costs.

4.5.3 Essentials of a good steam boiler

A good boiler should possess the following features:

   (i) The boiler should produce the maximum weight of steam of the required quality at minimum expenses.

  (ii) Steam production rate should be as per requirements.

 (iii) It should be absolutely reliable.

  (iv) It should occupy minimum space.

   (v) It should be light in weight.

  (vi) It should be capable of quick starting.

 (vii) There should be an easy access to the various parts of the boiler for repairs and inspection.

(viii) The boiler components should be transportable without difficulty.

  (ix) The installation of the boiler should be simple.

   (x) The tubes of the boiler should not accumulate soot or water deposits and should be sufficiently strong to allow for wear and corrosion.

  (xi) The water and gas circuits should be such as to allow minimum fluid velocity (for low frictional losses).

4.5.4 Uses of boilers

   (i) Industrial Boilers are made of steel and used for generation of steam.

  (ii) Vaporizes water, combustion of fuels takes place E.g. Coal, coke, oil, wood, saw dust and bagasse etc.

 (iii) Steam used for producing power, for industrial process work or for heating proposes.

4.6   Boiler Mounting and Accessories

4.6.1   Boiler mounting

Mountings are the devices necessary for the operation and safety of the boiler. The important Boiler Mountings are described below.

- (i)   Water level indicator
- (ii)   Pressure gauge
- (iii)   Dead safety valve
- (iv)   Lever safety valve
- (v)   Fusible plug
- (vi)   Blow off cock
- (vii)   Feed check valve

- **(i)**   ***Water level indicator***: The function of water level indicator is to indicate the level of water in the boiler constantly. It is also called water gauge. A is the front end plate of the boiler (Fig. 4.13). F is a very hard glass tube indicating water level and is connected to the boiler Plate through stuffing boxes in hollow gun metal casting (b, c) having flanges x, y for bolting the plate. For controlling the passage of stream and water, cocks D and E are provided. When these cocks are opened water stand in the same level as in the boiler. K is the drain cock to blow out water at interval so as not to allow any sediment to accumulate. Upper and lower stuffing boxes are connected by a hollow metal column G. Balls J and H rest in the position shown in the normal working of the gauge. When the glass tube breaks due to rush of water in the bottom passage the balls move to dotted position and shut off the water and stream. Then the cocks D and E can be safely closed and broken glass tube replaced. M, N, P and R are screwed caps for internal cleaning of the passage after dismantling. L is the gauge glass, it is tough and does not give splinters.

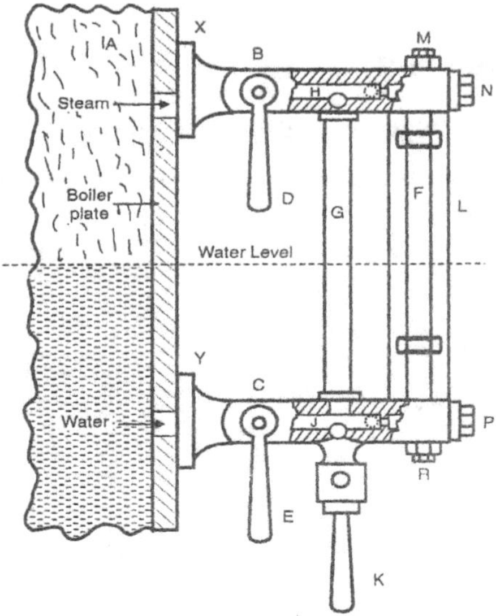

Fig. 4.13 Water level indicator

On breaking thus when the gauge glass breaks, and this guard glass which normally will hold flying pieces, also gives way, the piece will not hurt the attendant.

(ii) ***Pressure gauge***: The function of a pressure gauge is to measure the pressure exerted inside the vessel. A bourdon pressure gauge is a common type of pressure gauge used as shown in Fig. 4.14. The essential feature of this gauge is the elliptical spring tube which is made of a special quality of bronze and is solid drawn.

One end A is closed by a plug and the other is connected with a block C, the block is connected with a siphon Tube. The steam pressure force the water from the syphon tube into elliptical tube and this causes the tube to become circular is cross section. As the tube is fixed at C, the other end A moves out words. This outward movement is magnified by the rod R and transmitted to toothed sector T. This toothed sector is hinged at the point H and meshes with the pinion P fixed to the spindle of the pointer N. Thus the pointers moves and register the pressure on a graduated dial. The movement of the free end of the elliptical tube is proportional to the difference between external and internal pressure on the tube. Since the outside pressure on the tube is atmospheric, the movement of the free end is a measure of the boiler pressure above atmosphere

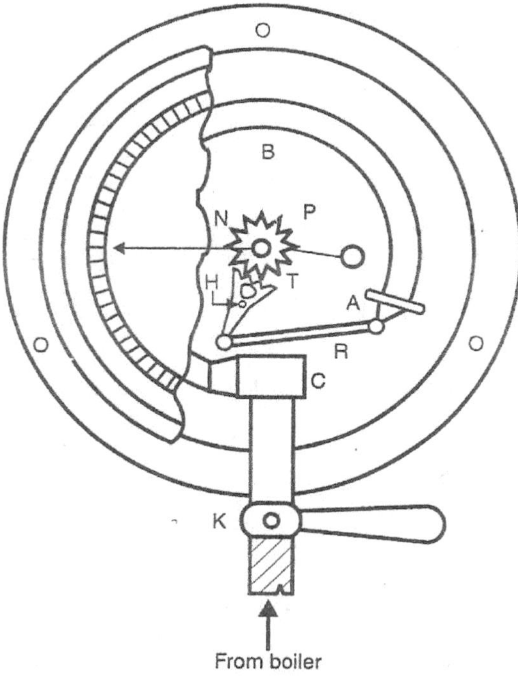

Fig. 4.14 Pressure gauge

(iii) ***Dead safety valve***: A is the vertical cast iron pipe through which steam pressure acts (Fig. 4.15). B is the bottom flange directly connected to seating block on the boiler shell communicating to the steam space. V is the gun metal valve and VS is the gun metal valve seat. D is another cast iron pipe for discharge of excess steam from the boiler. W are the weights in the form of cylindrical disc of cast iron. WC is the weight carrier carrying the weight W. The cover plate C cover these weights. The steam pressure acts in the upward direction and is balanced by the force of the dead weight W. The total dead weight consists of the sum of the weight W. Weight of the valve V, weight carrier and weight of the cover plate C.

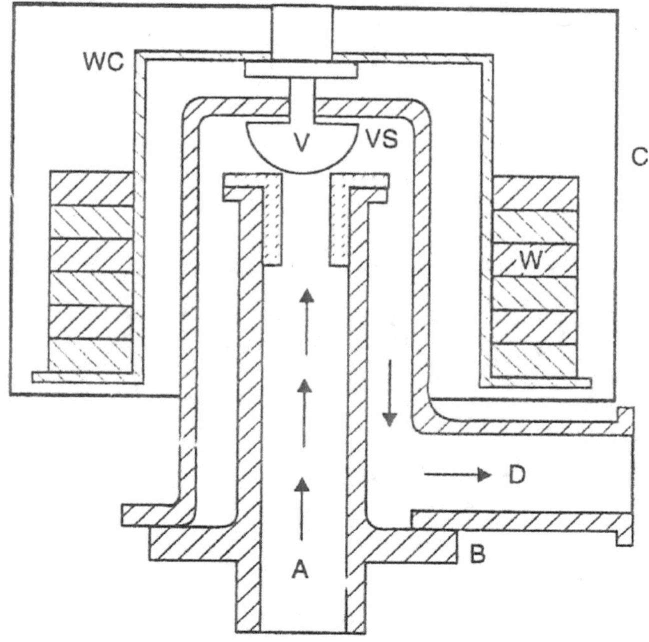

Fig. 4.15 Dead Safety Valve

(iv) *Lever safety valve*: It consists of a lever and weight W (Fig. 4.16). The valve rest on the valve seat which is screwed into the valve

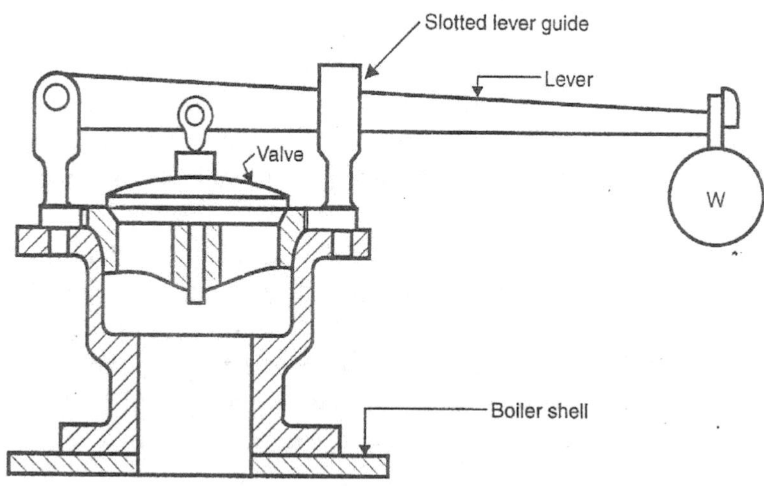

Fig. 4.16 Lever safety valve

body, the valve seat can be replaced if required. The valve body is fitted on the boiler shell. One end of the lever is hinged while at the other is suspended a weight W. The struck presses against the steam pressure below the valve. The slotted lever guide allows vertical movement to the lever. When the steam pressure become greater than the normal working pressure, the valve is lifted with the weight. Consequently, the steam escape through the passages between the valve and seat and the steam pressure decreases.

(v) *Fusible plug*: Function of the fusible plug is to protect the fire box crown plate , or the fire tube from burning when the level of the water in the boiler shell falls abnormally low. The fusible plug consists of a hollow gun metal body screwed into the fire box crown (Fig. 4.17). The body has hexagonal flange to tighten it into

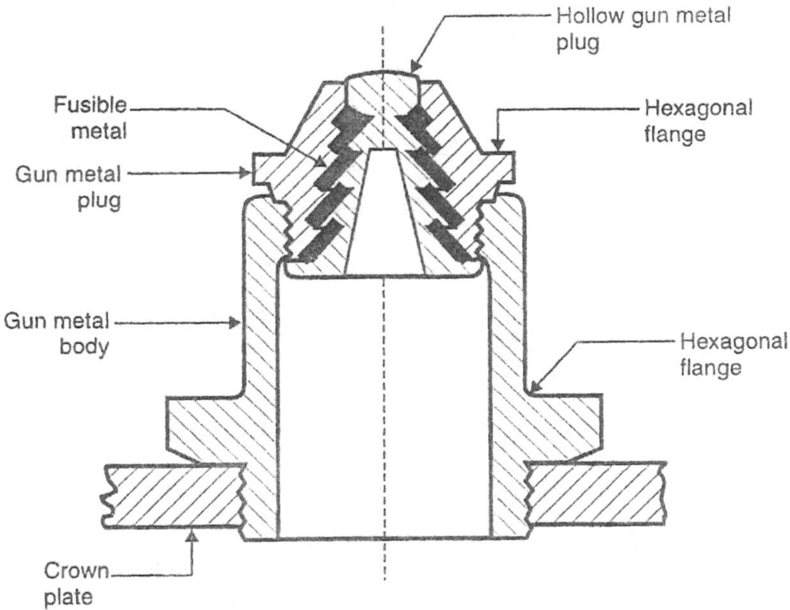

Fig. 4.17 Fusible Plug

the shell. There is another solid plug made of copper with conical copper plug and the gun metal plug together due to depressions provided at the mating surfaces. This plug under normal conditions is covered with water in the boiler which keeps the temperature of the fusible metal below its melting point. When the water level falls low enough to uncover the top of the plug, the fusible metal quickly melts, the plug drops out and the opening so made allows the steam to rush into the furnace. The steam thus puts out the fire or gives warning that the crown of furnace is in danger of being over-heated.

(vi) *Blow off cock*: It is used to empty the boiler whenever required and to discharge the mud scale or sediments which are accumulated at the bottom of the boiler. The blow cock is fitted to the bottom of a boiler drum and consists of a conical plug fitted to the body or casting (Fig. 4.18). When more than one boiler are working and they drain in the same waste pipe line, an isolating valve is necessary to prevent the discharge of one boiler, from entering into the other. Then plug P of the cock is conical and fits into the casting C which is packed with asbestos packing in grooves round the top and bottom of the plug. The shank S of the plug passes through a gland and stuffing box in the cover. The plug is held down by yoke Y and studs. A are the vertical slots for fixing the box spanner, on the top of the yoke. The plug spindle S

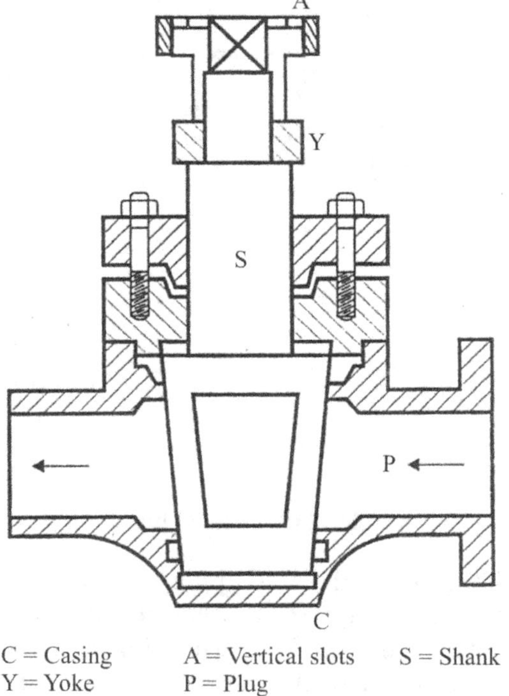

C = Casing   A = Vertical slots   S = Shank
Y = Yoke   P = Plug

Fig. 4.18 Blow-off Cock

is generally rotated by means of the box spanner. The plug P has a hole. When this is brought in line with the casting hole by rotating the spindle S, the water flows out of it and the water can not flow when the solid portion of the plug is in front of casting hole.

(vii)   *Feed check valve*: The function of feed check valve is to control the supply of water to the boiler and to prevent the escaping of water from the boiler when the pump pressure is less or pump is stopped. It consists of a check valve CV which moves automatically up and down under the pressure of water on its gun metal seat (Fig. 4.19). FV is the feed check valve which can be raised or lowered on its gun metal seat, there by opening the delivery passage and its opening control simultaneously lift of the check valve CV. F is the flange which is bolted to the front end of the point from which an internally perforated pipe leads the feed water and distributes it near the working level of the boiler. During normal operation the feed valve may be fed into the boiler. When the pump stops working, the pump pressure will be less than the boiler pressure, the valve may be closed to its seat automatically due to the pressure of water from the boiler. If the boiler is in operation it should be seen that there is sufficient water in the boiler

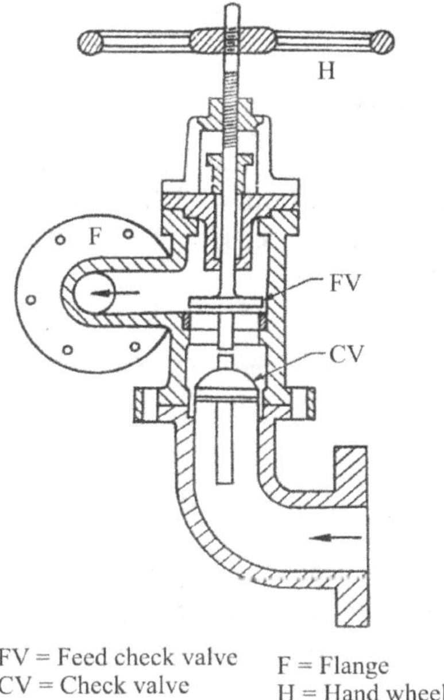

FV = Feed check valve   F = Flange
CV = Check valve   H = Hand wheel

Fig. 4.19 Feed check valve

***Merits of mounting*:**
 (i) Boiler mountings are primarily intended for safety of the boiler.
 (ii) A boiler cannot work without mounting.
 (iii) Mountings are mounted on the body of the boiler itself.

***Demerits of the mounting*:**
 (i) Unsuitable for use on any boiler where extensive vibration and movement are experienced.
 (ii) Dead weight safety valve is not suitable for high pressure boiler because a large amount of weight is required to balance the steam pressure.
 (iii) High pressure safety valve cannot be used on mobile boiler.
 (iv) Fusible plugs should generally be renewed after a period of about two years as they are liable to become defective over a long period of use.

***Uses of mounting*:**
 (i) Boiler of Lancashire and Cornish type should be fitted with a high pressure and low water safety valve.
 (ii) The entire land boiler should have a fusible plug in each furnace.
 (iii) The double-tube Bourdon gauge is widely used for Locomotive and Portable boilers.

4.6.2 Boilers Accessories

The accessories are mounted on the boiler to increase its efficiency. These are optional on efficient boilers. With addition of the accessories, the efficiency of the plant increases. The following accessories are normally used on a modern boiler:
 (i) Super heater
 (ii) Economiser
 (iii) Air Preheater
 (iv) Feed Water pump
 (v) Steam Injector

 (i) ***Superheater*:** It is a heat exchanger in which heat of combustion products is used to dry the wet steam and to increase its temperature. During superheating, the steam pressure remains constant, its specific volume and temperature increases. Fig. 4.20 shows schematics of Surgden's superheater. It consists of two mild steel headers. The U shaped steel tubes are connected to these heaters as shown. The steam generated into the boiler passes the valve C and enters the superheater U tubes through the intake header. The steam is made dry and superheated in these rubes by supplying heat and then it is taken for use through valve B via uptake header. The superheating of steam is controlled by controlling the quantity of flue gases by operating the dampers manually. If superheated steam is not needed or the superheater is under maintenance, the valve B and C are closed and steam is taken out through valve A.

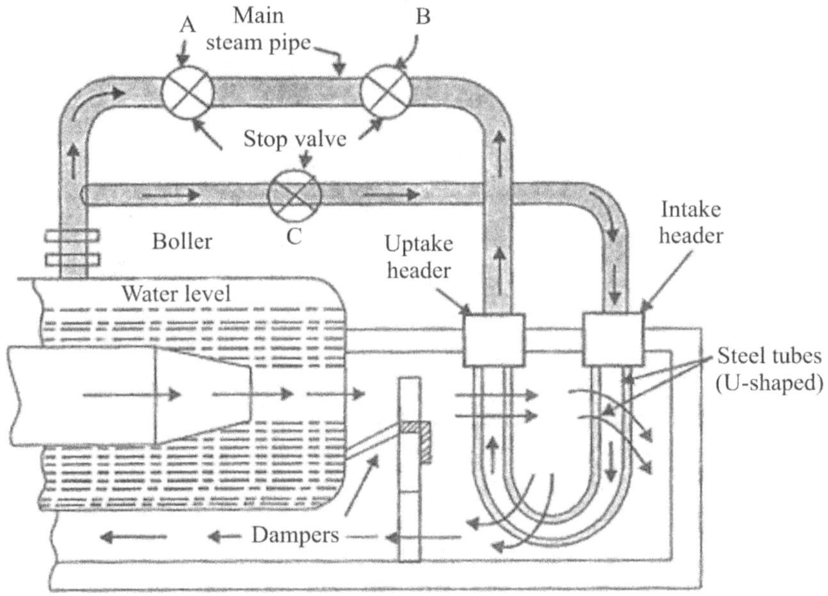

Fig. 4.20 Superheater

(ii) *Economizer*: An economiser is a heat exchanger, used for heating the feed water before it enters the boiler. The economizer recovers some of waste heat of flue gases going to chimney. It helps in improving the boiler efficiency. It is placed in the path of flue gases at the rear end of the boiler just before air preheater. The most popular economizer is Green's economizer as shown in the Fig. 4.21. The Green's economizer consist of a set of vertical cast iron pipes joined with horizontal lower and upper headers. The cold feed water flows through the vertical pipes via lower header. The hot flue gases pass over them and transferring heat to water. The heated water is supplied to boiler via upper header. The scrappers are provided on pipes, which move up and down slowly by means of chains and sprockets to avoid the soot deposition on the pipe surface. The soot collected in soot chamber can be removed from the door. Each economizer is equipped with a safety valve, a drain valve, a release valve, pressure gauge and thermometers.

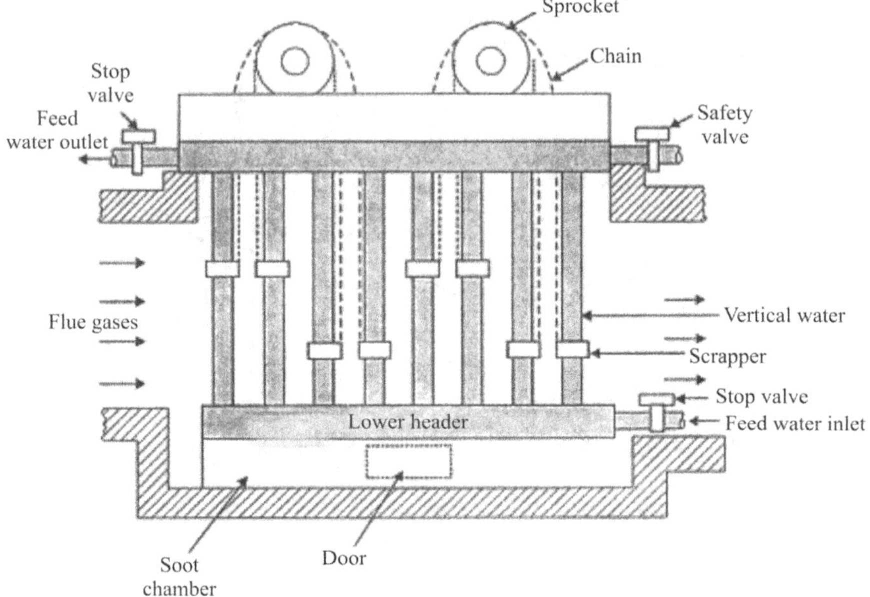

Fig. 4.21 Green's economizer

(iii) *Air preheater*: The function of an air preheater is similar to that of an economizer. It recovers some portion of the waste heat of flue gases going to chimney, and transfers same to the fresh air before it enters the combustion chamber. A tubular air preheater is shown in the Fig. 4.22. Due to preheating of air, the furnace temperature increases. It results in rapid combustion of fuel with less soot, smoke and ash. The high furnace temperature can permit low grade fuel with less atmospheric pollution. The air preheater is placed between economizer and chimney.

(iv) *Steam injector*: A steam injector lifts and forces the feed water into the boiler. It is usually used for vertical and locomotive boilers and can be accommodated in small space. It is less costly. It does not have any moving parts thus operation is salient. In a steam injector, the water is delivered to the boiler by steam pressure. The kinetic energy of the expanding steam is given to feed water. Thus, the pressure and velocity of water is increases.

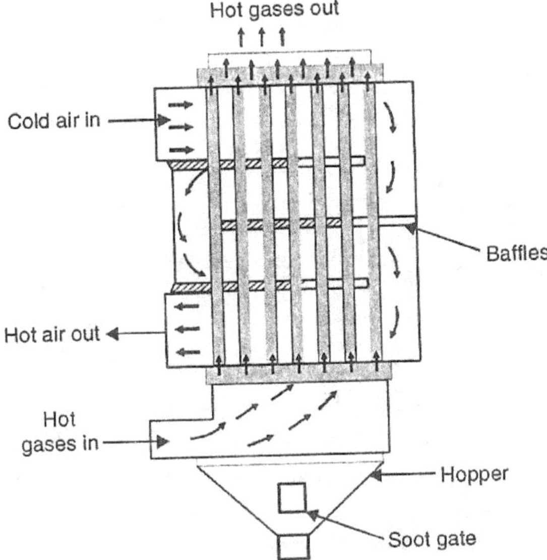

Fig. 4.22 Air preheater

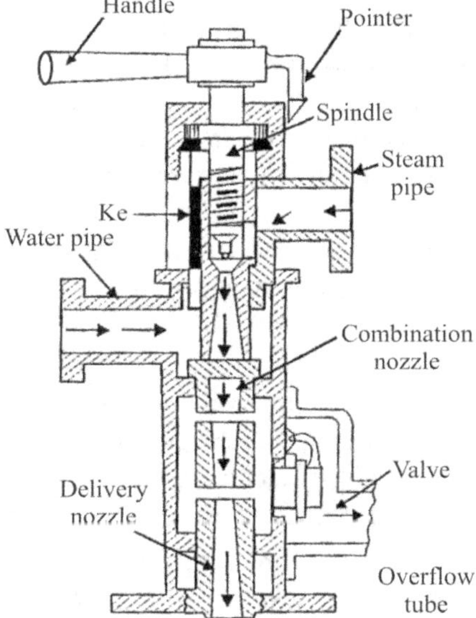

Fig. 4.23 Steam injector

Fig. 4.23 shows a Steam Injector. It consists of a set of nozzle, suction jet, delivery nozzle, steam and water inlet passages, and delivery passages, spindle, handle and a overflow valve. When steam valve is opened with the help of spindle and handle, the steam from the boiler enters the steam nozzles. The steam pressure drops in nozzles and steam velocity increases. The high velocity steam creates a partial vacuum in steam jet, thus the water from the feed tank is sucked and mixed with steam thus steam condenses. This mixture then enters the delivery tube, where kinetic energy of the mixture is converted to pressure energy. When the pressure of water delivered by nozzle is less than boiler pressure, the water leaves the injector is placed through the over flow valve to over flow pipe. But when sufficient pressure in the delivery tube is raised then the water is supplied to boiler.

(v) *Feed pump:* The feed pump delivers feed water at a pressure higher than that in boiler. The commonly used feed pumps are:

(a) *Rotary pump:* It is a high speed centrifugal pump and is used to deliver a large quantity of water in to the boiler. It consists of an impeller and a casting. The impeller shaft is driven by an electric motor or by a prime mover. It utilizes a centrifugal force of the impeller for pumping water

(b) *Reciprocating Pump*: These are positive displacement type pumps. Most popular type of reciprocating feed pump used in a boiler is a duplex feed pump. It consists of a steam engine and a water pump side by side. The reciprocating water pump is driven by reciprocating steam engine. The piston rods of the water pump and steam engines are coupled. So the steam engine and water pump operate simultaneously. The steam generated in the boiler is used to run the steam engine. The steam engine drives the water pump.

Merits of accessories:

1. In superheater steam consumption of engine is reduced and losses due to condensation in the cylinder and the steam pipes are reduced. Due to which efficiency of steam plant is increased and erosion of turbine blade is also increases.

2. In economizer the temperature range between various parts of the boiler is reduced which results in reduction of stresses due to unequal expansion, evaporative capacity of the boiler is increased.

If the boiler is fed with cold water it may result in chilling the boiler metal. Hot feed water checks it.

3. In injector, initial cost is low and thermal efficiency is very high.

Demerits of accessories:

1. In injector pumping efficiency is low and it cannot force very hot water.
2. Storage type air preheater cannot employ with small boiler.

Uses of accessories:

1. Boiler accessories are used to increase the efficiency of boiler.
2. Storage type air preheaters are employed widely in larger plants.
3. Sugden's superheater installed in a Lancashire boiler.
4. An injector is commonly employed for vertical and locomotive boiler.

4.7   Lancashire Boiler

Lancashire boiler is very widely used as a stationary boiler because of its good steaming quality & because it can also burn coal of inferior quality.

4.7.1   Specification of lancashire boiler

(i)   Horizontal smoke tube boiler; size range from a shell 2 m dia × 6 m long to 3 m dia × 10 m long.

(ii)   Working pressure range is up to 20 kgf/cm^2.

(iii)   Ratio of heating surface to grate area is 24-30.

(iv)   Efficiency is about 56% without economizer and 75% with economizer.

(v)   Two flue tubes are present in Lancashire Boiler.

4.7.2 Construction

The Lancashire boiler has cylindrical shell usually 2-3 meters in diameter & 8-9.5 m in length. This boiler is used for power generation at moderate steam pressure of 15 bar. The main features of the Lancashire boiler & its brickwork setting are shown in Fig. 4.24. The boiler consists of a cylindrical shell with flat ends & two furnace tubes pass right through this. There are one bottom flue & two side flues formed by brickwork setting. These flues provide part of the heating surface on the main external shell.

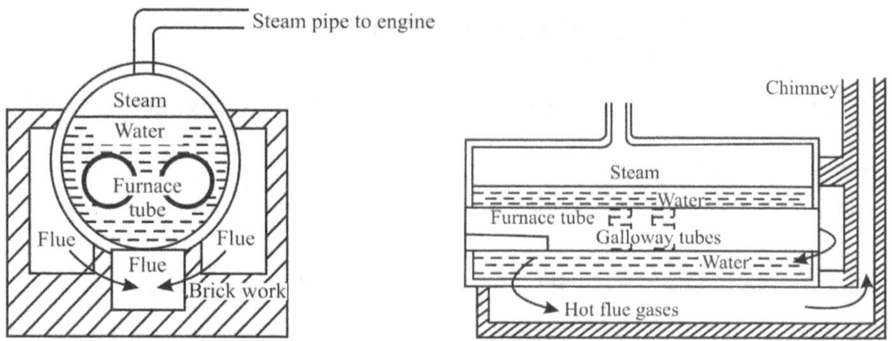

Fig. 4.24 Lancashire boiler

4.7.3   Principle parts and their functions

(i)   **Drum:** The boiler drum consists of a shell and end heads. The shell of the boiler consists of one or more steel plates bent into the cylindrical form and riveted or welded together. The ends of the shell are closed by means of the flat or curved boiler head.

(ii) **Setting:** It is also called foundation and constructed of bricks. It supports the boiler drum and other components. It forms the wall of the furnace, combustion chamber and passage to flue gases.

(iii) **Grate:** It is the space located below the furnace and consists of cast iron bars and can support the combustion process. The ashes can fall down through these spaces.

(iv) **Furnace:** It is the space above the grate and below the boiler shell, where the volatile matter and combustible fuel are burnt and flue gases are generated.

(v)   **Flue gases:** It is the hot mixture of products of combustion, generated in the furnace.

(vi) **Heating surface:** It is the surface of boiler, which is exposed to hot flue gases on one side, water on other.

(vii) **Water space:** The space of the boiler shell occupied by water is called water space. The level of water is maintained in the boiler and can be seen through water level indicator.

(viii) **Steam space:** The entire space of boiler shell which is not occupied by water is called steam space.

(ix) **Feed water:** The water supplied to the boiler is called feed water. The pump which supplies water is called feed pump.

(x) Economizer: The feed water supplied by the feed pump is heated by the waste hot gases in the economizer before they escape to the chimney. Therefore, some of the waste heat is recovered by feed water and plant efficiency improves.

(xi) Air preheater: Similar to feed water heating, the fresh air going to furnace is also preheated to improve combustion process. The device of air heating system is known as *preheater.*

(xii) Super heater: It is the device which heats the saturated steam generated in the boiler. The super heater is located above the furnace and it increases the heat content of steam without increasing its pressure.

4.7.4 Working

The gases from the furnace pass to back end of the boiler where they go down & enter the bottom flue end, travel through it to the front end of the boiler. The gases then divide themselves here and enter the side flues and travel through them to the back of the boiler and then to the chimney. The Lancashire boiler is often fitted with cross tubes known as Galloway tubes. They are fitted across the furnace tubes. They are provided with blow-off cock at the bottom of the front end, and the feed check valve with feed pipe on the front end plate. Some Lancashire boilers are fitted with shortened furnace and a number of smoke tubes. These smoke tubes increase the heating surface of the boiler. The Lancashire boilers are sometimes called economical boilers.

4.7.5 Advantages

(i)   Ease of operation.

(ii) Simple design.

(iii) The ratio of the volume of the boiler to its rated evaporative capacity is high.

4.7.6 Disadvantages

(i)   It occupies considerable floor area.

(ii) Brickwork setting is expensive in the initial cost and troublesome in maintenance.

(iii) The shell construction restricts the maximum working pressure to about 17.5 bars.

(iv) The grate area is restricted by the diameter of the internal furnace tubes.

(v) There is very large water capacity and little encouragement to water circulation, especially between the furnace tubes and the bottom of the shell, which is impossible in emergency to raise steam pressure rapidly from the cold water.

4.7.7 Application

It is commonly used in sugar mills and textile industries for sizing and bleaching and in chemical industries where along with the power steam and steam for the process work is also needed. In addition, this boiler is used where larger reserve of water and steam are needed.

4.8 Cochran boiler

4.8.1 Specifications of cochran boiler

(i)   Shell diameter: 2.75 m.

(ii)   Height: 45.79 m.

(iii)   Working pressure: 6.5 bar (max pressure: 15 bar).steam capacity: 3500 kg/hr.

(iv)   Max. Capacity: 4000 kg/hr.

(v)   Heating surface: 120 m^2.

(vi)   Efficiency: 70 to 75% (Depending on the fuel used).

4.8.2 Constructional Details

Cochran boiler is one of the best types of vertical multi-tubular boilers, and has a number of horizontal fire tubes (Fig. 4.25). It is made in different designs and sizes of evaporative capacities ranging from 150 to 3000 kg of water per hour and for working pressures up to 20 bar and is suitable for different types of fuels. Thermal efficiency of the Cochran boiler is about 70% with coal firing and about 75% with oil firing. Cochran boiler consists of a cylindrical shell with its crown having a hemispherical shape, such a shape of the crown plate gives enough strength to withstand the bulging effect of the inside steam pressure. The fire-box is made in one piece and has no joints. The fire-box also has a crown of hemispherical shape. The hemispherical shape of the crown is advantageous for the absorption of the radiant heat from the furnace. The convection heat surfaces are provided by a large number of horizontal smoke tubes. The fuel gases from the fire-box enter through the small fuel-pipe into the combustion chamber and strike on the boiler shell plate which forms the back of the combustion chamber. The back plate of the combustion chamber is lined with fire-bricks and can be conveniently dismantled and removed for cleaning smoke tubes. The back plate directs the fuel gases into the smoke tubes. The smoke tubes generally have 62.5 mm external diameter and are 165 in number. Most of the smoke tubes are fixed in the vertical tube plates by being expanded in the holes but some of them are fixed by screwing into the holes. The screwed tubes structure settle to the vertical tubes and prevent them from bulging out due to the inside steam pressure. A number of hand holes are provided around the outer shell for cleaning purposes. The flat top of the combustion chamber is strengthened by gusset stays. The gases passing through the fire tubes transfer a large portion of heat to the water by convection. The fuel gases coming out of the fire tubes are finally discharged to the atmosphere through a chimney.

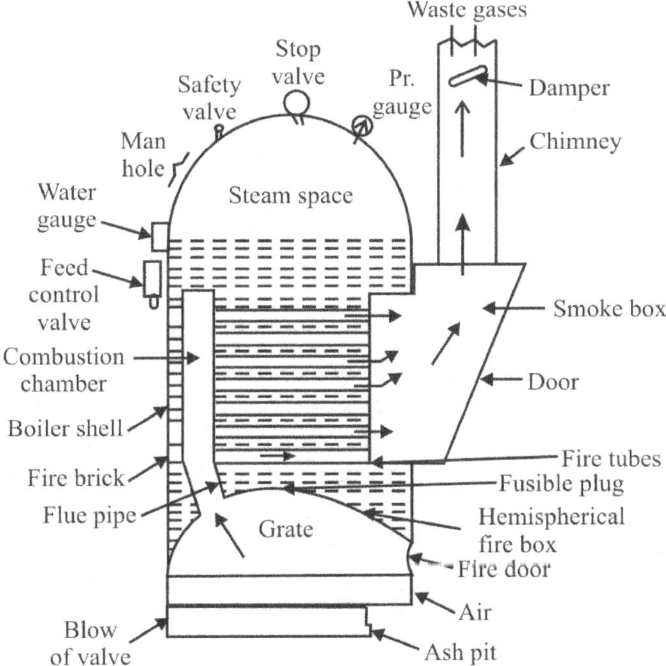

Fig. 4.25 Cochran boiler

4.8.3 Principle parts and their functions

The various parts of a boiler along with their functions are as follows:

   (i)   Cylinder shell
   (ii)   Combustion chamber
   (iii)   Fire brick
   (iv)   Furnace (dome shaped)
   (v)   Blow off cock
   (vi)   Steam stop valve
   (vii)   Anti priming valve
   (viii) Pressure gauge
   (ix)   Flue tube
   (x)   Smoke box
   (xi)   Chimney
   (xii)   Fire tube
   (xiii) Grate
   (xiv)   Ash pit
   (xv)   Safety valve
   (xvi)   Man hole
   (xvii) Water level gauge

   (i)   **Shell:** The shell of a boiler consists of one or more steel plates bent into a cylindrical form and riveted or welded together. The shell ends are closed with the end plates.

   (ii) **Combustion chamber:** It is the space, generally below the boiler shell, meant for burning fuel in order to produce steam from the water contained in the shell.

   (iii) **Grate:** It is the platform in the furnace upon which fuel is burnt and it is made of cast iron bars. The bars are so arranged that air may pass on to the fuel for combustion. The area of the grate on which the fire rests in a coal or wood fired boiler is called grate surface.

   (iv) **Furnace:** It is a chamber formed by the space above the grate and below the boiler shell, in which combustion takes place. It is also called a fire box.

   (v)   **Heating surface:** It is that part of the boiler surface, which is exposed to the fire (or hot gases from the fire).

(vi)   Water space and steam space: The volume of the shell that is occupied by the water is termed as water space while the entire shell volume less the water and tubes (if any) is called the steam space.

(vii)   Mountings: The items such as stop valve, safety valves, water level gauges, fusible plug, blow-off cock, pressure gauges, water level indicator etc., are termed as mountings and a boiler cannot work safely without them.

Cochran boiler mountings are:

(a)   Water level gauge: The function of a water level indicator is to indicate the level of water in the boiler constantly. It is also called water gauge. Normally two water level indicators are fitted at the front end of every boiler. For the proper working of boiler, the water must be kept at safe-level. If the water level falls below the safe level and the boiler goes on producing steam without the addition of feed water, great damage like crack and leak can occur to the boiler which gets uncovered from water. This can result in the stoppage of steam generation and boiler operations.

(b)   Safety valve: The function of a safety valve is to release the excess steam when the pressure of steam inside the boiler exceeds the rated pressure by opening the safety valve automatically and excess steam rushes out into the atmosphere till the pressure drops down to the normal value. A safety valve is generally mounted on the top of the shell.

(c)   Blow off cock: A blow off cock or valve performs the two functions:

    (i)   It may discharge a portion of water when the boiler is in operation to blow out mud, scale or sediments periodically.

    (ii)   It may empty the boiler when necessary for cleaning, inspection and repair. It is fitted on the boiler shell directly or to a short branch pipe at the lowest part of the water space.

(d)   Pressure gauge: The function of a pressure gauge is to measure the pressure exerted inside the vessel. The gauge is usually mounted on the front top of the shell or the drum. It is usually constructed to indicate up to double the maximum working pressure.

4.8.4 Working Principle

The function of a boiler is to evaporate water into steam at a pressure higher than the atmospheric pressure. Water free from impurities such as dissolved salts, gases and non soluble solids should be supplied to boilers. This is done by suitable water treatment. Steam is useful for running steam turbines in electrical power stations, ships and steam engines in railway locomotives. Cochran boiler is one of the best types of vertical multi-tubular boiler, and has a number of horizontal fire tubes. These tubes generally have 62.5 mm external diameter and are 165 in number. The gases from the smoke box pass to the atmosphere through a chimney. Cochran boiler consists of a cylindrical shell with a dome shaped top where the space is provided for steam. The furnace is one piece construction and is seamless. Its crown has a hemispherical shape and thus provides maximum volume of space. The fuel is burnt on the grate and ash is collected and disposed off from ash pit. The gases of combustion produced by burning of fuel enter the combustion chamber through number of horizontal tubes, being surrounded by water. After which the gases escape to the atmosphere through smoke box and chimney. A number of hand holes are provided around the outer shell for cleaning purposes.

4.8.5 Merits of cochran boiler

   (i)   It involves lesser risk of explosion due to lower pressure in the steam space.

  (ii)   It requires less skill for efficient and economic working.

 (iii)   It is very compact, hence requires lesser floor area.

  (iv)   It is quite portable.

   (v)   It does not require a strong foundation.

  (vi)   It can use any type of fuel.

4.8.6 Demerits of cochran boiler

   (i)   Steam production is less in this boiler.

  (ii)   It is not suitable for large power plants.

 (iii)   For a given power it occupies more floor area.

  (iv)   Its construction is difficult and expensive too.

   (v)   Various parts not so easily accessible for cleaning, repair and inspection.

4.8.7 Application of cochran boiler

The Cochran boiler may be employed for following purposes:
   (i)   For generating power in steam engines and steam turbines.
   (ii) In textiles industries for sizing and bleaching etc, and many other industries like sugar mills, chemical industries.
   (iii) For heating the buildings in cold weather and for producing hot water from water supply.
   (iv) In small and medium sized industries and in mines.

4.9 Locomotive Boiler

4.9.1 Specifications of Locomotive Boiler

Dimensions and the specifications of the Locomotives Boilers (Made at Chitranjan Works in India) are given below:
   (i)   Barrel diameter = 2.095 m
   (ii)   Length of the barrel = 5.206 m
   (iii) Size of the tubes (superheater) = 14 cm
   (iv)   No. of superheater tubes = 38
   (v)   Size of ordinary tubes = 5.72 cm
   (vi)   Steam capacity = 9000 kg/h
   (vii) Working pressure =14 bar
   (viii) Grate area = $4.27 m^2$
   (ix)   Coal burnt/h = 1600 kg
   (x)   Heating surface = $271 m^2$
   (xi)   Efficiency = 70%

4.9.2 Constructional Details

Fig. 4.26 shows all the constructional details of locomotive boilers and they are discussing as:

Boiler: The boiler consists of a steel shell, which includes the boiler barrel, the outer firebox wrapper plate, the inner firebox, boiler back plate, smoke box tube plate and throat plate.

Firebox Stays: The inner firebox is supported from the outer firebox by the foundation ring at the bottom, crown stays at the top and palm stays between the firebox tube plate and the boiler barrel. In addition, copper or steel stays run between the firebox and the outer firebox wrapper plate, the boiler back plate and throat plate.

Tubes: The tubes run between the firebox tube plate and the smokebox tube plate. In superheated boilers, large tubes contain the Superheater Elements.

Dome: The dome is positioned at the highest part of the boiler barrel where it forms a collector for steam above the surface of the water.

Main Steam Pipe: Steam collected in the dome is fed into the main steam pipe via the regulator valve to the header in the smokebox.

Smokebox: The smokebox is an extension of the front end of the boiler barrel. Exhaust steam from the cylinders passes through the blast pipe into the chimney and creates a partial vacuum. This causes hot gases to be drawn through the tubes which, in turn, draw air into the firebox through the grate and firehole door. The smokebox also contains the main steam pipes to the steam chests, blower, superheater header, tubes and exhaust for the vacuum Ejector where fitted.

Brick Arch: The brick arch serves several purposes. It protects the firebox tube plate from the direct flame of the fire, radiates heat to prevent rapid fluctuation of the tube plate temperature, and ensures thorough combustion of volatiles by lengthening their path from the fire to the tube plate.

Fire door: Fire hole doors vary from locomotive to locomotive. These give access for firing and can be adjusted to control the flow of secondary air.

Baffle Plate: The baffle plate placed in the fire hole is designed to direct the secondary air down towards the fire bed in order to mix it thoroughly with the hot gases and flames.

Fusible Plugs: Fusible plugs are screwed into the firebox crown. They are made of brass and have a lead core. If the water level in the boiler drops and uncovers the firebox crown, the lead melts allowing steam to escape into the firebox. This warns the Enginemen and helps to deaden the fire. Both injectors should be put on immediately if this occurs and steps taken to remove or deaden the fire.

Superheater: The superheater consists of a superheater header and superheater elements. Steam from the main steam pipe arrives at the saturated steam chamber of the superheater header and is fed into the superheater elements. Superheated steam arrives back at the superheated steam chamber of the superheater header and is fed into the steam pipe to the cylinders. Superheated steam is more expansive.

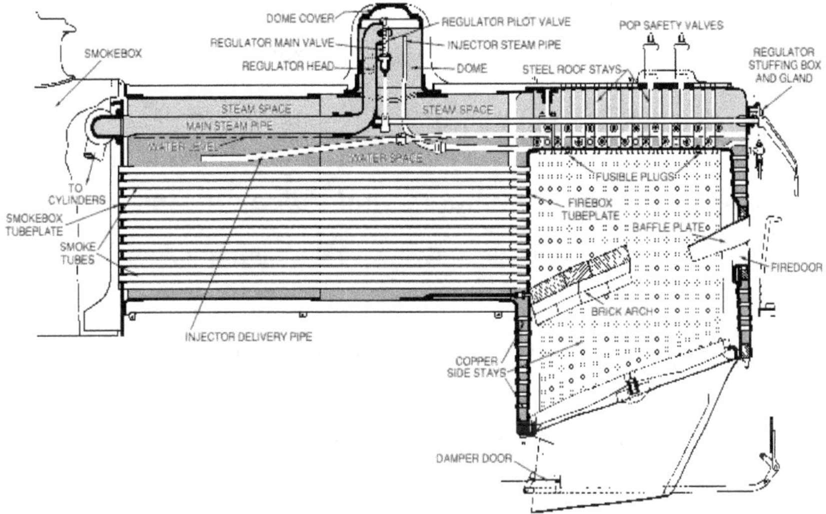

Fig. 4.26. Locomotive Boiler

4.9.3 Advantages

(i)   High steam capacity.

(ii) Low cost of construction.

(iii) Portability.

(iv) Low installation cost.

(v)   Compact.

4.9.4 Disadvantages

(i)   There are chances to corrosion and scale formation in the water legs due to the accumulation of sediments and the mud particles.

(ii) It is difficult to clean some water spaces.

(iii) Large flat surfaces need bracing.

(iv) It cannot carry high overloads without being damaged by overheating.

(v) There are practical constructional limits for pressure and capacity which do not meet requirements.

4.9.5 Application of Locomotive Boiler

(i)   Used in electric and diesel railway vehicles.

(ii) Used in Traction Engines.

(iii) Used in steam rollers and steam road vehicles.

4.10   Babcock and Wilcox boiler

4.10.1 Specifications of Babcock and Wilcox Boiler

The particulars (dimensions, capacity etc,) relating to this boiler are given below:

Diameter of the drum:	1.22 to 1.83 m
Length:	6.096 to 9.144 m
Size of the water tubes:	7.62 to 10.16 cm
Working pressure:	40 bar (max)
Steaming capacity:	40,000 kg/h (max)
Efficiency:	60 to 80%

4.10.2 Constructional Details

Boiler is built in two general classes; the longitudinal drum type and the cross drum type. The longitudinal drum boiler is the generally accepted standard of Babcock & Wilcox construction. The cross drum boiler, though originally designed to meet certain conditions of headroom, has become popular for numerous classes of work where low headroom is not a requirement which must be met.

Babcock and Wilcox boiler consist of a drum connected to a series of front end and rear end header by short riser tubes. To these headers are connected a series of inclined water tubes of solid drawn mild steel. The angle of inclination of the water tubes to the horizontal is about 15 or more. At the lowest point of boiler is provided a mud collector to remove the mud particles through a blow-down-cock. The entire boiler except the furnace is hung by the means of metallic slings or straps or wrought iron girders supported on pillars. This arrangement enables the drum and the tube to expand or contract freely. The brick work around the boiler encloses the furnace and the hot gases.

4.10.3 Working principle

The steam drum A at the top is supported by straps from steel columns set in walls lined with refractory bricks. The water tubes, from which the boiler gets its name, are mounted between two sets of headers. The rear set B nearly touch the boiler floor and are fed by large tubes from the bottom of the drum. The front set C connect with the front of the drum by a short manifold so that the group of sixty 4 inch diameter tubes slopes up towards the front over the fire grate F. This layout promotes good water circulation by thermo-syphon action. You can see the front headers on the unrestored boilers, "saturated" steam is taken direct from the top of the steam drum via a stop valve at the rear of the drum (Fig. 4.27). Steam for the engine, however, is "superheated" by being collected through two tubes high in the steam space and fed through twenty U-shaped tubes D below the drum where the steam picks up extra heat energy from the flue gases before passing to the main stop valve E. The front of the steam drum has a pressure gauge and two gauge glasses to monitor the water level. The original working pressure was 160 p.s.i. but we now operate at 100 p.s.i. (7 bar or 0.7 MPa). The water level is maintained by one or more of the engine driven feed pump, auxiliary steam feed pumps or an electric pump. The boiler feed water comes from condensed steam in the hot well, an additional water storage tank beside the chimney is recommended to hold treated boiler water whilst the boiler is in use or in a dry storage condition. Steam grate blowers keep the boiler airways free of ash and induced air is forced up through the grate. Any solids deposited in the water tubes during steaming tend to collect at the bottom of the "down comers", the headers at the rear of the tube stack. Here, there is a special drain called the "blow down" which, when operated during steaming, dumps water and deposited sludge in the blow down pit outside. Steam plus hot water at 175 °C is violent and dangerous. Control of the furnace is by a damper, a sliding trapdoor, in the passage from the boiler to the chimney G. The damper is counterbalanced and operated by a large weight hanging to the right of the fire doors. The operating chain passes over guide boiler frame.

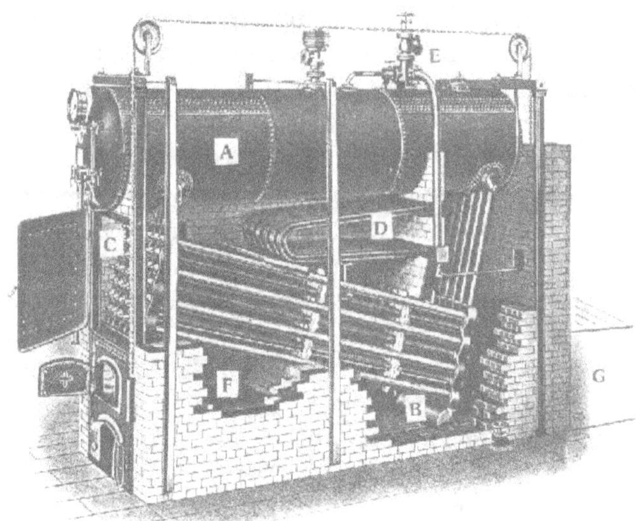

(a) Babcock & wilcox land boiler, fitted with superheater

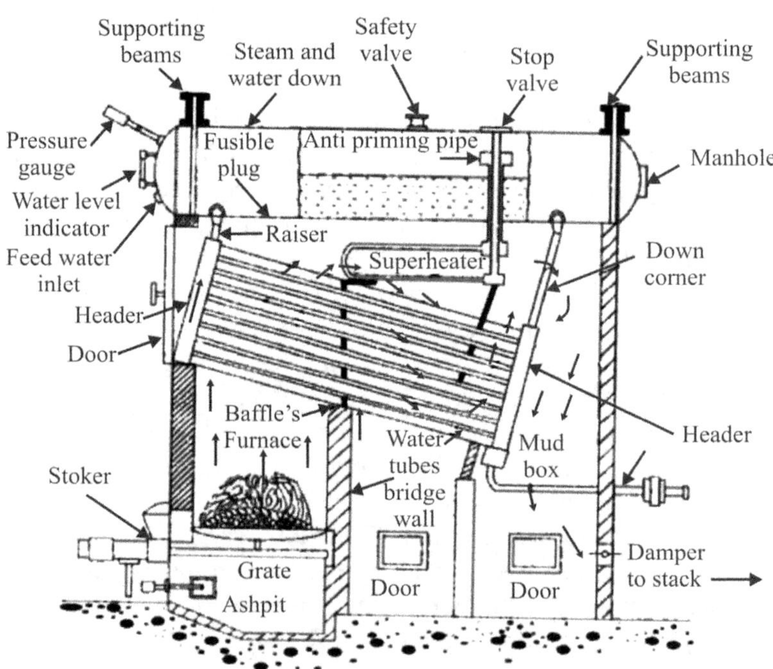

(b) Babcock & wilcox boiler

Fig. 4.27

4.10.4 Merits

The advantages of the Babcock & Wilcox boiler may perhaps be most clearly set forth by a consideration.

(i) *Safety*: The most important requirement of a steam boiler is that it shall be safe in so far as danger from explosion is concerned. The main feature to which the Babcock & Wilcox boiler owes its safety is the construction made possible by the use of headers, by which the water in each vertical row of tubes is separated from that in the adjacent rows. The headers in a Babcock & Wilcox boiler are practically free from any danger of explosion.

(ii) *Economy*: In economy, the construction made possible by the use of headers in Babcock & Wilcox boilers appears as a distinct advantage. The parts used in the Babcock & Wilcox boiler are more economical.

(iii) *Capacity*: Due to the generally more efficient circulation found in water-tube than in fire-tube boilers, rates of evaporation are possible with water-tube boilers that cannot be approached where fire-tube boilers are employed.

(iv) *Quick Steaming*: Another important result of the better circulation ordinarily found in water-tube boilers is in their ability to raise steam rapidly in starting and to meet the sudden demands that may be thrown on them.

(v) *Accessibility*: In the "requirements of a perfect steam boiler", Babcock demonstrates the necessity for complete accessibility to all portions of the boiler for cleaning, inspection and repair.

(vi) *Cleaning*: Water-tube boilers in general are from the nature of their design more readily accessible for cleaning than are fire-tube boilers.

(vii) *Inspection*: The objections given above in the consideration of the inability to properly clean fire-tube boilers hold as well for the inspection of such boilers.

(viii) *Repairs*: In the case of a rupture in a water-tube boiler, the loss will ordinarily be limited to one or two tubes which can be readily replaced. The lack of accessibility in fire-tube boilers further leads to difficulties where repairs are required. In fire tube boilers renewal are serious undertakings.

4.10.5 Demerits

(i)   Water circulates inside the tubes which are surrounded by hot gases from the furnace.

(ii) High operating cost.

(iii) The bursting chances are more.

(iv) If burst, does not cause much destruction to the whole boiler.

(v)   It is less suitable for impure and sedimentary water.

4.10.6 Applications

(i)   For small power generation.

(ii) For processing as in sizing, bleaching etc., in textile industry and sugar industry, chemical etc.

4.11 Boiler Draught

The small pressure difference which causes a flow of gas to take place is termed as a draught. The function of the draught, in case of a boiler, is to force air to the fire and to carry away the gaseous products of combustion. In a boiler furnace proper combustion takes place only when sufficient quantity of air is supplied to the burning fuel.

The draught may be classified as:

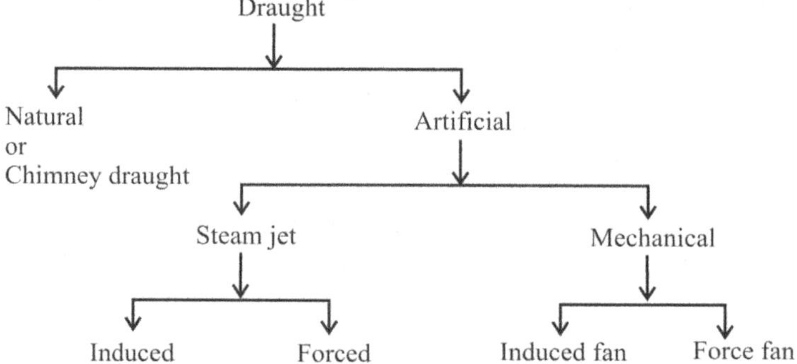

4.11.1 Natural draught – Chimney

Natural draught is obtained by the use of a chimney. The chimney in a boiler installation performs one or more of the functions:

(i)   It produces the draught whereby the air and gas forced though the fuel bed, furnace, boiler passes and settings.

(ii) It carries the products of combustion to such a height before discharging them that will not be objectionable or injurious to surroundings. A chimney is vertical tubular structure built either of masonry, concrete or steel. The draught produced by the chimney is due to the density difference between the column of hot gases inside the chimney and the cold air outside.

Fig.4.28 below shows a diagrammatic arrangement of a chimney of height 'H' metre above the grate.

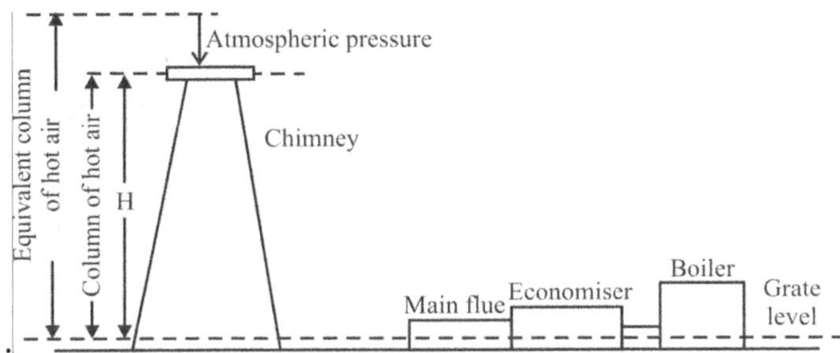

Fig. 4.28. Chimney height

We have,

$$P_1 = P_a + \rho_g\, g\, H$$

where, P_1 = pressure at the grate level (Chimney side)

P_a = atmospheric at the chimney top

$\rho_g\, g\, H$ = pressure due to the column of hot gas of height H metre, and

ρ_g = average mass density of hot gas.

Similarly, $P_2 = P_a + \rho_a\, g\, H$

where, P_2 = pressure acting on the grate on the open side.

$\rho_a\, g\, H$ = pressure exerted by the column of cold air outside the chimney of height H metre.

ρ_a = mass density of air outside the chimney.

Net pressure difference causing the flow through the combustion chamber,

$$\Delta P = P_2 - P_1 = (\rho_a - \rho_g)\, g\, H$$

This difference of pressure causing the flow of gases is known as 'static draught'. Its value is small and is generally measured by a water manometer.

It may be noted that this pressure difference in chimney is generally less than 12 mm of water.

Chimney Height and Diameter: The calculation of chimney height and its diameter requires the knowledge of chemical reaction occurring during combustion and pressure difference of cold and hot gas column.

When carbon burns the reaction is represented by the equation

$$C + O_2 = CO_2 \qquad \qquad .....(1)$$

1Vol. + 1 Vol. = 1 Vol.

when hydrogen burns the reaction is represented by the equation

$$2H_2 + O_2 = 2H_2O \qquad \qquad .....(2)$$

2 Vol.+ 1 Vol. = 2 Vol.

Therefore, one may assume that the volume of products of combustion (i.e., flue gases) in the chimney is equal to the volume of air supplied per combustion, both volumes being taken at the same temperature. It is to be observed that masses of air and the flue gases are different though the volumes are same.

If m_a kg be the mass of air supplied per kg of fuel then $(m_a + 1)$ kg will be the mass of the chimney gas

From the above analysis the volume of 1 kg of chimey gas at 0 °C or 273 K is equal to volume of 1 kg of air at 0 °C or 273 K. At NTP 1 kg of air occupies a volume equal to

$$v = \frac{RT}{p} = \frac{287 \times 273}{1.013 \times 10^5} = 0.7734 \ m^3$$

The difference in pressure at the furnace and the chimney base is very small, of the order of a maximum of 25 m, therefore the pressure may be assumed constant so that the volume of air flue gas at higher temperature may be calculated.

Let,

T_g = mean absolute temperature of the flue gases, K

T_a = absolute temperature of the outside air, K

Then, volume per kg of air at temperature T_a is

$$0.7734 \times \frac{T_a}{273}$$

Therefore, volume of m_a kg of air at T_a is

$$\frac{0.7734 \times m_a \times T_a}{273} \qquad .....(3)$$

Again volume of (m_a+1) kg of chimney gases at T_g is

$$\frac{0.7734 \times (m_a + 1) \times T_g}{273} \qquad .....(4)$$

Hence, the density of air at temperature T_a is

$$\rho_a = \frac{Mass}{Volume} = \frac{m_a \times 273}{0.7734 \times m_a \times T_a} = 1.293 \left(\frac{273}{T_a} \right) kg/m^3 \qquad .....(5)$$

Similarly,

Density of chimney gases at temperature T_g is

$$\rho_g = \frac{Mass}{Volume} = \frac{(m_a + 1) \times 273}{0.7734 \times m_a \times T_g}$$

$$= 1.293 \left(\frac{m_a + 1}{m_a} \right) \times \left(\frac{273}{T_g} \right) \ kg/m^3 \qquad .....(6)$$

Chimney Height: Let H be the height of the chimney required in metre measured from the grate level and h mm be column of water which will be balance by a draught.

Pressure exerted by a column of hot flue gas of H metre height is

$$p_g = \rho_g \times g \times H = 1.293 \times \left(\frac{m_a + 1}{m_a} \right) \times \frac{273}{T_g} \times g \times H \ / \ Nm^2$$

Pressure exerted by a column of cool air outside the chimney of H metre height is

$$p_a = \rho_a \times g \times H = 1.293 \times \frac{273}{T_a} \times g \times H \ / \ Nm^2$$

At the same height H, the pressure difference between hot flue gas and cool air outside the chimney is

$$p = p_a - p_g = g \times H \times \left(\rho_a - \rho_g\right) / Nm^2 \qquad\qquad .....(7(a))$$

$$p = p_a - p_g$$

$$= \left[1.293 \times \frac{273}{T_a} \times g \times H\right] - \left[1.293 \times \left(\frac{m_a + 1}{m_a}\right) \times \frac{273}{T_g} \times g \times H\right]$$

$$= 353 \times g \times H \times \left[\frac{1}{T_a} - \left(\frac{m_a + 1}{m_a}\right) \times \frac{1}{T_g}\right] / Nm^2$$

$$.....(7(b))$$

If h m be the draught measured in water column then

$$h = H \ (\rho_a - \rho_g), \ \text{mm of water} \qquad\qquad .....(8(a))$$

From Eqn. 7a and 8a

$$h = \frac{p_a - p_g}{g}$$

$$h = 353 \times H \times \left[\frac{1}{T_a} - \left(\frac{m_a + 1}{m_a}\right) \times \frac{1}{T_g}\right] \text{mm of water} \qquad\qquad .....(8(b))$$

If H_g is the height of a column of hot gas expressed in metre which would produce the draught pressure p in N/m^2,

Then,

$$H_g = \frac{p\left(N/m^2\right)}{\text{Density of hot gas}\left(kg/m^3\right) \times g} \ m$$

$$H_g = \frac{353 \times g \times H \times \left[\dfrac{1}{T_a} - \left(\dfrac{m_a+1}{m_a}\right) \times \dfrac{1}{T_g'}\right]}{353 \times \left(\dfrac{m_a+1}{m_a}\right) \times \left(\dfrac{1}{T_g}\right) \times g} \qquad .....(9)$$

$$= H\left[\left(\frac{m_a}{m_a+1}\right) \times \left(\frac{T_g'}{T_a}\right) - 1\right] metre\ of\ gas$$

Height of the chimney depends on temperature of hot flue gas and cool air outside the chimney.

Diameter of Chimney: Assuming no loss, the velocity of the hot flue gases passing through the chimney is given by

$$C = \sqrt{2gH_g} \qquad .....(10)$$

If the pressure loss in the chimney is equivalent to a hot gas column of ΔH_g meters, then velocity of the gas in the chimney is

$$C = \sqrt{2g\left(H_g - \Delta H_g\right)} = 4.429\sqrt{H_g\left(1 - \frac{\Delta H_g}{H_g}\right)}$$

$$C = 4.429\sqrt{\left(1 - \frac{\Delta H_g}{H_g}\right)} \times \sqrt{H_g} = k \times \sqrt{H_g} \qquad .....(11)$$

where, $k = 4.429\sqrt{\left(1 - \dfrac{\Delta H_g}{H_g}\right)}$

The value of K = 0.825 for brick chimneys, and K = 1.1 for steel chimneys.

(i)   Considering steel chimney

$$1.1 = 4.429\sqrt{\left(1 - \frac{\Delta H_g}{H_g}\right)}$$

(ii)   Considering Brick chimney

$$\frac{\Delta H_g}{H_g} = \frac{18.4}{19.61}$$

$$0.825 = 4.429 \sqrt{\left(1 - \frac{\Delta H_g}{H_g}\right)}$$

$$\left(\frac{0.825}{4.429}\right)^2 = 1 - \frac{\Delta H_g}{H_g}$$

$$\frac{\Delta H_g}{H_g} = 1 - \frac{0.680625}{19.616041}$$

$$\frac{\Delta H_g}{H_g} = \frac{18.93}{19.61}$$

The above calculations show that 18/19th of the draught is lost due to friction and only 1/19th is available for creating the flow velocity of the chimney.

The mass of the gases flowing through any cross-section of the chimney is given by

$$m_g = A \times C \times \rho_g$$

$$m_g = k \times \sqrt{H_g} \times \rho_g \times \frac{\pi D^2}{4}$$

$$D = 1.128 \times \sqrt{\frac{m_g}{\rho_g}} \times \sqrt{\frac{1}{k \times \sqrt{H_g}}} \qquad \ldots\ldots(12)$$

The diameter of the chimney can be calculated from Equ. 12.

Condition for maximum discharge through a chimney

The chimney draught is most effective when the maximum weight of hot gases is discharged in a given time, and it will be shown that this occurs when the absolute temperature of the chimney gases bears a certain relation to the absolute temperature of the outside air.

We know that the velocity of gas through the chimney, assuming the losses to be negligible, is given by

$$C = \sqrt{2gH_g}$$

Substituting the value of Hg from Equ. 9

$$C = \sqrt{2gH\left[\left(\frac{m_a}{m_a+1}\right)\times\left(\frac{T_g}{T_a}\right)-1\right]} \qquad .....(13)$$

we know that

$$\rho_g = 1.293\left(\frac{m_a+1}{m_a}\right)\times\left(\frac{273}{T_g}\right) \qquad .....(14)$$

$$= \frac{K}{T_g}$$

where K is the constant and we can say that density of gas is inversely proportional to its temperature.

Mass of the gases flowing through a cross sectional area A of the chimney is

$$m_g = A \times C \times \rho_g$$

$$m_g = A \times \sqrt{2gH\left[\left(\frac{m_a}{m_a+1}\right)\times\left(\frac{T_g}{T_a}\right)-1\right]} \times \frac{K}{T_g}$$

$$m_g = K_1 \times \sqrt{\left[\left(\frac{m_a}{m_a+1}\right)\times\left(\frac{1}{T_a\times T_g}\right)-\frac{1}{T_g^2}\right]} \qquad .....(15)$$

where $\quad K_1 = A \times K \times \sqrt{2gH}$

In Equ. 15 T_a and K_1 are constant and m_g and T_g are variables.

Hence, for the maximum discharge condition

$$\frac{dm_g}{dT_g} = 0$$

$$\frac{dm_g}{dT_g} = K_1 \times \frac{1}{2} \times \frac{-\left(\frac{m_a}{m_a+1}\right)\times\frac{1}{T_a}\times\frac{1}{T_g^2}+\frac{2}{T_g^3}}{\sqrt{\left[\left(\frac{m_a}{m_a+1}\right)\times\frac{1}{T_aT_g}-\frac{1}{T_g^2}\right]}} = 0$$

$$-\left(\frac{m_a}{m_a+1}\right)\times\frac{1}{T_a}\times\frac{1}{T_g^2}+\frac{2}{T_g^3}=0$$

$$-\left(\frac{m_a}{m_a+1}\right)\times\frac{1}{T_a}\times\frac{1}{T_g^2}=-\frac{2}{T_g^3}$$

$$T_g=2\times\left(\frac{m_a+1}{m_a}\right)\times T_a$$

$$\frac{T_g}{T_a}=2\times\left(\frac{m_a+1}{m_a}\right)\qquad\qquad.....(16)$$

The equ. 16 illustrate that for the maximum discharge through chimney, the absolute temperature of the chimney gases bears a certain ratio to the absolute temperature of outside air.

Now, substitute the value of Equ. 16 in Equ. 9

$$H_g=H\left[\left(\frac{m_a}{m_a+1}\right)\times2\times\left(\frac{m_a+1}{m_a}\right)-1\right]=H(2-1)$$

$$H_g=H_m$$

Thus for the maximum discharge the height of the gas column producing draught is equal to height of the chimney.

$$\left(m_g\right)_{max}=A\times\rho_g\times\sqrt{2gH}=A\times\frac{p}{RT_g}\times\sqrt{2gH}$$

$$\left(m_g\right)_{max}=\frac{Ap}{2RT_g}\left(\frac{m_a}{m_a+1}\right)\times\sqrt{2gH}\qquad.....(17)$$

For maximum discharge condition, draught in mm of water column can be obtain by substituting Equ 16 in Equ. 8b.

$$h=353\times H\times\left[\frac{1}{T_a}-\frac{1}{2T_a}\right]$$

$$h=176.5\times H\times\frac{1}{T_a}\ mm\ of\ water\qquad.....(18)$$

Efficiency of a Chimney: The temperature of the flue gases leaving a chimney, in case of natural draught, will be higher than that of flue gases leaving it in case of artificial draught system because a certain minimum temperature is needed to produce a given draught with the given height of chimney. As far as steam generation is concerned, in case of natural draught system the heat carried away by the flue gases is more due to higher flue gas temperature. This indicates that the draught is created at the cost of thermal efficiency of the boiler plant installation since a portion of the heat carried away by flue gases to produce the required draught could have been used either in heating the air going to furnace or in heating the feed water going to boiler; thereby improving the thermal efficiency of the installation.

Let,

T_g = absolute temperature of flue gases to create the natural draught.

T_g' = absolute temperature of flue gases in case of artificial draught.

c_p = mean specific heat flue gases.

The extra heat carried away by 1 kg of flue gas due to higher temperature required to produce the natural draught = $c_p(T_g - T_g')$ Since $T_g > T_g'$

The draught pressure produced by the natural draught system in height of hot gases column,

$$H_g = H\left[\left(\frac{m_a}{m_a + 1}\right) \times \left(\frac{T_g}{T_a}\right) - 1\right] \quad metre\ of\ gas$$

The maximum energy this head would give to 1 kg of flue gas which is at the expense of extra heat carried away from the boiler plant

$$= H\left[\left(\frac{m_a}{m_a + 1}\right) \times \left(\frac{T_g}{T_a}\right) - 1\right]$$

$$= \frac{H}{J}\left[\left(\frac{m_a}{m_a + 1}\right) \times \left(\frac{T_g}{T_a}\right) - 1\right] \qquad .....(19)$$

Efficiency of chimney,

$$\eta_{chimney} = \frac{H_g}{c_p\left(T_g - T_g'\right)}$$

$$\eta_{chimney} = \frac{\dfrac{H}{J}\left[\left(\dfrac{m_a}{m_a + 1}\right) \times \left(\dfrac{T_g}{T_a}\right) - 1\right]}{c_p\left(T_g - T_g'\right)} \ , \quad J = 1\ (S.I.\ units) \quad .....(20)$$

The efficiency of the chimney is proportional to the height but even for a very tall chimney the efficiency will be less than 1% and thus we see that chimney is very inefficient as an instrument for creating draught.

Draught losses: The loss of draught may be due to the reasons mentioned below:

(i)   The frictional resistance offered by the flues and gas passages to the flow of flue gases.

(ii)   Loss near the bends in the gas flow circuit. (1.20 to 1.25 mm of water).

(iii) Loss due to friction head in equipments like grate, economizer, superheater, etc.

(iv)   Loss due to imparting velocity to the flue gases.

(v)   The loss in draught in a chimney is 20 percent of the total draught produced by it.

Limitations of Chimney Height

(i)   As per I.B.R. (Indian boiler regulation), the chimney height minimum required is 100 ft.

(ii)   Self-supported or supported by wire ropes depends upon the type of chimney.

(iii) An undertaking regarding height of chimney should be submitted to the I.B.R.

(iv)   Chartered Engineers certificate that the chimney withstands a wind load of 207 kg per m^2 should be submitted to the I.B.R.

4.11.2 Artificial draught

In the boiler installations of today the total static draught required may carry from 30 to 350 mm of water column. It may not be possible to build a chimney high enough to produce draught of such a large magnitude. To meet this requirement artificial draught system should be used. It may be a mechanical draught or a steam jet draught. The former is used for central power stations and many other boiler installations while the latter is employed for small installations and in locomotives.

(i)   ***Forced draught***: In a mechanical draught system, the draught is produced by a fan. In a forced draught system, a blower or a fan is installed near or at the base of the boiler to force the air through the cool bed and other passages through the furnace, flues, air preheater, economizer, etc. it is a positive pressure draught. the closure for the furnace etc, has to be very highly sealed so that gases from the furnace do not leak out in the boiler house.

(ii) ***Induced draught*:** In this system a fan or blower is located at or near the base of the chimney. The pressure over the fuel bed is reduced below that of the atmosphere. By creating a partial vacuum in the furnace and flues, the products of combustion are drawn from the main flue and they pass up the chimney. This draught is used usually when economizers and air preheaters are incorporated in the system. The draught is similar in action to the natural draught.

(iii) ***Balanced draught*:** It is a combination of the forced and induced draught system. In this system the forced draught fan overcomes the resistance in the air preheater and chain grate stoker while the induced draught fan overcomes draught losses through boiler, economizer, air preheater and connecting flues.

(iv) ***Steam jet draught*:** Steam jet draught may be induced draught or forced draught depending upon where the steam jet is located. In a steam jet draught, the exhaust steam, from a non-condensing steam engine, is used for producing draught. It is mostly used in locomotive engine. In an induced steam jet draught, the steam jet issuing from the nozzle is placed in the chimney. In a forced steam jet draught, the steam jet issuing from the nozzle is placed in the ash pit under the grate of the furnace. Induced draught produced by steam jet is shown in Fig.4.29.

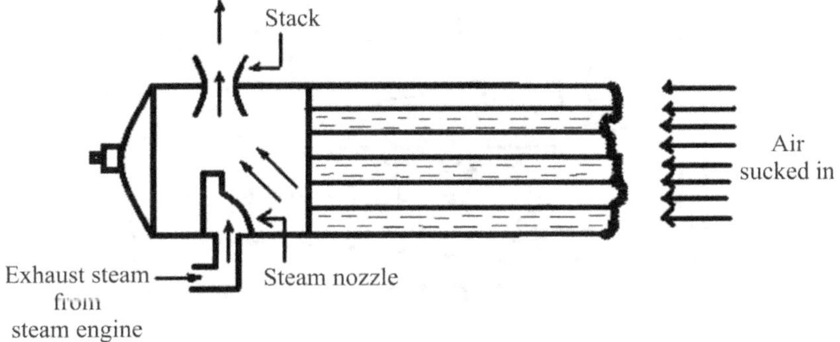

Fig. 4.29 Induced Draught by steam jet

This system is used in a boiler for a locomotive. When the locomotive is stationary, steam from the boiler may be lead to the smoke box through the nozzle to produce draught. When the locomotive is in motion, the air enters through the damper and forces its way through the grate and flues to the smoke box. But besides this, the exhaust steam from the cylinder can be lead to the nozzle in the smoke box, and induced draught created.

Forced draught produced by steam jet is shown in Fig.4.30. In a forced draught, steam from the boiler is throttled by a valve to a pressure of about 1.5 to 2 bar and passed through the nozzles projecting in a diffuser pipe. The steam thus emerges out with higher velocity and drags the column of air along with it thus allowing fresh outside air to enter. Mixture of steam and air having great kinetic energy passes in the diffuser pipes. The kinetic energy is converted to pressure energy and thus air is forced through the coal bed to the furnace. Normally, the natural draught or chimney draught is also provided in conjunction with steam jet draught. The steam jet draught is only an accelerating device helping to make the natural draught more effective.

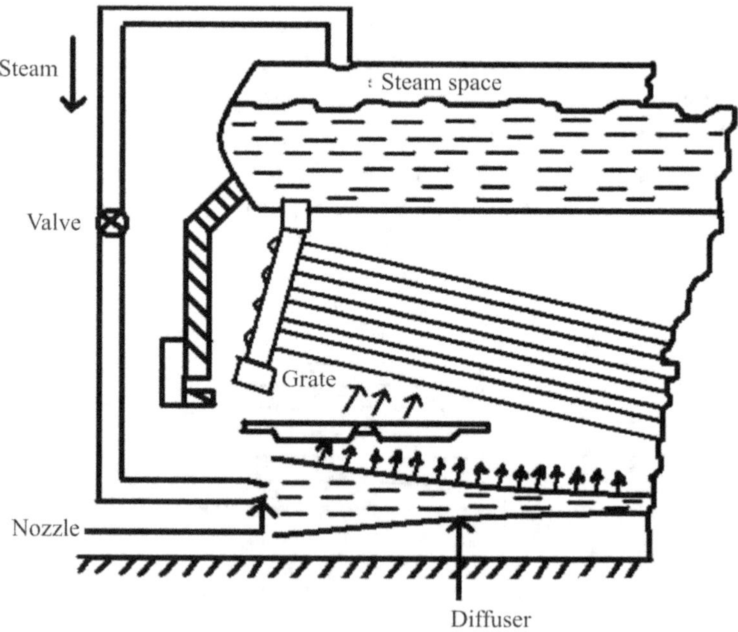

Fig. 4.30 Forced Draught by Steam Jet

4.11.3 The steam jet draught entails the following advantages

(i)   Very simple and economical.

(ii)   Occupies minimum space.

(iii)   Requires very little attention.

(iv)   In forced type steam jet draught, the steam keeps the fire bars cool and prevents the adhering of clinker to the fire bars.

(v)   Several classes of low grade fuels can be used with this system.

4.11.4 Advantages of mechanical draught

The mechanical draught possesses the following advantages:

(i)   Easy control of combustion and evaporation.

(ii)   Increase of evaporative power of a boiler.

(iii)   Improvement of efficiency of the plant.

(iv)   Reduced chimney height.

(v)   Prevention of smoke.

(vi)   Capability of consuming low grade fuel.

(vii)   Low grade fuel can be used as the intensity of artificial draught is high.

(viii) The fuel consumption per H.P. due to artificial draught is 15% less than that for natural draught.

(ix)   The fuel burning capacity of grate is 200 to 300 kg/m^2-h with mechanical draught whereas it is hardly 50 to 100 kg/m^2-h with natural draught.

4.11.5 The forced draught entails following advantages over induced draught

(i)   Forced draught fan does not require water cooled bearings.

(ii)   Tendency to air leak into the boiler furnace is reduced.

(iii)   No loss due to in rush of cold air through the furnace doors when they are opened for fire and cleaning fires.

(iv)   Fan size and power required for the same draught are 1/5 to 2/5 of that required from an induced draught fan installation because forced draught fan handles cold air.

4.12   Boiler Trial and Performance Analysis

The performance of the boiler is measured by evaporation capacity and its efficiency.

4.12.1 Equivalent evaporation

It is the quantity of steam generated per hour in Kg at full load. Equivalent evaporation is also defined as – quantity of steam generated by the boiler per unit time from water at 100°C to steam at 100°C. 2260KJ is taken as standard evaporation rate.

$$E = m_e (h\text{-}h_f)/ 2260$$

where m_e - Quantity of steam

h- Enthalpy of steam at working pressure

h_f- Enthalpy of feed water

4.12.2 Boiler Efficiency

It is defined as the ratio of heat absorbed by the steam to the heat supplied by the combustion of fuel.

Boiler Efficiency = Heat used to produce steam/ heat supplied by the furnace

$$= m_s (h\text{-}h_f)/ m_f \times C.V.$$

m_s - mass of steam generated (kg/hr)

m_f - mass of fuel burned (kg/hr)

h - Enthalpy of steam at working pressure

h_f - Enthalpy of feed water

4.12.3 Boiler Trail

In order to evaluate the efficiency and to determine the performance of the boiler, a test is conducted by operating the boiler for certain time and data is recorded, this procedure of testing is known as boiler trail

4.12.4   Objective of Boiler trail

Following are the objectives of boiler trial:

(i)   To determine the evaporative capacity of the boiler.

(ii)   To determine the thermal efficiency of the boiler.

(iii) To draw up heat balance sheet of the boiler.

4.12.5 Heat Balance Sheet

Heat balance sheet is a result of 'Boiler Trail'. It is a systematic representation of heat released and heat utilised in 'minute', 'hour' or per kg of fuel basis.

The typical Layout of heat balance sheet is shown below:

Heat Supplied	KJ	%	Heat Lost	KJ
Heat Supplied By fuel	$Q_s = m_f \times C.V.$	100%	(a) Heat carried by Dry flue gases. (b) Heat carried by steam in flue gases (c) Heat lost due to incomplete combustion (d) Heat lost due to unburnt fuel. (e) Unaccounted or Radiation Losses	$Q_1 = m_g C_{pg} (T_2 - T_1)$ $Q_2 = m_{s1} (hg_1 - h_{f1})$ $Q_3 = mf_1 \times C.V$ $Q_4 = mf_2 \times C.V$ $Q_5 = Q - (Q_1 + Q_2 + Q_3 + Q_4)$
	Qs	100		Q

4.13 Refrigeration

The term "refrigeration" is used to denote maintenance of a system or body at a temperature lower than that of its surroundings. The system maintained at the lower temperature is known as refrigerated system, while the equipment used to maintain this lower temperature is known as the refrigerating system or refrigerator. Refrigeration applications at the domestic, commercial and industrial levels are becoming an integral part of the present day living. The vapor compression refrigeration cycle is a common method for transferring heat from a low temperature to a high temperature.

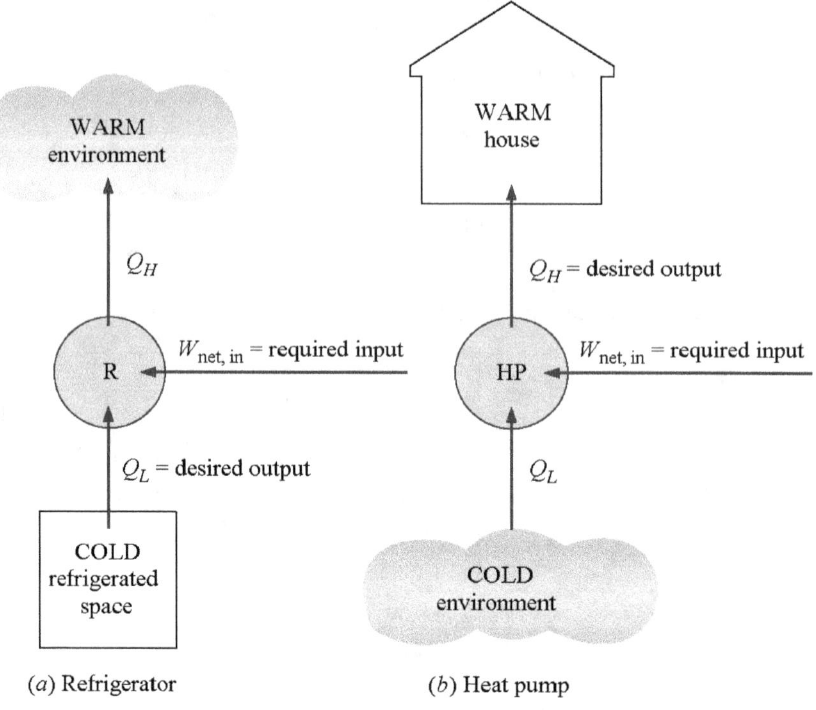

Fig. 4.31 Refrigerator & heat pump

Fig. 4.31 shows the objectives of refrigerators and heat pumps. The purpose of a refrigerator is the removal of heat, called the cooling load, from a low temperature medium. The purpose of a heat pump is the transfer of heat to a high temperature medium, called the heating load. When we are interested in the heat energy removed from a low temperature space, the device is called a refrigerator. When we are interested in the heat energy supplied to the high temperature space, the device is called a heat pump. In general, the term "heat pump" is used to describe the cycle as heat energy is removed from the low temperature space and rejected to the high temperature space.

The performance of refrigerators and heat pumps is expressed in terms of *coefficient of performance* (COP), defined as

$$COP_R = \frac{\text{Desired output}}{\text{Required input}} = \frac{\text{Cooling effect}}{\text{Work input}} = \frac{Q_L}{W_{net,in}}$$

$$COP_{HP} = \frac{\text{Desired output}}{\text{Required input}} = \frac{\text{Heating effect}}{\text{Work input}} = \frac{Q_H}{W_{net,in}}$$

Both COP_R and COP_{HP} can be larger than 1. Under the same operating conditions, the COPs are related by

$$COP_{HP} = COP_R + 1$$

4.13.1   Unit of Refrigeration

Capacity of refrigeration unit is generally defined in ton of refrigeration. A ton of Refrigeration is defined as the quantity of heat to be removed in order to form one ton (1000 kg) of ice at 0^0C in 24 hrs, from liquid water at 0^0C. This is equivalent to 3.5 kJ/s (3.5 kW) or 210 kJ/min

4.13.2 Refrigerant

A refrigerant is a fluid in a refrigerating system that by its evaporating takes the heat of the cooling coils and gives up heat by condensing the condenser. The refrigerant is a heat carrying medium, which undergoes the thermodynamic cycle of refrigeration (i.e., compression, condensation, expansion and evaporation). In refrigeration system, it absorbs the heat at low temperature medium and discards the absorbed heat to a high temperature environment. The two main uses of refrigerants are refrigerators/ freezers and air conditioners.

4.13.3 Identifying refrigerants by numbers

The present practice in the refrigeration industry is to identify refrigerants by numbers. The identification system of numbering has been standardized by the American society of heating, refrigerating and air conditioning engineers (ASHRAE), some refrigerants in common use are-

Classification	Denomination	Composition or chemical formula	Safety classification
(mass percentage)			
INORGANIC COMPOUND			
R717	ammonia	NH_3	B2
R718	water	H_2O	A1
R744	carbon dioxide	CO_2	A1
ORGANIC COMPOUND			
Hydrocarbons			
R170	ethane	CH_3CH_3	A3
R290	propane	$CH_3CH_2CH_3$	A3
R600a	isobutane	$CH(CH_3)_2CH_3$	A3

Halocarbons			
Chlorofluorocarbons (CFCs) and Bromofluorocarbons (BFCs)			
R11	trichlorofluoromethane	CCl_3F	A1
R12	dichlorodifluoromethane	CCl_2F_2	A1
Hydrochlorofluorocarbures (HCFC)			
R22	chlorodifluoromethane	$CHClF_2$	A1
R141b	1,1-dichloro-1-fluoroethane	CH_3CCl_2F	A2
R142b	1-chloro-1,1-difluoroethane	CH_3CClF_2	A2
Hydrofluorocarbons (HFCs)			
R32	difluoromethane	CH_2F_2	A2
R125	pentafluoroethane	CHF_2CF_3	A1
R134a	1,1,1,2-tetrafluoroethane	CH_2FCF_3	A1
R143a	1,1,1-trifluoroethane	CH_3CF_3	A2
R152a	1,1-difluoroethane	CH_3CHF_2	A2
Azeotropic mixtures			
R502		R22/R115 (48.8/51.2)	A1
R507		R125/R143a (50/50)	A1
Zeotropic mixtures			
R404A		R125/R143a/R134a (44/52/4)	A1
R407C		R32/R125/R134a (23/25/52)	A1

4.13.4 New Eco-friendly Refrigerant

The relative ability of a substance to deplete the ozone layer is called the Ozone Depletion Potential (ODP) R-11 (CFC-11) & R-12 (CFC-12) have the highest value, ODP = 1.0. According to the Environmental Protection Agency, most air conditioners use refrigerant R-22, which adds to the problem of global climate change. Most of refrigerant being used contain Chlorine which reacts rapidly with oxygen atom of ozone causing **ozone depletion.** So, now people are working on eco-friendly refrigerants to prevent ozone depletion.

Musicool: Musicool is a hydrocarbon refrigerant that can be used in automobile air-conditioning systems, split/window refrigeration units, water dispensers, cold storages and chillers. The refrigerant is commonly used in developed nations like Australia, Canada, Europe, Japan and the

United States since it is environmentally friendly and has non-toxic low global warming effect and can save up to 40 per cent of energy.

R-410A: R-410A is a new environmentally safe liquid coolant that makes air conditioning possible. It was developed as an alternative to R-22, also known as Freon, which will be phased out in the near future along with other ozone depleting chemicals. R-410A is economically and environmentally beneficial, which makes it a smart choice for future air conditioning units. Table below shows the ODP of refrigerants.

Sub Group	Ref. Designation	Chemical formula	Refrigerant Name	ODP
	R-11	CCl_3F	Trichlorofluoro methane	1
	R-12	CCl_2F_2	Dichlorodifluoro methane	1
CFCs	R-113	$CF_2Cl\text{-}CFCl_2$	Trifluoro-trichloro ethane	0.8
	R-114	$CF_2Cl\text{-}CF_2Cl$	Tetrafluoro-dichloro ethane	1
HCFCs	R-22	$CHClF_2$	Chlorodifluoro methane	0.05
	R-123	$CHCl_2F_3$	Dichlorotrifluoro ethane	0.02
HFC	R-134a	CH_2FCF_3	Tetrafluoro ethane	0
Azeotrope	R-500	R-12/R-152		0.75
	R-717	NH_3	Ammonia	0

4.13.5 Properties of an Ideal Refrigerant

A good refrigerant should have the following properties:

(i) *Saturation Pressure*: The saturation pressure of the refrigerant at desired low temperature should be above or equal to atmospheric pressure in order to avoid the leakage in the evaporator. The pressure at the condenser must not be excessively high for the same reasons.

(ii) *Latent Heat*: Latent heat of evaporation at low temperature should be as high as possible to give a reasonably low mass flow rate of refrigerant for given capacity.

(iii) *Specific Volume*: The size of compressor depends on the specific volume of refrigerant at evaporator pressure. Thus the specific volume of refrigerant at the compressor suction should be in order to avoid the large compressor for the required mass flow rate.

(iv) *Chemical Stability*: The refrigerant should be chemically stable & should not react with lubricant used in reciprocating compressor and should be miscible with oil.

(v) *Flammability*: It should not be non-flammable and explosive.

(vi) *Toxicity*: It should be non-toxic. If toxic, then to a limit, below the acceptable level.

(vii) *Specific Heat*: It should have low specific heat of liquid for better heat transfer in condenser.

(viii) *C.O.P.*: The refrigerant should give high value of C.O.P. with low power input per ton of refrigeration.

(ix) *Thermal Conductivity*: The refrigerant should have good thermal conductivity for better heat transfer in condenser and evaporator.

(x) *Freezing Point*: The refrigerant must have freezing point temperature well below the lowest temperature in the cycle.

(xi) *Eco-Friendly*: Other considerations are chemically stable, non corrosiveness, low cost and overall eco-friendly.

4.13.6 Simple Vapor Compression Refrigeration System

A simple vapor compression refrigeration system consists of the following equipments: (i) Compressor (ii) Condenser (iii) Expansion valve (iv) Evaporator.

The most widely used refrigeration cycle is the *vapor-compression refrigeration cycle*. In an ideal vapor-compression refrigeration cycle, the refrigerant enters the compressor as a saturated vapor and is cooled to the saturated liquid state in the condenser. It is then throttled to the evaporator pressure and vaporizes as it absorbs heat from the refrigerated space.

The ideal vapor compression cycle consists of four processes are shown in Fig. 4.32 and Fig. 4.33.

Process	Description
1-2	Isentropic compression
2-3	Constant pressure heat rejection in the condenser
3-4	Throttling in an expansion valve
4-1	Constant pressure heat addition in the evaporator

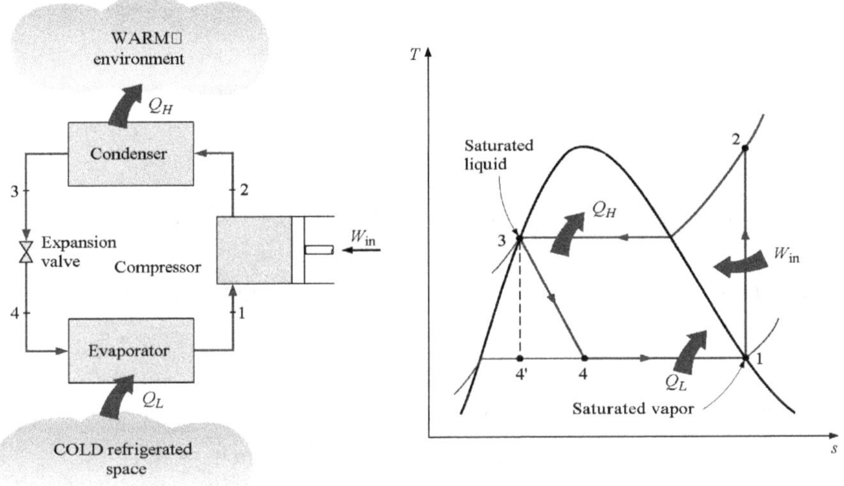

Fig. 4.32 Vapour Compression system with T-s Diagram

The p-h diagram is another convenient diagram often used to illustrate the refrigeration cycle.

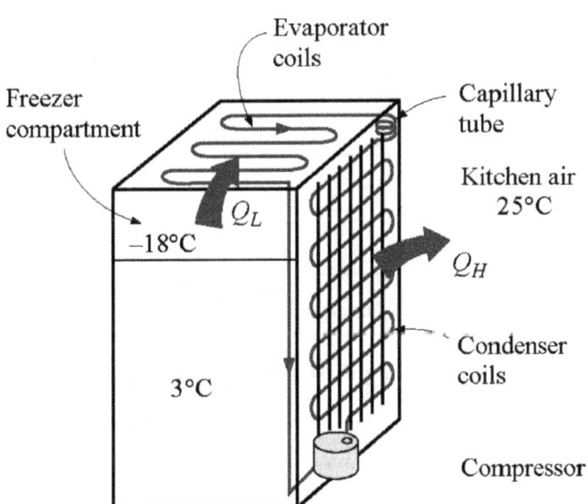

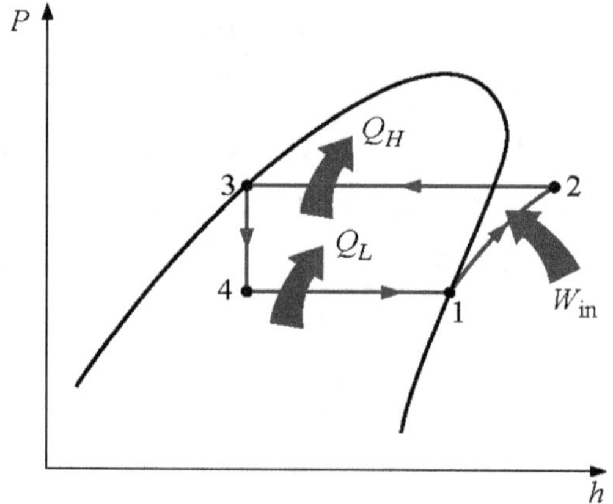

Fig. 4.33 Vapour Compression System with p-h Diagram

The ordinary household refrigerator is a good example of the application of this cycle.

Component	Process	Analysis
Compressor	$s = $ Const.	$\dot{W}_{in} = \dot{m}(h_2 - h_1)$
Condenser	$P = $ Const.	$\dot{Q}_H = \dot{m}(h_2 - h_3)$
Throttle Valve	$\Delta s > 0$	$h_4 = h_3$
	$\dot{W}_{net} = 0$	
	$\dot{Q}_{net} = 0$	
Evaporator	$P = $ Const.	$\dot{Q}_L = \dot{m}(h_1 - h_4)$

$$COP_R = \frac{\dot{Q}_L}{\dot{W}_{net,in}} = \frac{h_1 - h_4}{h_2 - h_1}$$

$$COP_{HP} = \frac{\dot{Q}_H}{\dot{W}_{net,in}} = \frac{h_2 - h_3}{h_2 - h_1}$$

Advantages of Vapor compression refrigeration

- Since the working cycle approaches closer to Carnot cycle, the C.O.P is quite high.

- Operational cost of vapour compression system is just above 1/4th of air refrigeration system.

- Since the heat removed consists of the latent heat of vapour, the amount of liquid circulated is less and as a result the size of the evaporator is smaller.

- Any desired temperature of the evaporator can be achieved just by adjusting the throttle valve.

Disadvantages of Vapour compression refrigeration system

- Initial investment is high

- Prevention of leakage of refrigerant is a major problem

4.13.7 Vapor Absorption Refrigeration System

Another form of refrigeration that becomes economically attractive when there is a source of inexpensive heat energy at a temperature of 100 to 200°C is *absorption refrigeration* .The absorption refrigeration system is a heat-operated unit which used a refrigerant that is alternately absorbed and liberated by the absorbent. The minimum numbers of primary units essential in an absorption system include an evaporation, absorber, generator and condenser, where the refrigerant is absorbed by a transport medium and compressed in liquid form. The most widely used absorption refrigeration system is the ammonia-water system, where ammonia serves as the refrigerant and water as the transport medium. The work input to the pump is usually very small, and the COP of absorption refrigeration systems is defined as

$$\text{COP}_R = \frac{\text{Desired output}}{\text{Required input}} = \frac{\text{Cooling effect}}{\text{Work input}} = \frac{Q_L}{Q_{gen} + W_{pump,in}} \cong \frac{Q_L}{Q_{gen}}$$

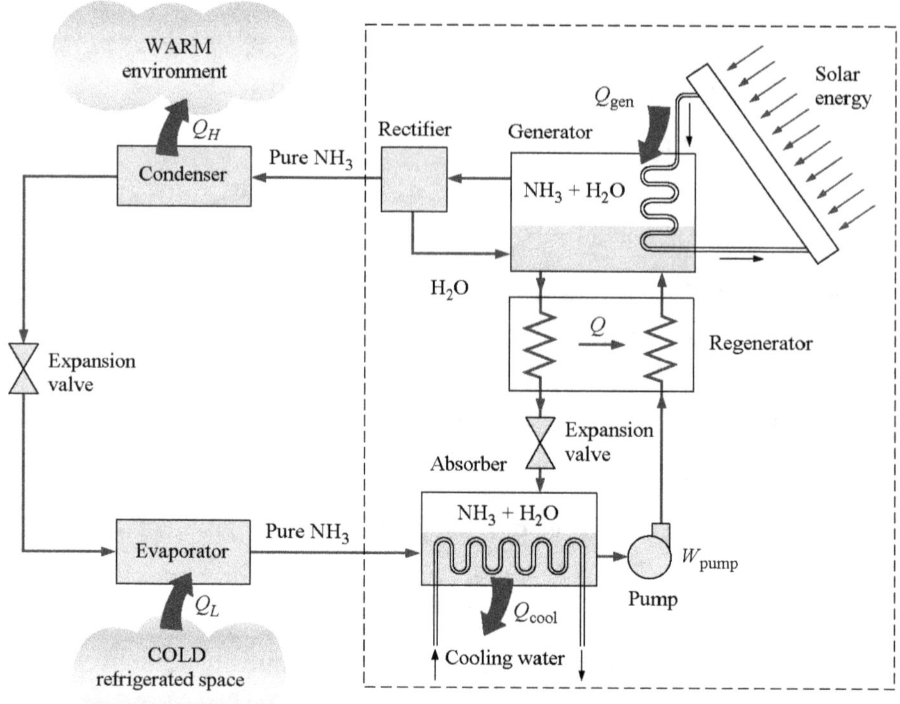

Fig. 4.34 Vapour Absorption System with T-S Diagram

A simple absorption cycle is shown in Fig. 4.34. This cycle differs from a vapour compression cycle by the substitution of an absorber, generator, pumps and reducing valve for the compressor. Various combinations of fluids may be used, but that of ammonia, a strong solution that contains about as much ammonia as possible; a weak solution contains considerably less ammonia. The weak solution containing very little ammonia is sprayed or otherwise exposed in the absorber and absorbs ammonia coming from the evaporation. Absorption of the ammonia lowers the pressure in the absorber, which in turn draws more ammonia vapour from the evaporator. Usually some forms of cooling is employed in the absorber to remove the heat of condensation and the heat of solution evolve there. The strong solution is then pumped into a generator, which is at higher pressure and is where heat is applied; the heat vaporizes the ammonia driving it out of solution and into the condenser, where it is liquefied. The liquid ammonia passes on to the receiver if a separate one is used, or through the expansion valve and into the evaporator. The weak solution left in the generator after the ammonia has been drive off flows through the reducing valve back to the absorber.

Comparison between Vapor Compression and Absorption system:

Compression System	Absorption system
Using high-grade energy like mechanical work.	Uses low grade energy like heat. Therefore, may be worked on exhaust systems from I.C engines, etc
Moving parts in the compressor are more.	Moving parts are only in the pump and less.
More wear, tear and noise.	Smooth operation.
COP decreases with decrease in evaporator pressure.	System can work on lower evaporator pressures also without affecting the COP.
Performance is adversely affected at partial loads.	No effect of reducing the load on performance.
Liquid traces in suction line may damage the compressor.	Liquid traces of refrigerant present in piping at the exit of evaporator.
It is difficult.	Automatic operation for controlling the capacity is easy.
High COP	Low COP
Work operated	Heat operated
Regular Maintenance is required	Very low Maintenance required
Higher noise and vibration	Less noise & vibration

4.13.8 Applications of refrigeration

(i) *Industrial Applications*:

    (a)   Processing of food products.

    (b)   Processing of farm crops.

    (c)   Processing of textiles, printing work, photographic materials, etc.

    (d)   Cooling of concrete for dams.

    (e)   Treatment of air for blast furnace.

    (f)   Processing of tobacco, petroleum and other chemical products.

(ii) *Preservation of Perishable Goods*:

    (a)   Manufacturing of ice.

    (b)   Freezing or chilling, storage and transportation of food stuffs including beverages, meat, poultry products, dairy products, fish, fruits, vegetables, fruit Juices, etc.

    (c)   Preservation of photographic films, archeological documents etc.

(iii) *Providing comfortable environment*:

    (a)   Industrial air-conditioning.

    (b)   Comfort air-conditioning of hospitals, residences, hotels, restaurants, theatres, offices, etc

Solved Numerical Examples

***Example* 1:** A system receives 10×10^6 J in the form of heat energy in a specified process and it produces work of 4×10^6 J. The system velocity changes form 10 m/s for 50 kg mass of the system. Determine the change in internal Energy.

Solution:

From first law of Thermodynamics

$$Q_{1-2} - W_{1-2} = m(g\, z_2 - g\, z_1) + m\left(\frac{v_2^2}{2} - \frac{v_1^2}{2}\right) + \left(u_2 - u_1\right)$$

$$10 \times 10^6 - 4 \times 10^6 = 50\,(0) + 50\left(\frac{25^2}{2} - \frac{10^2}{2}\right) + \left(u_2 - u_1\right)$$

$$u_2 - u_1 = 5986875 \text{ J/kg} = 598.6875 \text{ kJ/kg}$$

***Example* 2:** A machine working as a Carnot cycle operates between 305 k and 260 K. Determine the COP when it is operated as

    1. A refrigerating machine

    2. A heat pump

    3. A heat engine

Solution:

    1. COP of a refrigerating machine $= \dfrac{T_2}{T_1 - T_2} = \dfrac{260}{305 - 260} = 5.78$

2. COP of a heat pump $= \dfrac{T_1}{T_1 - T_2} = \dfrac{305}{305 - 260} = 6.78$

3. COP of a heat engine $= \dfrac{T_1 - T_2}{T_1} = \dfrac{305 - 260}{305} = 0.147$

Example 3: An investor claims to have developed an engine which receives 100 kJ at a temperature of 106 °C. It rejects heat at a temperature of 50 °C and delivers 0.12 kWh of mechanical work. Is this a valid claim?

Solution:

Work done = 0.12 kWh = 0.12 × 3600 = 432 kJ

Heat supplied = 1000 kJ.

$$\eta_{Engine} = \frac{Work\ done}{Heat\ supplied} = \frac{432}{1000} = 43.2\%$$

$$\eta_{Carnot} = \frac{T_1 - T_2}{T_1} = \frac{433 - 278}{433} = 35.79\%$$

$\eta_{Engine} > \eta_{carnot}$. Hence the claim is not possible (or) false.

Example 4: An Investor claims to have developed a refrigerating machine which operates between − 20 °C and 30 °C and consumes 1 kW of power. The machine gives a refrigerating effect of 21.6 kJ/hr. Comment on the claims of the Investor.

Solution:

$$Actual\ COP = \frac{Refrigerating\ Effect}{Work\ done},\ kJ/sec$$

$$= \frac{2.16 \times 1000}{1 \times 3600} = 6$$

$$(COP)_{max} = \frac{T_2}{T_1 - T_2} = \frac{273 - 20}{273 + 30 - (273 - 20)}$$

$$= 5.06.$$

Actual COP > COP$_{max}$. So the Investor's claim is false.

Example 5: Determine the quantity of heat required to produce 1 kg of steam at a pressure of 6 bar at a temperature of 25 °C under the following conditions

1. When the steam is wet having dryness fraction 0.9

2. When the steam is dry saturated

3. When it is superheated at a constant pressure of 250 °C.

   Assume $c_p = 2.3$ kJ/kg–K

Solution:

At 6 bar pressure: From steam table

$$h_f = 670.4 \text{ kJ/kg} \; ; \; h_{fg} = 2085 \text{ kJ/kg} \; ; \; t = 158.8°$$

1. Wet steam : $h = h_f + x \, h_{fg} = 670.4 + 0.9 \times 2085 = 2546.9$ kJ.
   Sensible heat $= m \times c_p \times (t_W - 0)$
   $$= 1 \times 2.3 \times 25 = 57.5 \text{ kJ.}$$
   $\therefore$ Actual Heat Supplied $= 2546.9 - 57.5 = 2489.4$ kJ

2. Dry saturated: $h_g = h_f + h_{fg} = 670.4 + 2085 = 2755.4$ kJ.
   Actual heat supplied $= 2755.4 - 57.5 = 2697.9$ kJ

3. Superheated Steam: $h_{sup} = h_g + c_p (t_{sup} - t)$
   $$= 2755.4 + 2.3 (250 - 158.8)$$
   $$= 2965.16 \text{ kJ}$$

Actual Heat supplied $= 2965.15 - 57.5 = 2907.26$ kJ

***Example* 6:** Calculate the internal energy of 1 kg of steam at a pressure of 10 bar, when the steam is

(i) 0.9 dry   (ii) dry saturated   (iii) Super heated to 250 °C

Solution:

At 10 bar pressure : From steam table

$h_f = 762.6$ kJ/kg ; $h_{fg} = 2013.6$ kJ/kg ; $v_g = 0.1943$ m^3 / kg

(i)   Dry steam $u = h_f + x \, h_{fg} - 100 \, p \, x \, v_g$
   $$= 762.6 + 0.9 \times 2013.6 - 100 \times 10 \times 0.9 \times 0.1943$$
   $$= 2574.8 - 174.8 = 2400 \text{ kJ}$$

(ii)   Dry saturated $u = h_f + h_{fg} - 100 \, p \, v_g$
   $$= 762.6 + 2013.6 - 100 \times 10 \times 0.1943$$
   $$= 2776.2 - 194.3 = 2581.9 \text{ kJ.}$$

(iii)   Superheated $u = h_g + c_p [t_{sup} - t] - 100 \, p \, v_{sup}$
   $$= 2776.2 + 2.1 [250 - 179.91] - 100 \times 10 \times 0.1943$$
   $$= 2729.089 \text{ kJ/kg}$$

***Example* 7:** A pressure cooker contains 2 kg of steam at a pressure of 8 bar. Find the quantity of heat which must be rejected to reduce the quality of steam to 70% dry.

Solution:

At 8 bar pressure : From steam tables

$$h_g = 2767.5 \text{ kJ / kg}; \; v_g = 0.24 \text{ m}^3/ \text{kg}$$

Volume of pressure cooker $= 2 \times 0.24 = 0.48 \text{ m}^3$

Initial Energy

$$U_1 = h_1 - p_1 \, v_{g1}$$
$$= (h_f + h_{fg}) - 100 \, p_1 \, v_{g1}$$
$$= 2767.5 - 100 \times 8 \times 0.24 = 2575.5 \text{ kJ/kg}$$

Volume remains constant

Volume occupied by steam = volume of pressure cooker

$$0.48 = 2 \times 0.7 \times v_{g2}$$
$$v_{g2} = 0.343 \text{ m}^3/\text{kg}$$

At 0.343 m^2 / kg Specific volume : From steam tables

$$P_2 = 5.5 \text{ bar} = h_f = 655.8 \text{ kJ/kg} ; h_{fg} = 2095.9 \text{ kJ/kg}$$

Final Internal Energy

$$u_2 = h_f + x_2 \, h_{fg} - 100 \, p_2 \, x_2 \, v_{g2}$$
$$= 655.8 + 0.7 \times 2095.9 - 100 \times 5.5 \times 0.7 \times 0.343$$
$$= 2123 - 132 = 1991 \text{ kJ / kg}$$

Heat Rejected $= u_1 - u_2 = 2575.5 - 1991 = 584.5 \text{ kJ/kg}.$

Heat Rejected by 2 kg of steam $= 2 \times 584.5 = 1169 \text{ kJ}$

***Example* 8:** Calculate the height of chimney required to produce a draught equivalent to 20 mm of water, if the flue gas temperature is 35 °C and ambient temperature is 27 °C and minimum amount of air per kg of fuel is 18 kg.

Solution:

Height of chimney

$$H_W = 353 \times H \left[\frac{1}{T_a} - \left(\frac{m_a + 1}{m_a} \right) \times \frac{1}{T_g} \right]$$

$$20 = 353 \times H \left[\frac{1}{300} - \left(\frac{18+1}{18} \right) \times \frac{1}{588} \right]$$

$$H = \frac{20}{0.53} = 38 \text{ m}$$

***Example* 9:** The following observations were made in a boiler that

Coal used = 200 kg/hr

Mass of steam = 2000 kg/hr

Steam pressure = 11.2 bar

Dryness fraction = 0.95

Feed water temp = 32.15 $^{\circ}$C

Calorific value = 28800 kJ/kg

Calculate the equivalent evaporation from and at 100 $^{\circ}$C per kg of coal and boiler efficiency.

From steam table

At 11.2 bar pressure and $x = 0.95$.

$$h_f = 784.89$$

$$h_{fg} = 1997.5$$

$$h = h_f + x \, h_{fg} = 784.89 + 0.95 \times 1997.5$$

$$= 2682.51 \text{ kJ / kg}$$

At 32.15°C Temperature

$$h_{f1} = 134.79 \text{ kJ/kg}$$

$$m_e = \frac{\text{Mass of Steam}}{\text{Mass of fuel burnt}} = \frac{m_s}{m_f} = \frac{2000}{200}$$

$$= 10 \text{ kg /kg of coal.}$$

Equivalent evaporation "From and at 100 $^{\circ}$C"

$$E = \frac{m_e \left(h - h_{f1} \right)}{2260}$$

$$= \frac{10 \times \left(2682.51 - 134.79 \right)}{2260}$$

$$= 11 \text{ kg/kg of coal.}$$

Boiler Efficiency

$$\eta_{Boiler} = \frac{m_g\left(h - h_{f_1}\right)}{m_f \times C_V}$$

$$= \frac{2000\left(2682.51 - 134.79\right)}{200 \times 28800}$$

$$= 0.084$$

$$= 88.46\%$$

Review Questions

1. State and explain First law of thermodynamics.

2. Write down first law of thermodynamics and enlist its short comings.

3. Explain the First Law of Thermodynamics with suitable examples.

4. State and explain second law of thermodynamics.

5. Write down Kelvin-Plank and Clausius equation of Second Law of Thermodynamics.

6. Give the two statements of Second Law of Thermodynamics.

7. Explain the various laws of thermodynamics with examples.

8. Show the various process of steam in a T-S Diagram.

9. How boilers are classified? Write down few names of mountings and accessories of a boiler.

10. How boilers are classified? Explain giving examples.

11. Explain the difference between water tube and fire tube boilers.

12. List Boilers mountings and accessories.

13. Differentiate clearly between mountings and accessories of a boiler with suitable examples.

14. Explain the construction and working of a Cochran boiler with the help of a neat sketch. Describe the classification, construction and working of a Cochran boiler.

15. What do you understand by the term "Equivalent Evaporation & Boiler Efficiency"? What is the purpose of Boiler trail?

16. Explain with a neat sketch the working of following boiler:
    (i)   Lancashire Boiler
    (ii)   Locomotive Boiler
    (iii) Cochran Boiler
    (iv)   Babcock and Wilcox Boiler

17. What is boiler draught? Derive an expression for chimney height.

18. Derive an expression for Chimney height and diameter.

19. Differentiate between:
    (a) Boiler Mountings and accessories
    (b) Vapour compression and vapour absorption refrigeration system.

20. What are the ideal properties of good refrigerant?

21. What do you understand by "Draught". What are the different methods of producing draught?

22. Explain the need of Eco-friendly refrigerant.

23. Define 1 ton of refrigeration.

24. What are the applications of refrigeration systems?

25. Mention the difference between heat pump and refrigerator.

26. Explain vapour compression refrigeration system with a neat sketch.

27. Explain vapour absorption refrigeration system with a neat sketch.

28. Explain the following:

    (i) COP (ii) Boiler efficiency (iii) Natural draught (iv) Entropy

29. Write short notes on the following:
    (i) Boiler classification   (ii)   Eco-friendly refrigerants.

Important Formulae

First law of Thermodynamics:

$$\Delta E = \delta Q - \delta W \text{ (or) } Q_{1-2} - W_{1-2} = E_2 - E_1.$$

$$Q_{1-2} - W_{1-2} = m\,(g\,Z_2 - g\,Z_1) + m\left(\frac{v_2^2}{2} - \frac{v_1^2}{2}\right) + (u_2 - u_1)$$

Second Law of Thermodynamics:

$$\eta_{Engine} = \frac{Q_1 - Q_2}{Q_1} = 1 - \frac{Q_2}{Q_1} = 1 - \frac{T_2}{T_1}.$$

$$(COP)_{Ref} = \frac{Q_2}{WR} = \frac{Q_2}{Q_1 - Q_2} = \frac{T_2}{T_1 - T_2}.$$

$$(COP)_{H.P} = \frac{Q_1}{WP} = \frac{Q_2}{Q_1 - Q_2} = \frac{T_1}{T_1 - T_2}.$$

$$(COP)_{H.P} = (COP)_{Ref} + 1.$$

Steam Properties:

$$\text{Dryness fraction (x)} = \frac{m_g}{m_g + m_f}$$

where m_g : Mass of dry steam

m_f : Mass of saturated liquid

Enthalpy (kJ/kg): Total Heat = Sensible heat + Latent Heat [kJ/kg]

    (i)   Wet Steam   : $h = h_f + x\, h_{f_g}$.

    (ii) Dry steam   : $h = h_f + h_{f_g} = h_g$.

    (iii) Superheated steam   : $h_{sup} = h_g + c_P\,(t_{sup} - t)$.

Specific volume [m^3 / kg]

    (i)   Wet steam   : $v = x\, v_g$.

    (ii) Dry steam   : $v = v_g$.

    (iii) Superheated Steam   : $v_{sup} = \dfrac{v_g T_{sup}}{T}$

Work done [kJ/kg]

    (i)   Wet Steam   : $W = 100\, p\, x\, v_g$.

    (ii)   Dry steam   : $W = 100\, p\, v_g$.

    (iii) Superheated steam   : $W = 100\, p\, v_{sup}$.

Internal Energy [kJ/kg]

(i)   Wet Steam   : $u = h_g - 100\, p\, v_g.$

(ii) Dry steam   : $u = h_f + x\, h_{f_g} - 100\, p\, v_g.$

(iii) Superheated steam   : $u = h_{sup} - 100\, p\, v_{sup}$

$$= [h_g + c_p\, (t_{sup} - t)] - 100\, p\, v_{sup}$$

Entropy [kJ/kg – K]

(i)   Dry steam   : $S = S_f + S_{f_g} = S_g.$

(ii) Wet Steam   : $S = S_f + x\, S_{f_g}.$

(iii) Superheated steam   : $S = S_g = 2.3 c_p\, \log_{10}\left(\dfrac{T_{sup}}{T}\right)$

Boiler chimney height

$$h = 353 \times H \times \left[\frac{1}{T_a} - \left(\frac{m_a + 1}{m_a}\right) \times \frac{1}{T_g}\right] mm\ of\ water$$

Boiler performance

Equivalent Evaporation

$$E = \frac{m_e\left(h - h_f\right)}{2260}$$

Boiler Efficiency

$$\eta_{boiler} = \frac{m_s\left(h - h_f\right)}{m_f\, CV}$$

CHAPTER 5

Reciprocating Machines

Reciprocating Machines: Steam Engines, Hypothetical and Actual indicator diagram; Carnot Cycle and Ideal efficiency; Otto and Diesel cycles; Working of two Stroke & four Stroke Petrol & Diesel IC Engines.

5.1 Reciprocating Engines

Reciprocating engines are those engines which convert pressure energy into rotating motion by the help of piston. They are also called **"piston engines"**.

Basically there are three types of reciprocating engines:

    (i)   Steam Engine

    (ii)   Internal Combustion Engine

    (iii) Stirling Engine

All the reciprocating engines have some of the properties in common with them. There may be one or more pistons. Each piston is inside a cylinder, into which a gas is introduced, either already hot and under pressure (steam engine), or heated inside the cylinder either by ignition of a fuel air mixture (internal combustion engine) or by contact with a hot heat exchanger in the cylinder (Stirling engine). The hot gases expand, pushing the piston to the bottom of the cylinder. The piston is returned to the cylinder top (Top Dead Centre) either by a flywheel or the power from other pistons connected to the same shaft. In most types the expanded or "exhausted" gases are removed from the cylinder by this stroke. The exception is the Stirling engine, which repeatedly heats and cools the same sealed quantity of gas.

5.2 Steam Engine

Thomas Savery was an English military engineer and inventor who in 1698 patented the first crude steam engine, based on Denis Papin's Digester or pressure cooker of 1679. Thomas Savery had been working on solving the problem of pumping water out of coal mines, his machine consisted of a closed vessel filled with water into which steam under pressure was introduced. This forced the water upwards and out of the mine shaft. Then a cold water sprinkler was used to condense the steam. This created a vacuum which sucked more water out of the

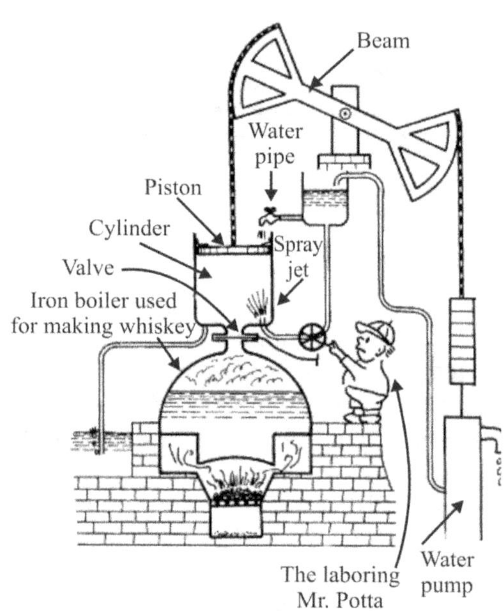

Fig. 5.1 Newcomen Steam Engine

mine shaft through a bottom valve. Thomas Newcomen was an English blacksmith, who invented the atmospheric steam engine, an improvement over Thomas Slavery's previous design. The Newcomen steam engine used the force of atmospheric pressure to do the work. Thomas Newcomen's engine pumped steam into a cylinder (Fig. 5.1).

The steam was then condensed by cold water which created a vacuum on the inside of the cylinder. The resulting atmospheric pressure operated a piston, creating downward strokes. James Watt was a Scottish inventor and mechanical engineer, born in Greenock, who was renowned for his improvements of the steam engine. In 1765, James Watt while working for the University of Glasgow was assigned the task of repairing a Newcomen engine, which was deemed inefficient but the best steam engine of its time. That started the inventor to work on several improvements to Newcomen's design. Most notable was Watt's 1769 patent for a separate condenser connected to a cylinder by a valve. Unlike Newcomen's engine, Watt's design had a condenser that could be cooled while the cylinder was hot. Watt's engine soon became the dominant design for all modern steam engines and helped bring about the Industrial Revolution.The working medium in steam engine is high pressure steam. After use of steam engines for a long time many difficulties were encountered like low efficiency, maintenance problem. This problem was cured to an extent by the use of **"Stirling Engines"**. And after that came into existence **"Internal Combustion Engine"**.

5.2.1 Construction of Steam Engine

(i)   Main and essential parts of steam engine are shown in Fig. 5.2.

(ii)   *Frame:* Frame is the main structure of engine which holds and supports all the moving parts of the engine. It is made of caste iron.

(iii)   *Cylinder:* Cylinder is the part of Steam Engine in which piston moves to and fro motion. It contains two valves from which high pressure steam is entered and act on piston. It is made of cast iron.

(iv)   *Steam Chest:* It is the part of machine which contains steam collected after removal from boiler. It supplies steam to the cylinder through valves. It is made of cast iron.

(v)   *Piston:* It is the main part of steam engine which performs the reciprocating motion. It is placed inside the cylinder. The steam with high pressure exerts force on the piston making it move in linear direction. It is connected with crank shaft via piston rod and connecting rod, by which it transfer the force to the crank shaft. It is made of grey cast iron.

(vi)   *Piston Ring:* They are used to prevent the leakage of steam. It is made of special grade cast iron.

(vii)   *Piston Rod:* It connects the piston to crank shaft via connection rod to transfer the force and convert the motion.

(viii) *Crosshead Bearing:* It connects the piston rod with connecting rod and guides the motion. It prevents the bending of piston rod and allows the transverse motion to convert into rotating motion. Its main function is to convert the transverse motion of piston rod into rotating motion of connecting rod and preventing the piston rod to come into defect.

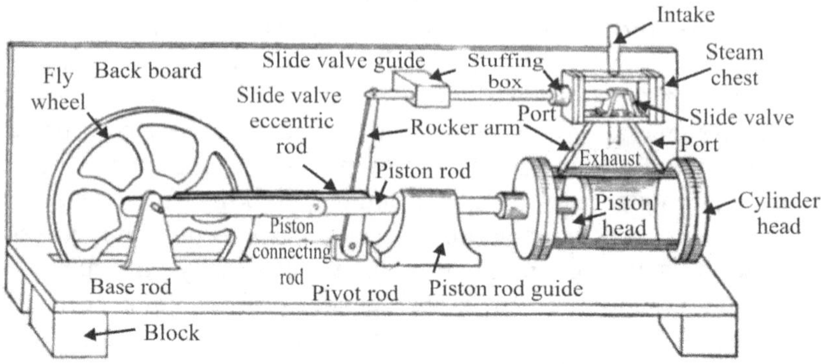

Fig. 5.2 Construction of steam engine

(i) *Connecting Rod:* It guides the crank circle to rotate in a particular fashion and transfer the force provided by the piston to the crank shaft.

(ii) *Crank:* It is that part of engine which govern crank shaft to rotate in particular fashion. It is its special shape which helps in the rotating motion.

(iii) *Eccentric valve motion:* It is mounted on crank shaft and connected with valve rod. Its particular motion governs the sliding motion of D-valves.

(iv) *Sliding Valve:* Sliding valve controls the flow of steam into the cylinder from steam chest. Due to its shape it is also called D-slide valve.

(v) *Centrifugal governor:* When the load is applied on the engine its speed is reduced according to the amount of load applied. To remove this problem centrifugal governor is used. The function of governor is to control the pressure of steam so that load applied can be adjusted accordingly.

(vi) *Flywheel:* It is a big wheel having heavy mass i.e. large inertia. It is mounted on crank shaft. The pressure exerted by the steam on piston is not uniform and resulting in non-uniform force and non-uniform motion. This problem is solved by flywheel as it stores the energy in the form of inertia when provided by the engine and returns back whenever required. This helps in producing uniform torque.

5.2.2 Terminologies

(i)   *Bore*: It is the diameter of the cylinder and is denoted by "D".

(ii)   *Stroke*: It is the length of the cylinder between TDC (Top Dead Center) and BDC (Bottom Dead Center). It is denoted by "L" and is equal to twice the radius of crank circle.

(iii)   *Dead Centers*: It is the position of piston inside the cylinder. When the cylinder is at the top position in cylinder then it is called Top Dead Center and when it is at the bottom position then it is called Bottom Dead Center.

(iv)   *Swept Volume*: It is the volume between the TDC and BDC. It is given by:

$$V_s = \frac{\pi D^2 L}{4}$$

(v)   *Clearance Volume*: The volume between the cylinder cover and the piston when it is at Top Dead Center is called clearance volume.

(vi)   *Compression Ratio*: It is the ratio of total volume to clearance volume.

(vii)   *Mean Effective Pressure*: It is that constant pressure which produces same work for the same piston displacement with the variable condition of temperature, pressure and volume. It is denoted by Pm.

5.2.3 Classification of Steam Engine

These steam engines can be classified according to its construction and operating features. Detail classification is covered as below.

(A)   *Position of Cylinder*: According to the position of the cylinders steam engines can be classified as:

1. Vertical Engine   2. Horizontal Engine   3. Inclined Engine

(B)   Working Stroke: Steam Engine is classified according to number of working strokes per revolution as:

1. Single Acting   2. Double Acting

In single acting engine there is only one stroke per revolution of the crankshaft as steam acts on one side of the piston. But in double acting engine two working stroke per revolution are obtained as steam acts on both sides of the piston.

(C)   *Speed of Engine*: According to the running speed of the engine it can be classified as:

1.   Low Speed Engine, where engine runs at 100 or less revolution per minute.

2.   Medium Speed Engine, where engine runs at 110 to 200 revolutions per minute.

3.   High Speed Engine, where engine runs above 200 or more revolution per minute.

(D)   Field of Application: Steam engines are classified according to its applications as:

1.   Locomotive or Portable Engines

2.   Stationary Engines

3.   Marine Engines

(E)   Exhaust of Steam: Steam engines are classified by exhaust of steam as:

1.   Condensing Engines, where exhaust steam passes into a condenser at lower pressure than atmosphere.

2.   Non Condensing Engines, where exhaust steam passes directly to the atmosphere.

(F)   Governing Methods: A classification is frequently made according to the governing method employed as:

1.   Throttle Governing, in this method the speed of engine is controlled by means of valve in the steam pipe which regulates the pressure of the steam entering the engine.

2.   Automatic Cut-Off Governing, in this method governor controls the quantity of steam admitted to the cylinder while the pressure of entering steam is remain constant.

(G)   Expansion of Steam: Steam engines are classified according to expansion of steam in number of stages as:

1.   Simple Steam Engine, where conversion of heat energy of steam into mechanical work in one stage only with one cylinder and piston.

2.   Compound Steam Engine, where the conversion of heat energy of steam into mechanical work occurs in two stages with high pressure and low pressure cylinder.

3.   Triple Expansion Steam Engine, where the steam expands successively in three cylinders.

4. Quadruple Expansion Steam Engine, where four cylinders and pistons are used for expansion of steam.

Besides above classification, steam engine can also be classified according its length of working stroke i.e., short stroke or long stroke engines and position of crank i.e., side crank or centre crank engines.

5.2.4 Working of Steam Engine

Fig. 5.3 shows the working of a vertical, double acting non-condensing D-slide valve type steam engine.

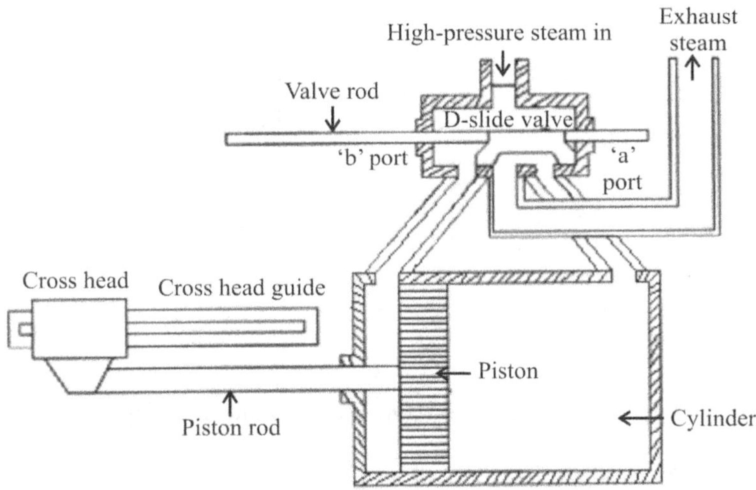

Fig. 5.3 Working of Steam Engine

The superheated steam at a high pressure (about 20 atm) from the boiler is led into the steam chest. After that the steam makes its way into the cylinder through any of the ports 'a' or 'b' depending upon the position of the D-slide valve. When port 'a' is open, the steam rushes to the left side of the piston and forces it to the right. At this stage, the slide valve covers the exhaust port and the other steam port 'b'. Since the pressure of the steam is greater on the left side than on the right side, the piston moves to the right.

When the piston reaches near the end of the cylinder, it closes the steam port 'a' and exhaust port. The steam port 'b' is now open, and the steam rushes to the right side of the piston. This forces the piston to the left and at the same time the exhaust steam goes out through the exhaust pipe, and thus completes the cycle of operation. The same process is repeated in other cycles of operation, and as such the engine works. At the end of each stroke, the piston changes its direction of motion.

5.2.5 Theoretical or Hypothetical Indicator Diagram

The theoretical or hypothetical indicator diagram without clearance and with clearance is shown. If there is no steam in the cylinder (or zero volume of steam at point 1), the indicator diagram will be as shown in Fig 5.4(a). Similarly, if there is some steam in the cylinder at point 1, the indicator diagram will be as shown in Fig 5.4(b).

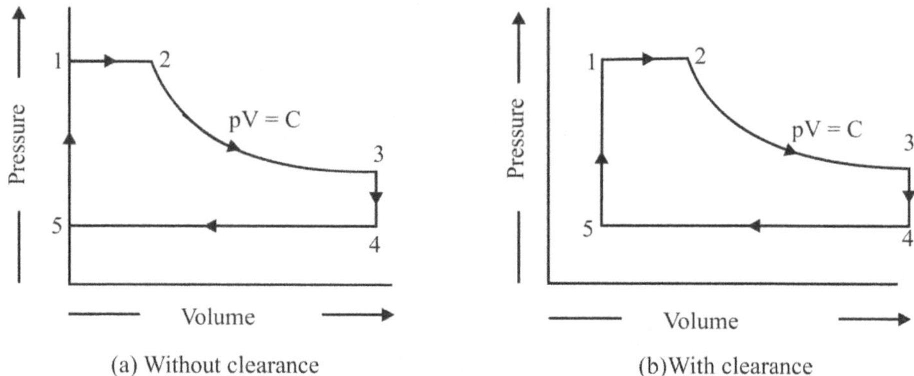

(a) Without clearance   (b) With clearance

Fig. 5.4 Theoretical or Hypothetical Indicator Diagram

The sequence of processes is given below:

(i)   *Process 1-2:* at point 1, the steam is admitted into the cylinder through the inlet port. Here steam comes at constant pressure i.e., atm. pressure as inlet port is open. Since the supply of steam is cut off at point 2, this point is known as cut-off point.

(ii) *Process 2-3:* at point 2, expansion of steam, in the cylinder, starts with the movement of the piston till it reaches the dead end. This expansion takes place hyperbolically (pV = C) and pressure falls considerably.

(iii) *Process 3-4:* at point 3, the exhaust port opens and steam is released from the cylinder to the exhaust. As a result of steam exhaust, pressure in the cylinder falls suddenly (without change in volume). Point 3 is known as release point.

(iv) *Process 4-5:* at point 4 return journey of the piston starts. The steam is exhausted at constant pressure, till the exhaust port is closed, and the inlet port is open.

(v) *Process 5-1:* at point 5 the inlet port is opened and some steam suddenly enters into the cylinder, which increases the pressure of steam (without change in volume). This process continues till the original position is restored.

5.2.6 Mean Effective Pressure (m.e.p.)

Mean effective pressure, pm, is defined as a hypothetical pressure, which is acting on the piston throughout the power stroke. It can be taken as the average height of the pv diagram of the cycle or indicator diagram of any actual engine.

Case 1. Theoretical indicator diagram without clearance and compression: The work done per cycle by an engine is equal to the area of the p-V diagram (Fig. 5.5).

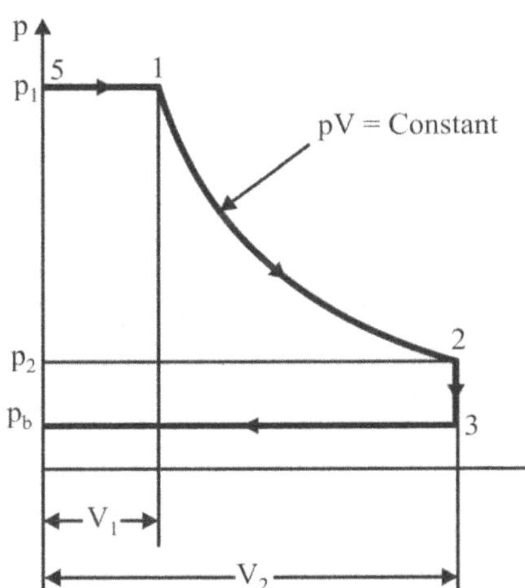

Fig. 5.5 Theoretical p-V Diagram without Clearance and Compression

∴ Work done,   W = Area of the theoretical indicator diagram

$$= p_1 V_1 + p_1 V_1 \log_e V_2/V_1 - p_b V_2$$

$$\text{Theoretical mean effective pressure} = \frac{\text{Theoretical work done}}{\text{Stroke volume}}$$

$$= \frac{p_1 V_1 + p_1 V_1 \log_e V_2 / V_1 - p_b V_2}{V_2}$$

$$= \frac{p_1 V_1}{V_2} (1 + \log_e V_2 / V_1) - P_b$$

The expansion ration, r, is defined as V_2 / V_1 therefore,

Theoretical / Hypothetical m.e.p., $p_m = \dfrac{P_1}{r}$ $(1+\log_e r) - p_b$

Note. The ratio $V_1 / V_2 = 1/r =$ cut-off ratio.

"Cut-off ratio is the ratio of volume between the points of admission and cut-off to the swept volume".

Case 2. Theoretical indicator diagram with clearance (Fig. 5.6).

Work done,   W = Area of indicator diagram = Area '512345'

$$= \text{Area '51ba5'} + \text{Area '12cb1'} - \text{Area 43ca4}$$

$$= p_1 (V_1 - V_c) + p_1 V_1 \log_e V_2/V_1 - p_b (V_2 - V_c)$$

$V_c = V_5 = V_4 =$ clearance volume

$$= p_1 (V_1 - V_c) + p_1 V_1 \log_e V_2/V_1 - p_b V_s \qquad .....(1)$$

$\therefore$ $V_2 - V_c = V_s =$ swept volume

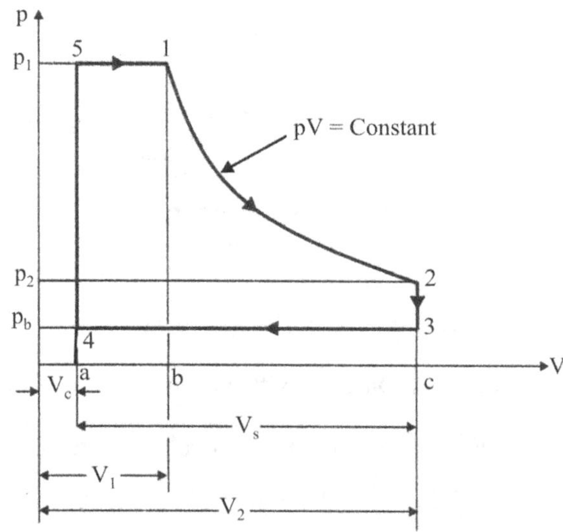

Fig. 5.6 Theoretical p-V Diagram with Clearance

Let,   c = ratio of clearance volume to swept volume

$$= V_c / V_s \qquad .....(2)$$

Also, $V_s = V_2 - V_c$ $\therefore V_2 = V_s + V_c$

Cut-off ratio, $\dfrac{1}{r} = \dfrac{V_1 - V_c}{V_s}$ $\qquad \therefore V_1 = \dfrac{V_s}{r} + V_c$

Inserting the value of V_c from equation (2), we get

$$V_1 = \dfrac{V_s}{r} + c.\,V_s$$

or $\ V_1 = V_s \left(c + \dfrac{1}{r} \right)$

Inserting the values of V_c, V_2 and V_1 in the equation (1), we get

$$W = p_1 \left(V_1 - V_c \right) + p_1 V_1 \, \log_e \dfrac{V_2}{V_1} - p_b V_s$$

$$W = p_1 \left[V_s \left(c + \dfrac{1}{r} \right) - c\,V_s \right] + p_1 V_s \left(c + \dfrac{1}{r} \right) \log_e \left(\dfrac{V_s(1+c)}{V_s \left(c + \dfrac{1}{r} \right)} \right) - p_b V_s$$

$$W = p_1 V_s \left(c + \dfrac{1}{r} - c \right) + p_1 V_s \left(c + \dfrac{1}{r} \right) \log_e \left(\dfrac{1+c}{c + \dfrac{1}{r}} \right) - p_b V_s$$

$$W = \dfrac{p_1 V_s}{r} + p_1 V_s \left(c + \dfrac{1}{r} \right) \log_e \left(\dfrac{1+c}{c + \dfrac{1}{r}} \right) - p_b V_s$$

$$p_m = \dfrac{\text{Work done}}{\text{Stroke volume}}$$

$$p_m = \dfrac{\dfrac{p_1 V_s}{r} + p_1 V_s \left(c + \dfrac{1}{r} \right) \log_e \left(\dfrac{1+c}{c + \dfrac{1}{r}} \right) - p_b V_s}{V_s}$$

$$p_m = \frac{p_1}{r} + p_1\left(c + \frac{1}{r}\right)\log_e\left(\frac{1+c}{c+\frac{1}{r}}\right) - p_b$$

$$p_m = p_1\left[\frac{1}{r} + \left(c + \frac{1}{r}\right)\log_e\left(\frac{1+c}{c+\frac{1}{r}}\right)\right] - p_b$$

Case 3. Theoretical indicator diagram with clearance and compression (Fig. 5.7)

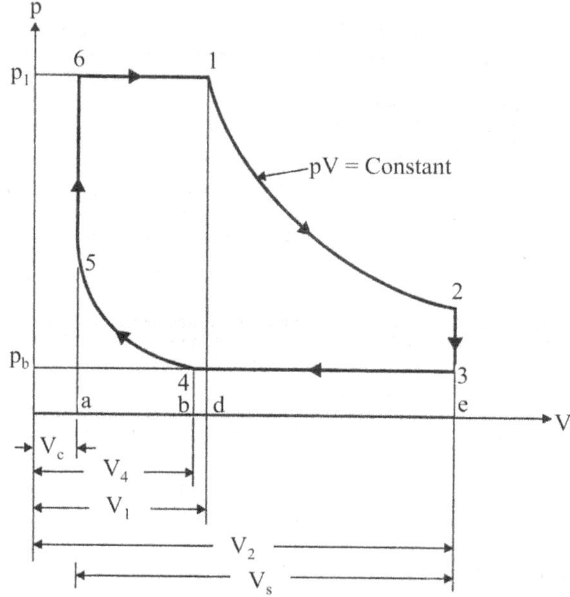

Fig. 5.7 Theoretical p-V Diagram with Clearance and Compression

Work done,

W = Area of the indicator diagram

= Area '6123456'

= Area '61da6' + Area '12ed1' – Area 34be3 – Area '45ab4'

$= p_1 (V_1 - V_c) + p_1 V_1 \log_e V_2/V_1 - p_b (V_2 - V_4) - p_b V_4 \log_e V_4/V_c.$ ....(1)

We know,

$$c = V_c / V_s \qquad \qquad .....(2)$$

Also, $\qquad V_s = V_2 - V_c \qquad \qquad \therefore V_2 = V_s + V_c$

Cut-off ratio,

$$\frac{1}{r} = \frac{V_1 - V_c}{V_s}$$

$$\Rightarrow V_1 = \frac{V_s}{r} + c\, V_s = V_s\left(\frac{1}{r} + c\right)$$

and α = ratio of the volume between points of compression and admission to the swept volume V_s.

$$\alpha = \frac{V_4 - V_c}{V_s} = \frac{V_4}{V_s} - c$$

$$\Rightarrow V_4 = (\alpha + c)\, V_s$$

Inserting the above values in equation (1), we get

$$W = p_1\left[V_s\left(c + \frac{1}{r}\right) - c\, V_s\right] + p_1 V_s\left(c + \frac{1}{r}\right)\log_e\left(\frac{V_s(1+c)}{V_s(c + \frac{1}{r})}\right)$$

$$- p_b V_s(1-\alpha) - p_b V_s(c+\alpha)\log_e\left[\frac{V_s(\alpha+c)}{V_s\, c}\right]$$

$$W = p_1\left[V_s\left(c + \frac{1}{r}\right) - c\, V_s\right] + p_1 V_s\left(c + \frac{1}{r}\right)\log_e\left(\frac{(1+c)}{(c + \frac{1}{r})}\right)$$

$$- p_b V_s(1-\alpha) - p_b V_s(c+\alpha)\log_e\left[\frac{(\alpha+c)}{c}\right]$$

$$p_m = \frac{\text{Work done}}{\text{Stroke volume}}$$

$$p_1\left[V_s\left(c+\frac{1}{r}\right)-cV_s\right]+p_1V_s\left(c+\frac{1}{r}\right)\log_e\left(\frac{(1+c)}{\left(c+\frac{1}{r}\right)}\right)-$$

$$p_m=\frac{p_bV_s(1-\alpha)-p_bV_s(c+\alpha)\log_e\left[\dfrac{(\alpha+c)}{c}\right]}{V_s}$$

$$p_m=p_1\left[\frac{1}{r}+\left(c+\frac{1}{r}\right)\log_e\left(\frac{(1+c)}{\left(c+\frac{1}{r}\right)}\right)\right]-p_b\left[(1-\alpha)-(c+\alpha)\log_e\left[\frac{(\alpha+c)}{c}\right]\right]$$

5.2.7 Actual Indicator Diagram

Hypothetical indicator diagram is the theoretical representation of the working cycle and all practical losses are assumed to be negligible. But the actual indicator diagram is as shown in Fig. 5.8:

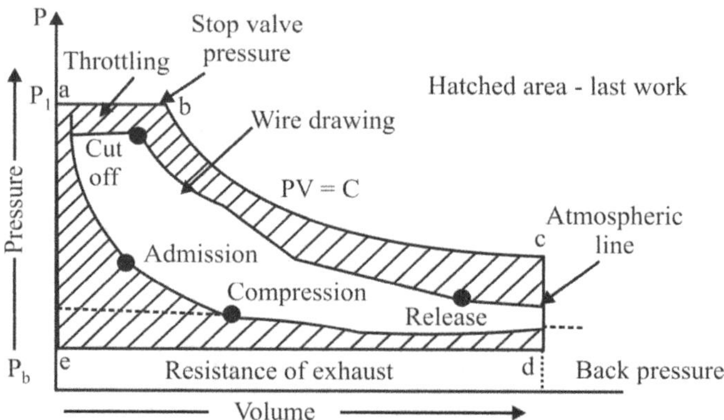

Fig. 5.8 Actual Indicator Diagram

5.2.8 Diagram Factor

The ratio of the areas of the two diagrams is known as the diagram factor, which is always less than unity.

$$\text{Diagram factor}=\frac{\text{Area of actual indicator diagram}}{\text{Area of hypothetical indicator diagram}}=\frac{p_m(\text{actual})}{p_m(\text{theoretical})}$$

The actual diagram differs from the hypothetical diagram due to the following reasons:

(i) The steam pressure drops considerably between the boiler and the engine cylinder. This is due to condensation caused by heat loss in pipes.

(ii) There is gradual drop in steam pressure before the point of cut off is reached. The reason is the condensation of steam in the cylinder upon contact with cylinder walls, which are colder than incoming steam.

(iii) The ports are not opened or closed instantaneously so there is rounding off of the diagram.

(iv) The actual expansion is not a hyperbolic curve. This is due to the varying interchange of heat between the steam and the cylinder walls.

(v) Release takes place before the end of the expansion stroke.

(vi) The exhaust pressure is slightly above the condenser pressure, as the steam has to be forced out from the cylinder.

(vii) Admission of steam occurs just before the end of compression of entrained steam.

5.2.9 Indicated Power (I.P.)

The indicated power is the power developed in the engine cylinder. The I.P. can be calculated if the following data are known :- (i) mean effective pressure active on the engine piston during the stroke, (ii) area of the piston, (iii) length of the stroke and (iv) number of working / min.

Let,

p_{m1} = actual mean effective pressure (bar) acting on the piston for cover end side of the engine.

p_{m2} = actual mean effective pressure (bar) acting on the piston for crank end side of the engine.

A_1 = piston area on the cover and side (m^2)

A_2 = piston area on the crank and side (m^2)

L = length of the stroke (m)

N = speed of the engine (r.p.m.).

Work obtained per stroke for cover end side = force × distance moved

$$= (\text{pressure} \times \text{area}) \times \text{distance}$$

$$= p_{m1} \, (10^5 \text{ N/m}^2) \times A_1 \, (\text{m}^2) \, L \, (\text{m}) \ \text{N-m}$$

Therefore, indicated power (cover end side)

$$= p_{m1} \, (10^5 \text{ N/m}^2) \times A_1 \, (\text{m}^2) \, L \, (\text{m}) \, N \, (\text{r.p.m}) \ \text{N-m/min}$$

$$= p_{m1} \, LA_1 N \times 10^5 \text{ N-m/min} = \frac{p_{m1} \, L \, A_1 \, N \times 10^5}{60} \ \text{N-m/s}$$

$$= p_{m1} \, L \, A_1 \, N \times \frac{10}{6} \text{ kJ/s} = \frac{10 p_{m1} \, L \, A_1 \, N}{6} \ \text{kW}$$

In case of a double acting steam engine the total indicated power developed is the sum of indicated power on cover end side and indicated power on crank end side.

$$\text{Indicated power for crank end} = \frac{10 p_{m2} \, L \, A_2 \, N}{6} \ \text{kW}$$

where $A_2 = A_1 - \text{area of piston rod}$

If D = diameter of piston, d = diameter of piston rod

Then, $A_2 = \dfrac{\pi}{4} D^2 - \dfrac{\pi}{4} d^2 = \dfrac{\pi}{4}\left(D^2 - d^2\right)$

Total indicated power developed by a double acting steam engine,

$$\text{I.P.} = \frac{10 \, p_{m1} \, L \, A_1 \, N}{6} + \frac{10 \, p_{m2} \, L \, A_2 \, N}{6}$$

If the area of piston rod is neglected, then

$$A_1 = A_2 = \frac{\pi}{4} D^2 = A$$

and $p_{m1} = p_{m2} = p_m$

Then $\text{I.P.} = 2 \times \dfrac{10 p_m \, LAN}{6}$

$$= \frac{10 p_m \, LAN}{3} \ \text{kW}$$

5.2.10 Brake Power (B.P.)

Brake power is the actual power available from the engine for doing the useful work. Brake power (B.P.) is always less than indicated (I.P.) since a part of power developed in the engine cylinder is used to overcome the frictional losses at different moving parts of the engine. The difference between I.P. and B.P. is called F.P. (frictional power).

$$\therefore \qquad \text{I.P.} = \text{B.P.} + \text{F.P.}$$

Brake power of an engine can be determined by a brake of some kind applied to the brake pulley of the engine. This arrangement for determination of B.P. of the engine is known as dynamometer. Usually, rope brake dynamometer is used for the purpose.

5.2.11 Rope Brake Dynamometer

It is very suitable for measuring B.P. of an engine of moderate size. It consists of rope wrapped round the brake drum or flywheel keyed to crankshaft of an engine whose B.P. is to be determined (Fig. 5.9). One end of the rope is connected to the spring balance while at the other end is hung a weight W. Wooden blocks are incorporated to check the rope slipping off the brake drum/flywheel. Since a lot of heat is produced in this arrangement, the rim of the wheel is water-cooled.

W opposes the rotation of wheel whereas spring balance S acts otherwise, so the net brake load (which opposes the rotation) is $(W - S)$.

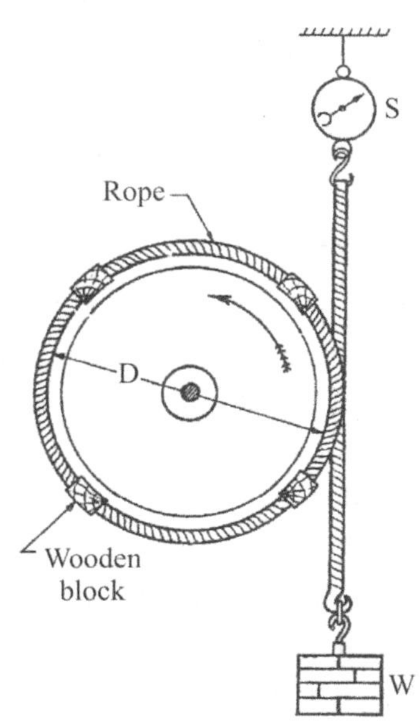

Fig. 5.9 Rope Brake Dynamometer

Let, W = dead weight on the rope (Newton),

 S = spring balance reading (Newton),

 D = diameter of the brake drum / wheel (m),

 d = diameter of the rope (m), and

 N = speed of the engine (r.p.m.)

Net load or frictional force acting on the drum/wheel

$$= (W - S) \text{ Newton}$$

The effective radius at which net load acts $= \dfrac{D+d}{2}$

$\therefore$ Braking torque, $T = \text{frictional force} \times \text{radius} = (W\text{-}S) \left(\dfrac{D+d}{2} \right) \text{ N-m}$

$\therefore$ Power absorbed $= \text{frictional torque} \times \text{angle turned in one minute}$

$$= T \times 2\pi N$$

$$= (W - S) \left(\dfrac{D+d}{2} \right) \times 2\pi N \quad \text{N-m/min}$$

$$= \dfrac{(W - S)\, \pi \, (D+d)\, N}{60} \quad \text{N-m/s or J.}$$

$\therefore$ Brake power, B.P. $= \dfrac{(W - S)\, \pi \, (D+d)\, N}{60 \times 10^3} \text{ kW}$

If diameter of the rope 'd' is neglected, then

$$\text{B.P.} = \dfrac{(W - S)\, \pi \, DN}{60 \times 10^3} \text{ kW} \quad (d = 0)$$

Also if T, braking torque is given,

Then,   B.P. $= \dfrac{2\pi \, NT}{60 \times 10^3} \text{ kW}$

5.2.12   Efficiency of Steam Engine

The efficiency of an engine can be calculated by dividing the number of joules of mechanical work that the engine produces by the number of joules of energy input to the engine by the burning fuel. The rest of the energy is dumped into the environment as heat. No pure heat engine can be more efficient than the Carnot cycle, in which heat is moved from a high temperature reservoir to one at a low temperature, and the efficiency depends on the temperature difference. Hence, steam engines should ideally be operated at the highest steam temperature possible (superheated steam), and release the waste heat at the lowest temperature possible.

In practice, a steam engine exhausting the steam to atmosphere will have an efficiency (including the boiler) of 1% to 8%, but with the addition of a condenser and multiple expansion engines the efficiency may be greatly improved to 25% or better. A power station with steam reheats, etc., will achieve 30% to 42% efficiency. Combined cycle in which the burning material is first used to drive a gas turbine can produce 50% to 60% efficiency. It is also possible to capture the waste heat using cogeneration in which the residual steam is used for heating. It is therefore possible to use as much as 90% of the energy produced by burning fuel—only 10% of the energy produced by the combustion of the fuel goes wasted into the atmosphere.

5.2.13 Advantages of Steam Engine

(i)   Since combustion of fuel takes place outside the engine cylinder, these engines are much smooth and silent running than internal combustion engines.

(ii) The working pressure and temperature inside the engine cylinder is low.

(iii) Because of low pressure and temperature, ordinary alloys are used for the manufacture of engine cylinder and its parts. So cost is low.

5.2.14 Disadvantages

(i)   Steam engines have less efficiency than I.C. engines.

(ii) It cannot be started instantaneously.

(iii) A steam engine requires a boiler and other components to transfer energy. Thus it is more cumbersome than I.C. engine.

5.2.15 Application

Uses of steam engines include engines used in thermal power stations and those that were used in mills, factories and to power cable railways and cable tramways before the widespread use of electric power. Very low power engines are used to power model ships and specialty applications such as the steam clock. Steam engines were used in locomotives too.

5.3 Thermodynamic Cycles

5.3.1 Carnot Cycle

Nicolas Leonard Sadi Carnot was the first to provide a thermodynamic model of a heat engine, abstracting from the only available heat engine, the steam engine, to pinpoint the fundamentals: the idea of a generic working fluid, performing a generic cyclic process, interacting with generic heat reservoirs. In that his only publication, Carnot concluded that all heat engines where limited in their energy-conversion efficiency by the operating temperatures, and that the maximum efficiency is obtained when the working fluid is assumed to follow four ideal processes (Fig. 5.10):

- An isentropic compression (1 to 2), to change temperature without heat transfer.

- An isothermal heat input (2 to 3), from the hot source, at the hot-source temperature.

- An isentropic expansion (3 to 4), to change temperature without heat transfer.

- An isothermal heat rejection to the cold source (usually the environment), at the cold-source temperature.

Carnot reached those conclusion by a set of rational deductions, namely: any engine with friction would have less efficiency than one without, among all engines exchanging heat at different temperatures, the one with highest efficiency only exchanges heat at the two extreme temperatures (the hottest and the coldest), and all reversible engines working with the same couple of temperature extremes have the same efficiency.

The energy can be easily deduced by establishing the overall energy conservation, $\Delta E_{univ} = Q_{hot} - W - Q_{cold} = 0$, and the overall entropy balance, $\Delta S_{univ} = - Q_{hot}/T_{hot} + Q_{cold}/T_{cold}, \geq 0$, the latter being zero in the ideal case of a Carnot cycle, what yields:

$$\eta_e \equiv \frac{W_{net}}{Q_{pos}} \rightarrow \eta_{e\ Carnot} = 1 - \frac{T_1}{T_2}$$

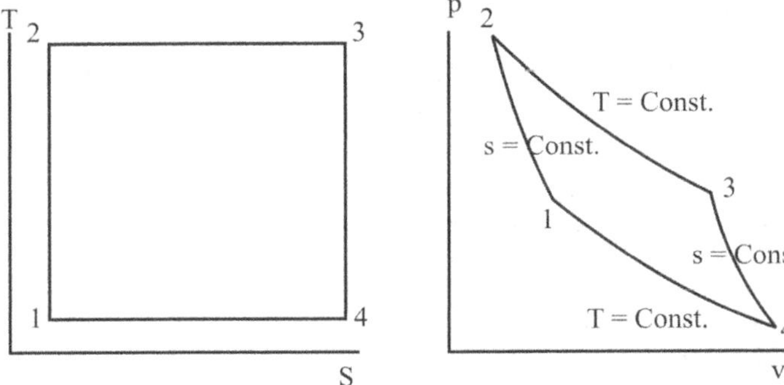

Fig. 5.10 The Carnot Cycle in T-s and P-v Diagrams.

Carnot cycle, and any other conceivable power cycle, is to be run clockwise in both the *T-s* and *p-V* thermodynamic diagrams, since the heat engine receives net heat, that in a reversible process is $Q_{net} = \int T dS$ (recall the area interpretation of integrals), and delivers net work, that in a reversible process is $W_{net} = \int p dV$; notice that in a cycle

$$\oint dU = \oint T dS - \oint p dV = 0.$$

The Carnot cycle is not practical, not only because of unavoidable frictional losses (could be minimised with appropriate lubrication), but because of the heat transfer with negligible temperature jump, that would render the heat transfer rate infinitesimal for a finite size engine with finite thermal transmittance with the heat sources.

5.3.2 Otto Cycle

The Otto cycle is a first approximation to model the operation of a spark-ignition engine, first built by Nikolaus Otto in 1876, and used in many cars, small planes and small power systems.

In the ideal air-standard Otto cycle, the working fluid is just air, which is assumed to follow four processes: isentropic compression, constant-volume heat input from the hot source, isentropic expansion, and constant-volume heat rejection to the environment (Fig. 5.11).

The main parameters of ideal and real Otto cycles are:

(i)   Size, measured by the displacement volume (the volume swept by the piston, $V_1 - V_2$), usually less than 0.5 litres per cylinder, to avoid self-ignition.

(ii)   Speed, more precisely crankshaft speed, n, with a typical operation range n = 1000-7000 rpm (n = 20-120 Hz). The maximum value may be in the range n_{max} = 6000-8000 rpm for four-stroke engines. Two-stroke motorcycle engines run quicker (n_{max} = 13000 rpm), the quickest (n_{max} = 20000 rpm) being the smallest engines (two stroke), used in aircraft modelling.

(iii)   Compression ratio, $r = V_1/V_2$, with a typical range of $r = 8\text{-}10$ (up to 14 in direct-injection spark-ignition engines), limited by the knock or self-ignition problem.

(iv)   Mean effective pressure, p_{me}, defined as the unit work divided by the displacement, with a typical range of 0.2-1.5 MPa (the full-load value may range from p_{me} = 1.2 MPa in two-stroke motorcycle engines, to p_{me} = 1.7 MPa in the largest turbocharged engines). Maximum pressure may have a typical range of 4-10 MPa. Performance maps of reciprocating engines are usually presented on a p_{me}–n diagram, i.e., mean effective pressure versus engine speed.

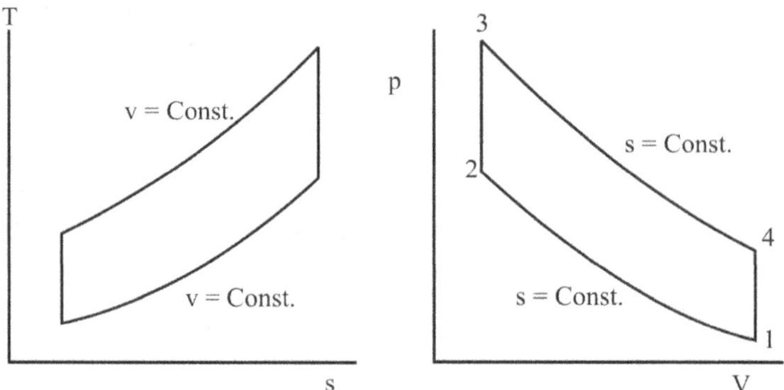

Fig. 5.11 The Ideal Otto Cycle in the *T-s* and *p-V* diagram

The cold-air-standard model takes as working fluid air with constant properties (those at the inlet, i.e. cold), what renders the analysis simple. The energy exchanges for the trapped control mass, m, are $W_{12}/m = c_v (T_2 - T_1)$, $Q_{12} = 0$, $W_{23} = 0$, $Q_{23}/m = c_v (T_3 - T_2)$, $W_{34}/m = c_v (T_4 - T_3)$, $Q_{34} = 0$, $W_{41} = 0$, $Q_{41}/m = c_v (T_1 - T_4)$, and the energy efficiencies is:

$$\eta_{e,Otto} = \frac{W_{34} - W_{12}}{Q_{23}} = \frac{\left(T_3 - T_4\right) - \left(T_2 - T_1\right)}{T_3 - T_2} = 1 - \frac{T_1}{T_2} = 1 - \frac{1}{r^{\gamma-1}}$$

Where, $\gamma = c_p/c_v$, 1.4 approx (for air)

Typical energy efficiencies of these engines are low, 25% to 35% when running at nominal power, and much lower at partial load (the main air flow is strangulated), but the engine is light, very powerful and responsive (accelerates very quickly).

One of the key advantages of Otto engines, is that cheap materials can be used in their construction (e.g. cast iron, against very expensive nickel alloys in gas turbines), because peak temperatures (up to 3000 K) are only realised within the burning gases and for short times, with an average temperature during the whole cycle of some 700 K, which would be the quasi-steady temperature level at the wall. Pollutant emission is higher because premixed fuel cannot burn close to the walls (the effect of walls is smaller in larger diesel engines, and most of the nearby gas is just non-premixed air), making the exhaust catalyser an environmental need.

5.3.3 Diesel Cycle

The Diesel cycle is a first approximation to model the operation of a compression-ignition engine, first built by Rudolf Diesel in 1893, and used in most cars, nearly all trucks, nearly all boats, many locomotives, some small airplanes, and many large electric power systems and cogeneration systems. It is the reference engine from 50 kW to 50 MW, due to the fuel used (cheaper and safer than gasoline) and the higher efficiency.

In the ideal air-standard Diesel cycle, the working fluid is just air, which is assumed to follow four processes (Fig. 5.12): isentropic compression, constant-pressure heat input from the hot source, isentropic expansion, and constant-volume heat rejection to the environment.

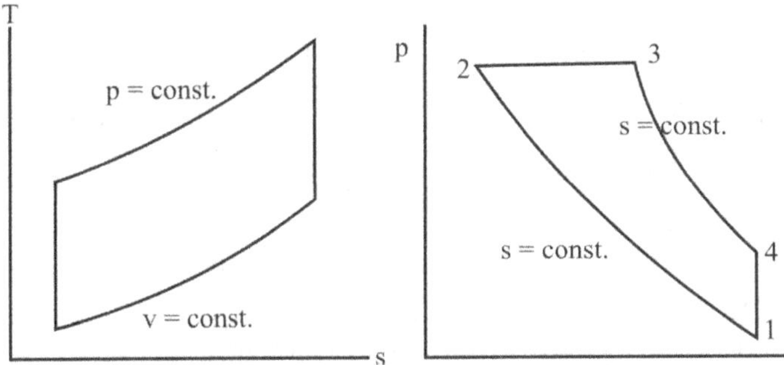

Fig. 5.12 The Ideal Diesel Cycle in the *T-S* and *p-V* Diagram

Similarly to the Otto cycle, the main parameters of ideal and real Diesel cycles are also the size, measured by the displacement volume (that may reach more than 1 m^3 per cylinder in large marine engines), the compression ratio, $r = V_1/V_2$ (with a typical range of 16-22, limited just by strength), the cut-off ratio, r_c, or the mean effective pressure (in the range 1-2 MPa), or the maximum pressure (in the range of 3 MPa to 20 MPa), and the speed (with a typical range of 100-6000 rpm). The energy efficiency can be expressed as:

$$\eta_{e,Diesel} = 1 - \frac{1}{r^{\gamma-1}} \frac{r_c^{\gamma} - 1}{\gamma(r_c - 1)}$$

Typical energy efficiencies are from 30% to 54% (based on the lower heating value of the fuel). Real compression-ignition engines take ambient air (often after a first stage compression) and compress it (inside the cylinder) so much, rising the temperature accordingly, that the fuel burns as it is injected (after a small initial delay due to vaporisation and combustion kinetics).

One of its key advantages compared to Otto engines is the great load increase per cylinder associated to the higher pressures allowed (the mixture of fuel and air would detonate in Otto engines at high compressions), and the further load increase associated to charging previously-compressed air (turbocharging). Another advantage is the much better performance at part load, since there it is achieved by injecting less fuel instead of by throttling, and the torque (power divided per angular speed), changes less with angular speed.

Table 5.1 Comparison between Otto and Diesel Cycle

Otto Cycle	**Diesel Cycle**
Petrol engine works on Otto cycle	Diesel engine works on Diesel cycle
During suction stroke, air petrol mixture is sucked into the cylinder.	During suction stroke, only air is sucked into the cylinder.
Heat is added at constant volume.	Heat is added at constant pressure.
Efficiency is higher, for the same compression ratio	Efficiency is low, for the same compression ratio
Compression ratio has to be kept below 12 due to knocking	No limitation
Thermal efficiency is lower due to lower CR.	Thermal efficiency is higher due to higher CR.

5.4 Internal Combustion Engine

The internal combustion engine is a heat engine that converts chemical energy in a fuel into mechanical energy, usually made available on a rotating output shaft.

Chemical energy of the fuel is first converted to thermal energy by means of combustion or oxidation with air inside the engine. This thermal energy raises the temperature and pressure of the gases within the engine and the high-pressure gas then expands against the mechanical mechanisms of the engine. This expansion is converted by the mechanical linkages of the engine to a rotating crankshaft, which is the output of the engine. Schematic view of IC Engine is shown in Fig. 5.13.

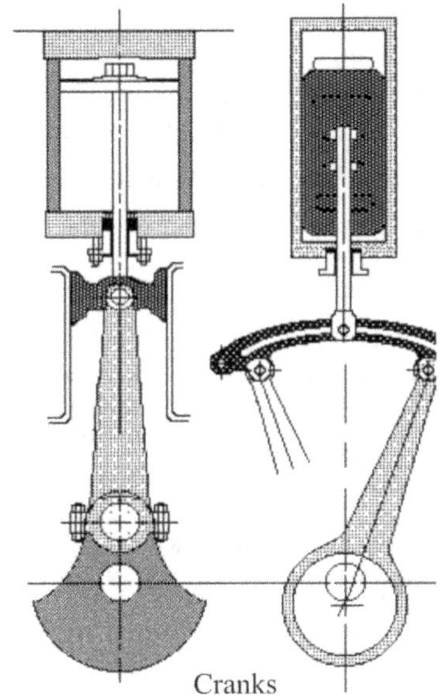

Fig. 5.13 Schematic View of IC Engine

The crankshaft, in turn, is connected to a transmission and/or power train to transmit the rotating mechanical energy to the desired final use. For engines this will often be the propulsion of a vehicle (i.e., automobile, truck, locomotive, marine vessel, or airplane). Other applications include stationary engines to drive generators or pumps, and portable engines for things like chain saws and lawn mowers.

Most internal combustion engines are reciprocating engines having pistons that reciprocate back and forth in cylinders internally within the engine.

5.4.1 Classification of Internal Combustion Engine

Internal combustion engines can be classified in a number of different ways:

(i)   Types of Ignition

    (a) *Spark Ignition (SI):* An SI engine starts the combustion process in each cycle by use of a spark plug. The spark plug gives a high-voltage electrical discharge between two electrodes which ignites the air-fuel mixture in the combustion chamber surrounding the plug. In early engine development, before the invention of the electric spark plug, many forms of torch holes were used to initiate combustion from an external flame.

    (b) *Compression Ignition (CI):* The combustion process in a CI engine starts when the air-fuel mixture self-ignites due to high temperature in the combustion chamber caused by high compression.

(ii)   *Engine Cycle*

    (a) *Four-Stroke Cycle*: A four-stroke cycle experiences four piston movements over two engine revolutions for each cycle.

    (b) *Two-Stroke Cycle*: A two-stroke cycle has two piston movements over one revolution for each cycle.

    Three-stroke cycles and six-stroke cycles were also tried in early engine development.

(iii)   Valve Location

    (a) Valves in head (overhead valve), also called I Head engine.

    (b) Valves in block (flat head), also called L Head engine. Some historic engines with valves in block had the intake valve on one side of the cylinder and the exhaust valve on the other side. These were called T Head engines.

    (c) One valve in head (usually intake) and one in block, also called F Head engine; this is much less common.

(iv)   *Basic Design*

    (a) *Reciprocating*: Engine has one or more cylinders in which pistons reciprocate back and forth. The combustion chamber is located in the closed end of each cylinder. Power is delivered to a rotating output crankshaft by mechanical linkage with the pistons.

(b) *Rotary*: Engine is made of a block (stator) built around a large non-concentric rotor and crankshaft. The combustion chambers are built into the nonrotating block.

(v) Position and Number of Cylinders of Reciprocating Engines

(a) *Single Cylinder*: Engine has one cylinder and piston connected to the crankshaft.

(b) *In-Line*: Cylinders are positioned in a straight line, one behind the other along the length of the crankshaft. They can consist of 2 to 11 cylinders or possibly more. In-line four-cylinder engines are very common for automobile and other applications. In-line six and eight cylinders are historically common automobile engines. In-line engines are sometimes called straight (e.g., straight six or straight eight).

(c) *V Engine*: Two banks of cylinders at an angle with each other along a single crankshaft. The angle between the banks of cylinders can be anywhere from 15° to 120°, with 60°-90° being common. V engines have even numbers of cylinders from 2 to 20 or more. V6s and V8s are common automobile engines, with V12s and V16s (historic) found in some luxury and high-performance vehicles.

(d) *Opposed Cylinder Engine*: Two banks of cylinders opposite each other on a single crankshaft (a V engine with a 180°V). These are common on small aircraft and some automobiles with an even number of cylinders from two to eight or more. These engines are often called flat engines (e.g., flat four).

(e) *W Engine*: Same as a V engine except with three banks of cylinders on the same crankshaft. Not common, but some have been developed for racing automobiles, both modern and historic. Usually 12 cylinders with about a 60° angle between each bank.

(f) *Opposed Piston Engine*: Two pistons in each cylinder with the combustion chamber in the center between the pistons. A single-combustion process causes two power strokes at the same time, with each piston being pushed away from the center and delivering power to a separate crankshaft at each end of the cylinder. Engine output is either on two rotating crankshafts or on one crankshaft incorporating complex mechanical linkage.

(g) *Radial Engine*: Engine with pistons positioned in a circular plane around the central crankshaft. The connecting rods of the pistons are connected to a master rod which, in turn, is connected to the crankshaft. A bank of cylinders on a radial engine always has an odd number of cylinders ranging from 3 to 13 or more. Operating on a four-stroke cycle, every other cylinder fires and has a power stroke as the crankshaft rotates, giving a smooth operation. Many medium- and large-size propeller-driven aircraft use radial engines. For large aircraft, two or more banks of cylinders are mounted together, one behind the other on a single crankshaft, making one powerful, smooth engine. Very large ship engines exist with up to 54 cylinders, six banks of 9 cylinders each.

(vi) *Air Intake Process*

(a) *Naturally Aspirated*: No intake air pressure boost system exists.

(b) *Supercharged*: Intake air pressure increased with the compressor driven off of the engine crankshaft.

(c) *Turbocharged*: Intake air pressure increased with the turbine-compressor driven by the engine exhaust gases.

(d) *Crankcase Compressed*: Two-stroke cycle engine uses the crankcase as the intake air compressor. Limited development work has also been done on design and construction of four-stroke cycle engines with crankcase compression.

(vii) *Method of Fuel Input for SI Engines*

(a) Carbureted.

(b) Multipoint Port Fuel Injection: One or more injectors at each cylinder intake.

(c) Throttle Body Fuel Injection: Injectors upstream in intake manifold.

(viii) Fuel Used

(a) Gasoline.

(b) Diesel Oil or Fuel Oil.

    (c) Gas, Natural Gas, Methane.

    (d) LPG.

    (e) Alcohol-Ethyl, Methyl.

    (f) *Dual Fuel*: There are a number of engines that use a combination of two or more fuels. Some, usually large, CI engines use a combination of methane and diesel fuel. These are attractive in developing third-world countries because of the high cost of diesel fuel. e.g. Combined gasoline-alcohol fuels

    (g) *Gasohol*: Common fuel is consisting of 90% gasoline and 10% alcohol.

(ix) *Application*

    (a) Automobile, Truck, Bus   (b) Locomotive.

    (c) Stationary.   (d) Marine.

    (e) Aircraft. (f) Small Portable, Chain Saw, Model Airplane.

(x) *Type of Cooling*

    (a) Air Cooled   (b) Liquid Cooled, Water Cooled.

5.4.2 Terminologies

The following terms are commonly used in IC engine technology literature and also shown in Fig. 5.14.

Spark Ignition (SI): An engine in which the combustion process in each cycle is started by use of a spark plug.

Compression Ignition (CI): An engine in which the combustion process starts when the air-fuel mixture self-ignites due to high temperature in the combustion chamber caused by high compression. CI engines are often called Diesel engines.

Top-Dead-Center (TDC): Position of the piston when it stops at the furthest point away from the crankshaft. Top, because this position is at the top of most engines (not always), and dead because the piston stops at this point. Because in some engines top dead center is not at the top of the engine (e.g., horizontally opposed engines, radial engines, etc.), some Sources call this position Head-End-Dead-Center (HEDC).

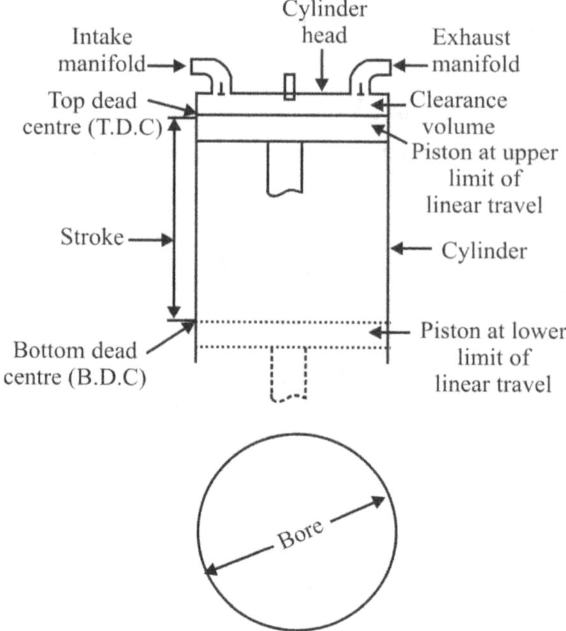

Fig. 5.14 Term relating in IC engine

Bottom-Dead-Center (BDC): Position of the piston when it stops at the point closest to the crankshaft. Some sources call this Crank-End-Dead-Center (CEDC) because it is not always at the bottom of the engine.

Bore (D): Diameter of the cylinder or diameter of the piston face, which is the same minus a very small clearance.

Stroke (L): Movement distance of the piston from one extreme position to the other i.e., TDC to BDC or BDC to TDC. It is equal to twice the radius of crank circle.

Clearance Volume: Minimum volume in the combustion chamber with piston at TDC.

Displacement or Displacement Volume: Volume displaced by the piston as it travels through one stroke. Displacement can be given for one cylinder or for the entire engine (one cylinder times number of cylinders). Some literature calls this *swept volume*.

$$V_s = \frac{\pi D^2 L}{4}$$

Compression Ratio (CR): It is the ratio of total volume to clearance volume.

5.4.3 Parts of Internal Combustion Engine

The following are the major components found in most reciprocating internal combustion engines (Fig. 5.15).

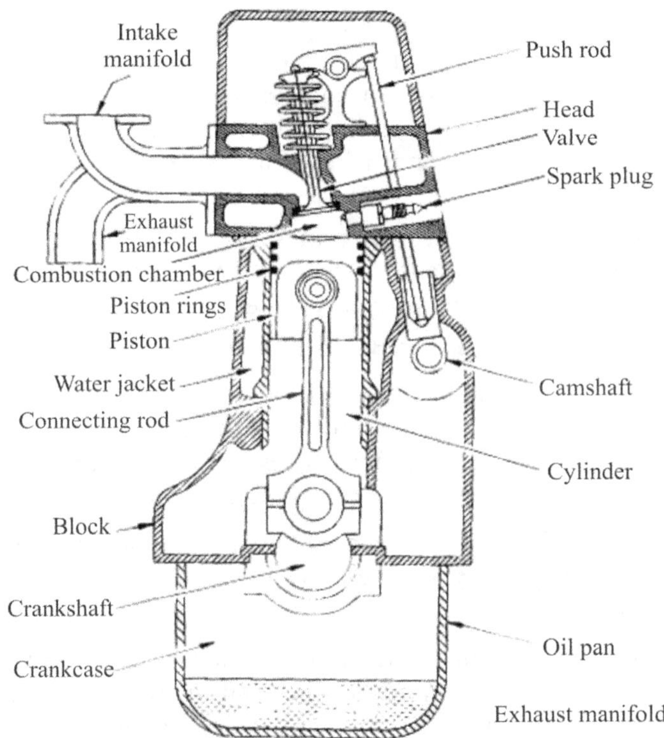

Fig. 5.15 Cross-Sections of Four-Stroke Cycle SI Engine

(i)   *Block*: Body of engine containing the cylinders, made of cast iron or aluminum. The block of water-cooled engines includes a water jacket cast around the cylinders. On air-cooled engines, the exterior surface of the block has cooling fins.

(ii) *Camshaft*: Rotating shaft used to push open valves at the proper time in the engine cycle, either directly or through mechanical or hydraulic linkage (push rods, rocker arms, tappets). Most modern automobile engines have one or more camshafts mounted in the engine head (overhead cam). Camshafts are generally made of forged steel or cast iron and are driven off the crankshaft by means of a belt or chain (timing chain).

(iii)   *Carburetor*: Venturi flow device which meters the proper amount of fuel into the air flow by means of a pressure differential.

(iv)   Catalytic converter Chamber mounted in exhaust flow containing catalytic material that promotes reduction of emissions by chemical reaction.

(v)   *Combustion chamber*: The end of the cylinder between the head and the piston face where combustion occurs. The size of the combustion chamber continuously changes from a minimum volume when the piston is at TDC to a maximum when the piston is at BDC.

(vi)   *Connecting rod*: Rod connecting the piston with the rotating crankshaft, usually made of steel or alloy forging in most engines but may be aluminum in some small engines.

(vii)   *Connecting rod bearing*: Bearing where connecting rod fastens to crankshaft.

(viii)   *Cooling fins*: Metal fins on the outside surfaces of cylinders and head of an air-cooled engine. These extended surfaces cool the cylinders by conduction and convection.

(ix)   *Crankcase*: Part of the engine block surrounding the rotating crankshaft. In many engines, the oil pan makes up part of the crankcase housing.

(x)   *Crankshaft*: Rotating shaft through which engine work output is supplied to external systems. The crankshaft is connected to the engine block with the main bearings. It is rotated by the reciprocating pistons through connecting rods connected to the crankshaft. Most crankshafts are made of forged steel, while some are made of cast iron.

(xi)   *Cylinders*: The circular cylinders in the engine block in which the pistons reciprocate back and forth. The walls of the cylinder have highly polished hard surfaces. Cylinders may be machined directly in the engine block, or a hard metal (drawn steel) sleeve may be pressed into the softer metal block.

(xii)   *Exhaust manifold*: Piping system which carries exhaust gases away from the engine cylinders, usually made of cast iron.

(xiii) *Exhaust system*: Flow system for removing exhaust gases from the cylinders, treating them, and exhausting them to the surroundings. It consists of an exhaust manifold which carries the exhaust gases away from the engine, a thermal or catalytic converter to reduce emissions, a muffler to reduce engine noise.

(xiv) *Flywheel*: Rotating mass with a large moment of inertia connected to the crank-shaft of the engine. The purpose of the flywheel is to store energy and furnish a large angular momentum that keeps the engine rotating between power strokes and smoothes out engine operation.

(xv) *Fuel injector*: A pressurized nozzle that sprays fuel into the incoming air on SI engines or into the cylinder on CI engines. On SI engines, fuel injectors are located at the intake valve ports on multipoint port injector systems and upstream at the intake manifold inlet on throttle body injector systems. In a few SI engines, injectors spray directly into the combustion chamber.

(xvi) *Fuel pump*: Electrically or mechanically driven pump to supply fuel from the fuel tank (reservoir) to the engine. Many modern automobiles have an electric fuel pump mounted submerged in the fuel tank. Some small engines and early automobiles had no fuel pump, relying on gravity feed.

(xvii) *Head*: The piece which closes the end of the cylinders, usually containing part of the clearance volume of the combustion chamber. The head is usually cast iron or aluminum, and bolts to the engine block. In some engines, the head is one piece with the block. The head contains the spark plugs in SI engines and the fuel injectors in CI engines. Most modern engines have the valves in the head.

(xviii) *Head gasket*: Gasket which serves as a sealant between the engine block and head where they bolt together. They are usually made in sandwich construction of metal and composite materials.

(xix) *Intake manifold*: Piping system which delivers incoming air to the cylinders usually made of cast metal, plastic, or composite material. In most SI engines, fuel is added to the air in the intake manifold system either by fuel injectors or with a carburetor.

(xx) *Oil pump*: Pump used to distribute oil from the oil sump to required lubrication points. The oil pump can be electrically driven, but is most commonly mechanically driven by the engine. Some small engines do not have an oil pump and are lubricated by splash distribution.

(xxi)   *Oil sump*: Reservoir for the oil system of the engine, commonly part of the crankcase.

(xxii)   *Piston*: The cylindrical-shaped mass that reciprocates back and forth in the cylinder, transmitting the pressure forces in the combustion chamber to the rotating crankshaft. The top of the piston is called the crown and the sides are called the skirt. The face on the crown makes up one wall of the combustion chamber and may be a flat or highly contoured surface. Pistons are made of cast iron, steel, or aluminum. Iron and steel pistons can have sharper corners because of their higher strength. Aluminum pistons are lighter and have less mass inertia. Sometimes synthetic or composite materials are used for the body of the piston. Some pistons have a ceramic coating on the face.

(xxiii)   *Piston rings*: Metal rings that fit into circumferential grooves around the piston and form a sliding surface against the cylinder walls. Near the top of the piston are usually two or more compression rings made of highly polished hard chrome steel. The purpose of these is to form a seal between the piston and cylinder walls and to restrict the high-pressure gases in the combustion chamber from leaking past the piston into the crankcase. Below the compression rings on the piston is at least one oil ring, which assists in lubricating the cylinder walls and scrapes away excess oil to reduce oil consumption.

(xxiv)   *Push rods*: Mechanical linkage between the camshaft and valves on overhead valve engines with the camshaft in the crankcase.

(xxv)   *Spark plug*: Electrical device used to initiate combustion in an SI engine by creating a high-voltage discharge across an electrode gap. Spark plugs are usually made of metal surrounded with ceramic insulation.

(xxvi)   *Supercharger:* Mechanical compressor powered off of the crankshaft, used to compress incoming air of the engine.

(xxvii)   *Throttle*: Butterfly valve mounted at the upstream end of the intake system, used to control the amount of air flow into an SI engine. Some small engines and stationary constant-speed engines have no throttle.

(xxviii) *Turbocharger*: Turbine-compressor used to compress incoming air into the engine.

The turbine is powered by the exhaust flow of the engine and thus takes very little useful work from the engine.

(xxix) *Valves*: Used to allow flow into and out of the cylinder at the proper time in the cycle. Most engines use poppet valves, which are spring loaded closed and pushed open by camshaft action. Valves are mostly made of forged steel. Many two-stroke cycle engines have ports (slots) in the side of the cylinder walls instead of mechanical valves.

(xxx) *Wrist pin*: Pin fastening the connecting rod to the piston (also called the piston pin).

5.4.4 Spark Ignition Engine (Petrol Engine)

Working of 2-stroke SI engine

(i) *Combustion*: With the piston at TDC combustion occurs very quickly, raising the temperature and pressure to peak values, almost at constant volume (Fig. 5.16 e).

(ii) *First Stroke*: Expansion Stroke or Power Stroke. Very high pressure created by the combustion process forces the piston down in the power stroke. The expanding volume of the combustion chamber causes pressure and temperature to decrease as the piston travels towards BDC (Fig. 5.16 a).

(iii) *Exhaust Blowdown*: At about BDC, the exhaust valve opens and blowdown occurs. The exhaust valve may be a poppet valve in the cylinder head, or it may be a slot in the side of the cylinder which is uncovered as the piston approaches BDC. After blowdown the cylinder remains filled with exhaust gas at lower pressure (Fig. 5.16 b).

(iv) *Intake and Scavenging*: When blowdown is nearly complete, at about BDC, the intake slot on the side of the cylinder is uncovered and intake air-fuel enters under pressure. Fuel is added to the air with either a carburetor or fuel injection. This incoming mixture pushes much of the remaining exhaust gases out the open exhaust valve and fills the cylinder with a combustible air-fuel mixture, a process called scavenging. The piston passes BDC and very quickly covers the intake port and then the exhaust port (or the exhaust valve closes). The higher pressure at which the air enters the cylinder is established in one of two ways. Large two-stroke cycle engines generally have a supercharger, while small engines will intake the air through the crankcase. On these engines the crankcase is designed to serve as a compressor in addition to serving its normal function (Fig. 5.16 c).

(v)   *Second Stroke*: *Compression Stroke:* With all valves (or ports) closed, the piston travels towards TDC and compresses the air-fuel mixture to a higher pressure and temperature. Near the end of the compression stroke, the spark plug is fired; by the time the piston gets to IDC, combustion occurs and the next engine cycle begins (Fig. 5.16 d).

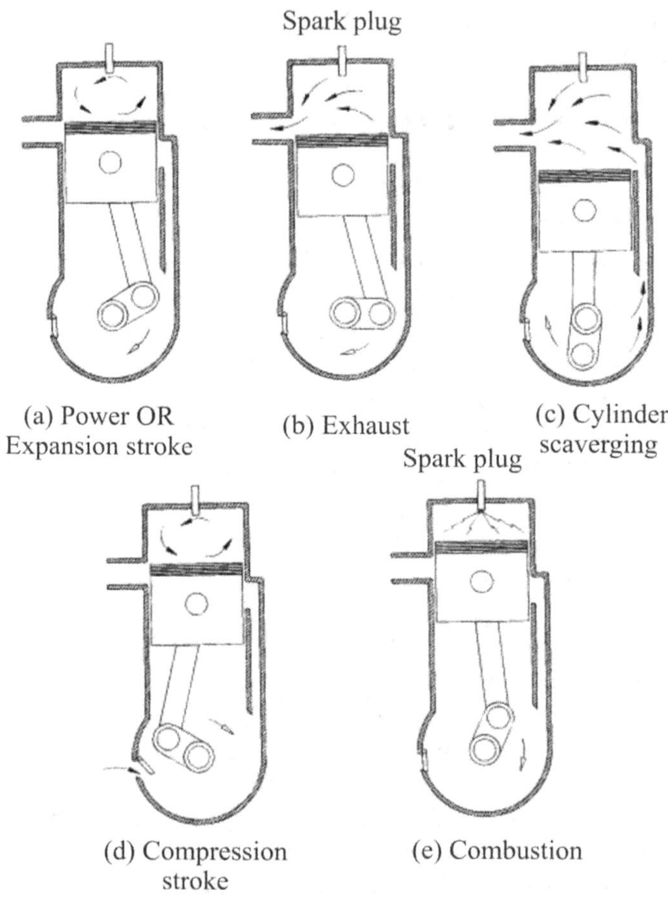

Fig. 5.16 2-Stroke SI Engine Operating Cycle

Figure 5.17 shows the following processes:

Process 1-2: isentropic compression of the charge in the cylinder. The charge inside the cylinder is compressed and the piston moves from bottom dead center to top dead center. This comprises first stroke of the engine.

Process 2-3: combustion at constant volume. This process takes place as process of constant volume and increase in pressure. In this process spark is ignited by spark plug inside the cylinder and fuel is burnt.

Process 3-4: isentropic expansion. The burnt fuel exerts pressure and moves the piston to bottom dead center. The gas expands in this process. This comprises the second stroke and the power stroke of the engine.

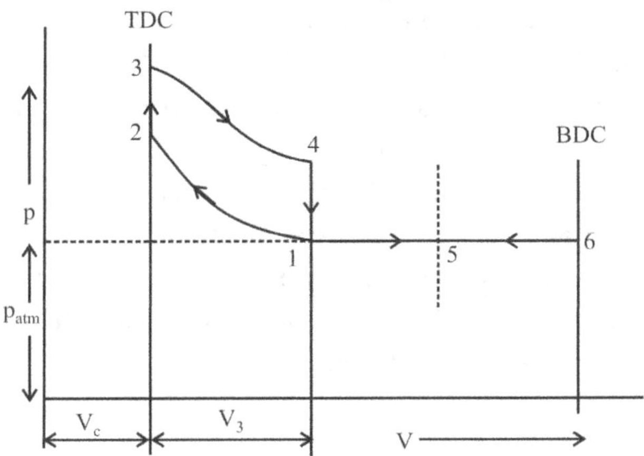

Fig. 5.17 Theoretical p-V Diagram of a Two Stroke Otto Cycle Engine

Process 4-1: sudden release of burnt gases to the atmosphere as the exhaust port opens. This process takes place as process of constant volume and decrease in pressure. In this process the burnt gas is exhausted outside the cylinder.

Process 1-6 & 6-1: sweeping out the exhaust gases to the atmosphere at the atmospheric pressure and charging the cylinder with charge.

Process 5-6 & 6-5: charging the cylinder with charge through the transfer ports and scavenging through exhaust ports.

The point "5" indicates the opening of the inlet or transfer ports.

Also the compression of the charge starts   from point "1" instead of point "6". Therefore, the effective stroke (compression stroke) is less then the actual stroke.

Here, P_{atm} = atmospheric pressure

The actual indicator diagram for a two stroke cycle petrol engine is shown in Fig. 5.18. The section is shown by the line 1-2-3 (from the instant transfer port opens and transfer port closes). During the section stage the exhaust port is also open. In the first half of suction stage, the volume of fuel-air mixture and burnt gases increases. This happens as the piston moves from 1 to 2. In the second half of suction stage, the volume of charge and burnt gases decreases. This happens as the piston moves upwards from 2 to 3. A little beyond 3, the exhaust port closes at 4. Now the charge inside the engine cylinder is compressed which is shown by the line 4-5. At the end of compression, there is an increase in pressure inside the engine cylinder. Shortly before the end of compression, the charge is ignited with the help of spark plug as shown in Fig. 5.16 (e). The sparking suddenly increases temperature and pressure of the product of combustion. But the volume, practically, remains constant as shown by the line 5-6. The expansion is shown by the line 6-7. Now the exhaust port opens at 7, and the burnt gases are exhausted into the atmosphere through the exhaust port. It reduces pressure. As the piston moving towards BDC, therefore volume of burnt gases increases from 7 to 1. At 1 the transfer port opens and suction starts.

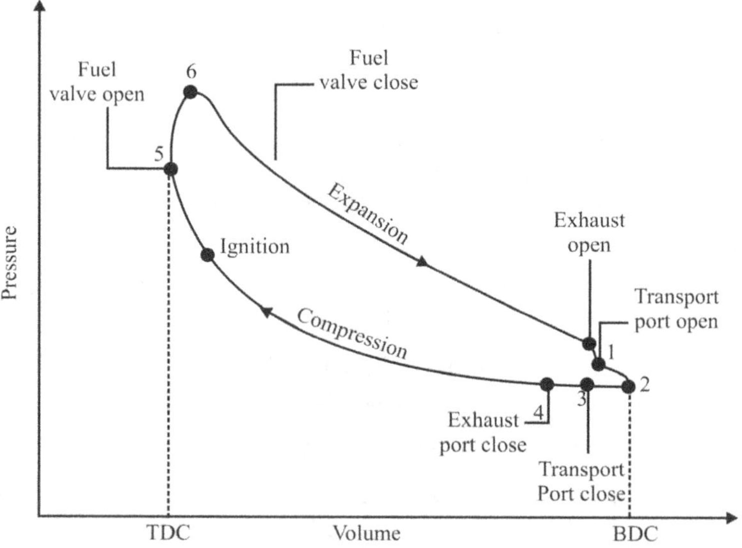

Fig. 5.18 Actual P-V Diagram of a Two Stroke Otto Cycle Engine

Working of 4-Stroke SI Engine:

(i)   *First Stroke*: Intake Stroke or Induction. The piston travels from TDC to BDC with the intake valve open and exhaust valve closed. This creates an increasing volume in the combustion chamber, which in turn creates a vacuum. The resulting pressure differential through the intake system from atmospheric pressure on the outside to the vacuum on the inside causes air to be pushed into the cylinder. As the air passes through the intake system, fuel is added to it in the desired amount by means of fuel injectors or a carburetor (Fig. 5.19 a).

(ii) *Second Stroke*: Compression Stroke. When the piston reaches BDC, the intake valve closes and the piston travels back to TDC with all valves closed. This compresses the air-fuel mixture, raising both the pressure and temperature in the cylinder. The finite time required to close the intake valve. Near the end of the compression stroke, the spark plug is fired and combustion is initiated (Fig. 5.19 b).

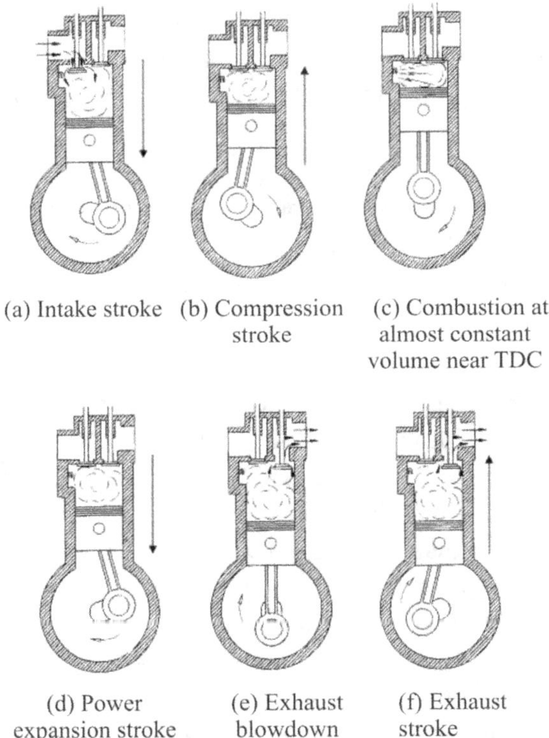

(a) Intake stroke   (b) Compression stroke   (c) Combustion at almost constant volume near TDC

(d) Power expansion stroke   (e) Exhaust blowdown   (f) Exhaust stroke

Fig. 5.19 4-Stroke SI Engine Operating Cycle

(iii) *Combustion*: Combustion of the air-fuel mixture occurs in a very short but finite length of time with the piston near TDC (i.e., nearly constant-volume combustion). It starts near the end of the compression stroke. Combustion changes the composition of the gas mixture to that of exhaust products and increases the temperature in the cylinder to a very high peak value. This, in turn, raises the pressure in the cylinder to a very high peak value (Fig. 5.19 c).

(iv) *Third Stroke*: Expansion Stroke or Power Stroke. With all valves closed, the high pressure created by the combustion process pushes the piston away from TDC. This is the stroke which produces the work output of the engine cycle. As the piston travels from TDC to BDC, cylinder volume is increased, causing pressure and temperature to drop (Fig. 5.19 d).

(v) *Exhaust Blowdown*: Late in the power stroke, the exhaust valve is opened and exhaust blowdown occurs. Pressure and temperature in the cylinder are still high relative to the surroundings at this point, and a pressure differential is created through the exhaust system which is open to atmospheric pressure. This pressure differential causes much of the hot exhaust gas to be pushed out of the cylinder and through the exhaust system when the piston is near BDC. This exhaust gas carries away a high amount of enthalpy, which lowers the cycle thermal efficiency. Opening the exhaust valve before BDC reduces the work obtained during the power stroke but is required because of the finite time needed for exhaust blowdown (Fig. 5.19 e).

(vi) *Fourth Stroke*: Exhaust Stroke. By the time the piston reaches BDC, exhaust blowdown is complete, but the cylinder is still full of exhaust gases at approximately atmospheric pressure. With the exhaust valve remaining open, the piston now travels from BDC to TDC in the exhaust stroke. This pushes most of the remaining exhaust gases out of the cylinder into the exhaust system at about atmospheric pressure, leaving only that trapped in the clearance volume when the piston reaches TDC. Near the end of the exhaust stroke, the intake valve starts to open, so that it is fully open by TDC when the new intake stroke starts the next cycle. Near TDC the exhaust valve starts to close (Fig. 5.19 f).

Theoretical indicator diagram for a four–stroke cycle petrol engine: In the above operation, the following assumptions were made:

(i)   Suction and exhaust take place at atmospheric pressure.

(ii)   Suction and exhaust take place at $180°$ rotation of crank.

(iii) Compression and expansion also take place at $180°$ rotation of crank.

(iv)   Compression and expansion are isentropic.

(v)   The combustion takes place instantaneously at constant volume at the end of compression stroke.

(vi)   Pressure suddenly falls to the atmospheric pressure at end of expansion stroke.

With these assumptions the working of 4-stroke Otto cycle engine on p-v diagram as shown in Fig. 5.20:

Process 5-1: Suction stroke. In this process fresh air and fuel mixture i.e., charge is passed inside the cylinder. The piston moves from top dead center to bottom dead center. This comprises the first stroke of the engine.

Process 1-2: Compression stroke. In this process the charge is compressed and piston is moved to top dead center. This comprises the second stroke of the engine.

Process 2-3: Instantaneous-Combustion. In this process spark plug ignites the spark and the fuel is burnt. This process is of constant volume and increase in pressure.

Process 3-4: Expansion Stroke. In this process the burnt fuel expands itself and exerts pressure on the piston. The piston moves from top dead center to bottom dead center. This comprises the third stroke of the engine and the power stroke.

Process 4-1: Sudden Fall in pressure. In this process the burnt gas is exhausted out and the pressure decreases with constant volume.

Process 1-5: Exhaust stroke. In this process the burnt gas is completely moved out of the cylinder by the action of piston. Piston moves from bottom dead center to top dead center. This comprises the fourth stroke of the engine.

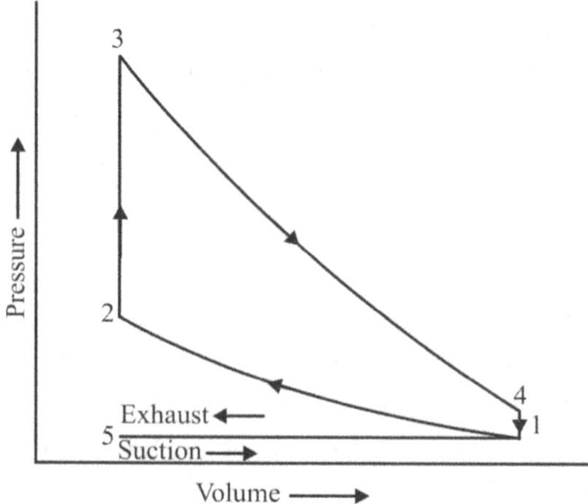

Fig. 5.20 Theoretical p-V Diagram of a Four Stroke Otto Cycle Engine

Actual indicator (p-v) diagram for a four stroke cycle petrol cycle: In the above operations, all the ideal conditions are assumed but in practice, the actual conditions differ from the ideal as described below (Fig. 5.21).

(i)   The suction of mixture in the cylinder is possible only if the pressure inside the cylinder is below atmospheric pressure.

(ii)   The burnt gases can be pushed out into the atmosphere only if the pressure of the exhaust gases is above atmospheric pressure.

(iii)   The combustion and expansion do not follow the isentropic law, as there will be heat exchange during process.

(iv)   Sudden pressure rise is not possible after ignition as combustion take sometime for completion and actual pressure rise is less than theoretical considered. The pressure increase takes place through some crank rotation, or increase in volume.

(v)   Sudden pressure release after the opening of expansion valve is not possible and also takes place through some crank rotation.

It may be noted that line 5-1 is below the atmospheric pressure line. This is due to the fact that owing to restricted area of the inlet passage, the entering fuel mixture cannot cope with the speed of the piston. The exhaust line 4-5 is slightly above the atmospheric pressure line. This is due to the fact that owing to restricted area of the exhaust passage, which does not allow exhaust gases to leave the engine cylinder quickly.

The loop, which has area 4-5-1is, called negative loop. It gives the pumping loss due to admission of fuel air mixture and removal of exhaust gases. The area 1234 is the total or gross negative work from the area 1234 that is gross work (Fig. 5.21)

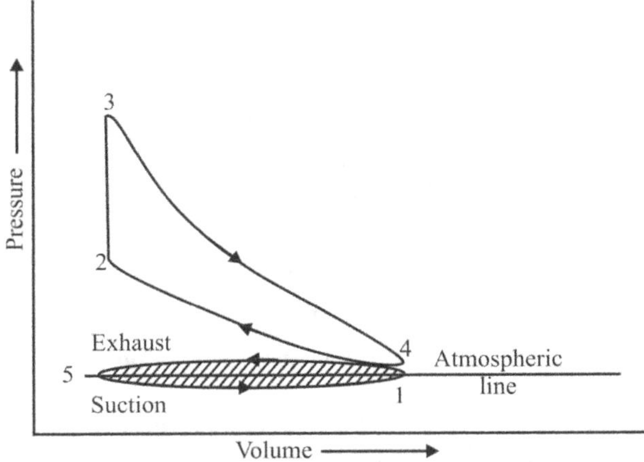

Fig. 5.21 Actual p-V Diagram of a Four Stroke Otto Cycle Engine

5.4.5 Compression Ignition Engine (Diesel Engine)

Working of 2-Stroke CI Engine:

A two-stroke diesel engine (Fig. 5.22) shares the same operating principles as other internal combustion engines. It has all of the advantages that other diesel engines have over gasoline engines. A two-stroke diesel engine does not produce as much power as a four-stroke diesel engine; however, it runs smoother than the four-stroke diesel. This is because it generates a power stroke each time the piston moves downward; that is, once for each crankshaft revolution. The two-stroke diesel engine has a less complicated valve train because it does not use intake valves. Instead, it requires a supercharger to force air into the cylinder and force exhaust gases out, because the piston cannot do this naturally as in four-stroke engines. The two-stroke diesel takes in air and discharges exhaust through a system called scavenging. Scavenging begins with the piston at bottom dead center. At this point, the intake ports are uncovered in the cylinder wall and the exhaust valve is open. The supercharger forces air into the cylinder, and, as the air is forced in, the burned gases from the previous operating cycle are forced out (Fig. 5.23).

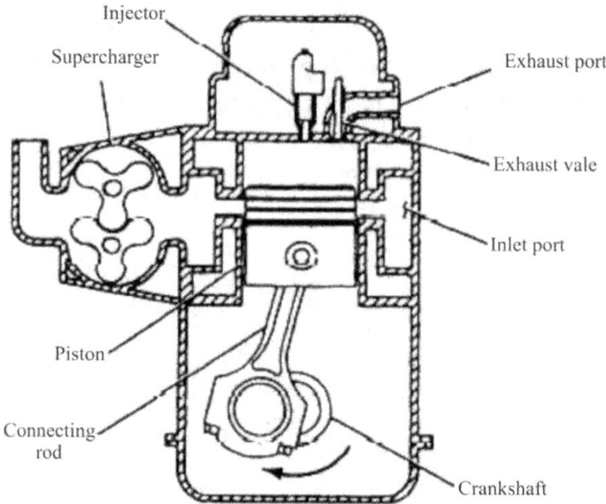

Fig. 5.22 Two-Stroke Cycle Diesel Engine

Compression Stroke: As the piston moves towards top dead center, it covers the intake ports. The exhaust valves close at this point and seals the upper cylinder. As the piston continues upward, the air in the cylinder is tightly compressed (Fig. 5.23). As in the four-stroke cycle diesel, a tremendous amount of heat is generated by the compression.

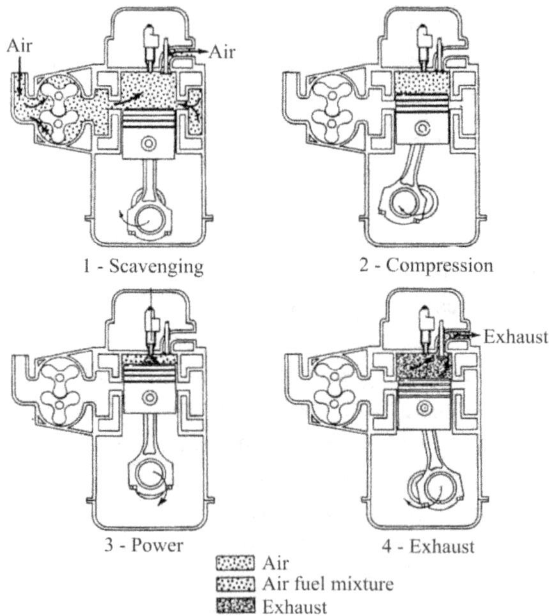

Fig. 5.23 Two-Stroke CI Engine Operating Cycle

Power Stroke: As the piston reaches top dead center, the compression stroke ends. Fuel is injected at this point and the intense heat of the compression causes the fuel to ignite. The burning fuel pushes the piston down, giving power to the crankshaft. The power stroke ends when the piston gets down to the point where the intake ports are uncovered. At about this point, the exhaust valve opens and scavenging begins again, as shown in Fig. 5.23. The operation of the valves in a timed sequence is critical. If the exhaust valve opened in the middle of the intake stroke, the piston would draw burnt gases into the combustion chamber with a fresh mixture of fuel and air. As the piston continues the power stroke, there would be nothing in the combustion chamber that would burn.

Working of the two stroke CI engine cycle on p-V diagram as shown in Fig. 5.24.

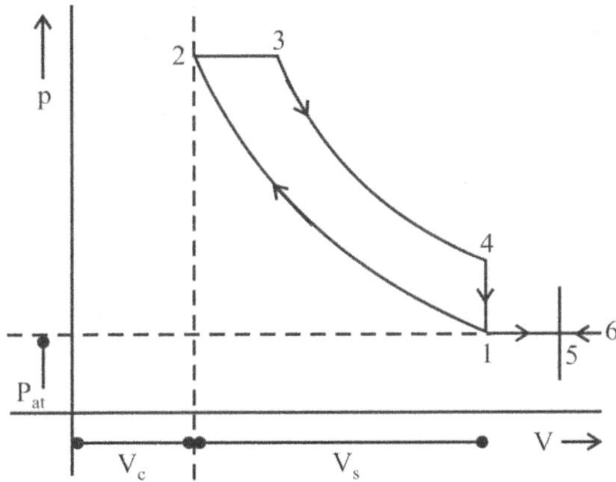

Fig. 5.24 Theoretical p-V Diagram of a Two Stroke Diesel Cycle Engine

Process 1-2: Isentropic compression of the air in the cylinder. The air inside the cylinder is compressed and the piston moves from bottom dead center to top dead center. This comprises the first stroke of the engine.

Process 2-3: Combustion at constant Pressure. In this process diesel is pumped inside the cylinder by the help of fuel pump and fuel is burnt.

Process 3-4: Isentropic Expansion of the burnt fuel in the cylinder. The burnt fuel exerts pressure and moves the piston to bottom dead center. The gas expands in this process. This comprises the second stroke and the power stroke of the engine.

Process 4-1: Sudden release of burnt gases to the atmosphere as the exhaust port opens. This process takes place as process of constant volume and decrease in pressure. In this process the burnt gas is exhausted outside the cylinder.

Process 1-6 & 6-1: Sweeping out the exhaust gases to the atmosphere.

Process 5-6 & 6-5: Charging the cylinder with fresh charge through the transfer ports and scavenging action takes place.

Fig. 5.25 shows the actual indicator diagram for a two stroke cycle diesel engine. The suction is shown by the line 1-2-3, from the instant transfer ports opens and trandfer port closes. During the suction stage, the exhaust port is opens. In the first half of suction stage, the volume of air and burnt gases increases. This happens as the piston moves from 1-2. In the second half of the suction stage, the volume of air and burnt gases decreases. This happens as the piston moves upwards from 2-3. A little beyond 3, the exhaust port closes at 4. Now the air inside the engine cylinder is compressed which is shown by the line 4-5. At the end of compression there is an increase in pressure inside the engine cylinder, shortly before the end of compression, the fuel valve opens and the fuel is injected into the engine cylinder. The fuel is ignited by high temperature of the compressed air. The ignition suddenly increases volume and temperature of the products of combustion. But the pressure, practically, remains constant as shown by the line 5-6. The expansion is shown by the line 6-7. Now the exhaust port opens at 7, and the burnt gases are exhausted into the atmosphere through the exhaust port. It reduces pressure. As the piston moving towards BDC, therefore volume of burnt gases increases from 7 to 1. At 1 the transfer port opens (TPO) and suction starts.

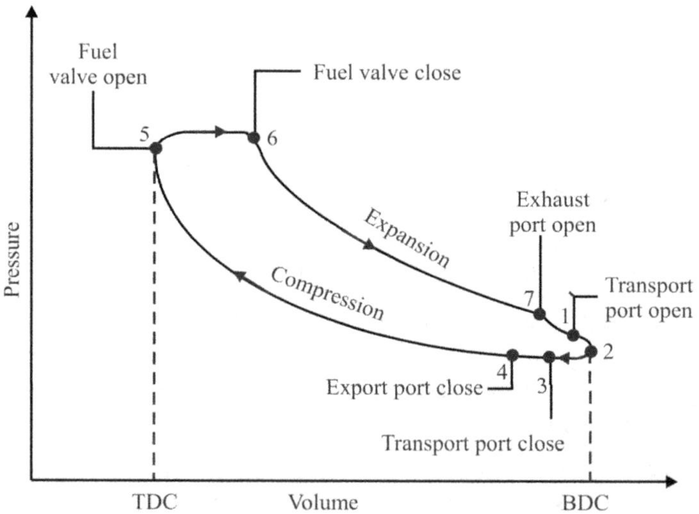

Fig. 5.25 Actual p-V Diagram of a Two Stroke Diesel Cycle Engine

Working of 4- Strokes Diesel Engine

Intake Stroke: The piston is at top dead center at the beginning of the intake stroke, and, as the piston moves downward, the intake valve opens. The downward movement of the piston draws air into the cylinder, and, as the piston reaches bottom dead center, the intake valve closes (Figure 5.26 A).

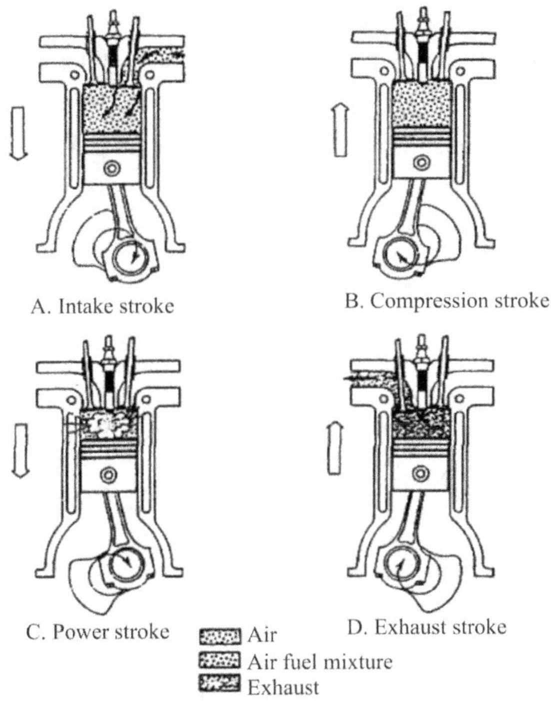

Fig. 5.26 Four Stroke CI Engine Operating Cycle.

Compression Stroke: The piston is at bottom dead center at the beginning of the compression stroke, and, as the piston moves upward, the air compresses. As the piston reaches top dead center, the compression stroke ends (Figure 5.26 B).

Power Stroke: The piston begins the power stroke at top dead center. The air is compressed to as much as 35 bar and at a compressed temperature of approximately 500 °C. At this point, fuel is injected into the combustion chamber and is ignited by the heat of the compression. This begins the power stroke. The expanding force of the burning gases pushes the piston downward, providing power to the crankshaft. The diesel fuel will continue to burn through the entire power stroke (a more complete burning of the fuel). The gasoline engine has a power stroke with rapid combustion in the beginning, but little to no combustion at the end (Figure 5.26 C).

Exhaust Stroke: As the piston reaches bottom dead center on the power stroke, the power stroke ends and the exhaust stroke begins. The exhaust valve opens, and, as the piston rises towards top dead center, the burnt gases are pushed out through the exhaust port. As the piston reaches top dead center, the exhaust valve closes and the intake valve opens. The engine is now ready to begin another operating cycle (Figure 5.29D).

Working of the four stroke CI engine cycle on p-V diagram as shown in Fig. 5.27.

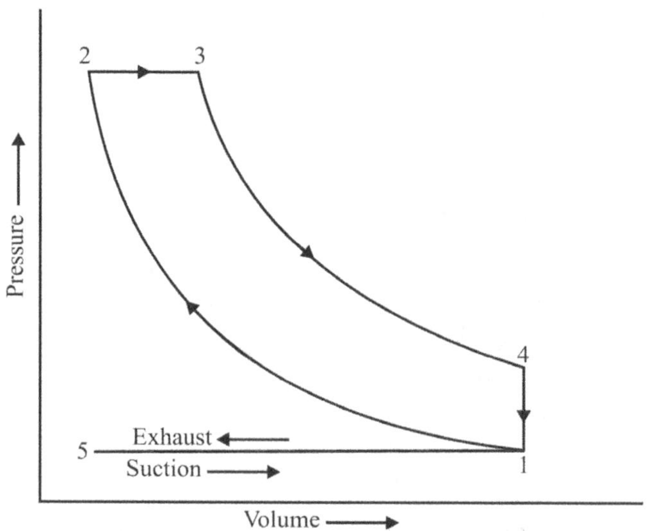

Fig. 5.27 Theoretical p-V Diagram of a Four Stroke Diesel Cycle Engine

Process 5-1: In this process fresh air is passed inside the cylinder. The piston moves from top dead center to bottom dead center. This comprises the first stroke of the engine.

Process 1-2: In this process the air is compressed and piston is moved to top dead center. This comprises the second stroke of the engine.

Process 2-3: In this process diesel is pumped into the cylinder through fuel pump and the fuel is burnt. This process is of constant volume and increase in pressure.

Process 3-4: In this process the burnt fuel expands itself and exerts pressure on the piston. The piston moves from top dead center to bottom dead center. This comprises the third stroke of the engine and the power stroke.

Process 4-1: In this process the burnt gas is exhausted out and the pressure decreases with constant volume.

Process 1-5: In this process the burnt gas is completely moved out of the cylinder by the action of piston. Piston moves from bottom dead center to top dead center. This comprises the fourth stroke of the engine.

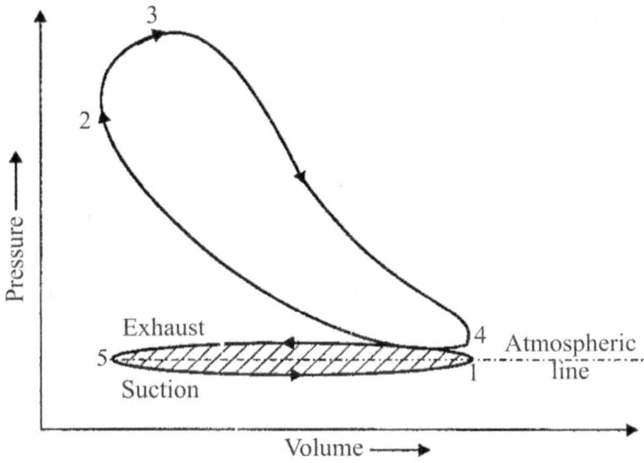

Fig. 5.28 Actual p-V Diagram of a Four Stroke Diesel Cycle Engine

The actual indicator diagram of a 4-stroke diesel engine has been shown in Fig. 5.28. The suction stroke is shown by line 1-2 which lies below the atmospheric pressure line. It is thus pressure difference, which makes the fresh air to flow into the engine cylinder. The inlet valve offers some resistance to the incoming air. That is why, the air can not enter suddenly into the engine cylinder. As a result of this, pressure inside the cylinder remains somewhat below the atmospheric pressure during the suction stroke. The compression stroke is shown by the line 2-3, which shows that the inlet valve closes a little beyond 2 (i.e., B.D.C.). At the end of this stroke, there is an increase of pressure inside the engine cylinder, shortly before the end of the compression stroke (i.e., T.D.C.), fuel valve opens and the fuel is injected into the engine cylinder. The fuel is ignited by the temperature of the compressed air. The ignition suddenly increases volume and temperature of the products of combustion. But the pressure, practically remains constant. The expansion stroke is shown by the line 4-5 in which the exit valve opens a little before 5 (i.e. B.D.C.).

Now the burnt gases are exhausted into the atmosphere through the exhaust valve, shown by 5-1, lies above atmospheric pressure line. The exhaust valve offers some resistance to the outgoing burnt gases. That is why, the burnt gases can not escape suddenly from the engine cylinder. As a result of this pressure inside the cylinder remains above the atmospheric pressure during the exhaust stroke.

Table 5.2 Comparison of Spark Ignition (S.I.) and Compression Ignition (C.I.) Engines

S. No	Features	S.I. engines	C.I. engines
1.	Thermodynamic Cycle	Otto cycle	Diesel cycle for slow speed engines Dual cycle for high speed engines
2.	Fuel used	Petrol (Gasoline)	Diesel
3.	Air fuel (A/F) ratio	10:1 to 20:1	18:1to 100:1
4.	Compression Ratio	Up to 11; Average value 7 to 9 Upper limit fixed by anti-knock quality of fuel.	Up to12 to 24; Average value 15 to 18; Upper limit is limited by thermal and mechanical stresses.
5.	Combustion	Spark Ignition	Compression Ignition
6.	Fuel Supply	By Carburettor (Low Cost)	By injection (High Cost)
7.	Operating pressure 1.Compression pressure 2. Maximum Pressure	7 bar to 15 bar 45 bar to 60 bar	30 bar to 50 bar 60 bar to 120 bar
8.	Operating Speed	High speed:2000 to 6000 r.p.m	Low speed: 400 r.p.m Medium speed: 400 to 1200 r.p.m. High speed: 1200 to 3500 r.p.m
9.	Power Control	Throttle governed the quantity	Rack governed the quantity
10.	Calorific value	44 MJ/kg	42 MJ/kg
11.	Cost of running	High	Low
12	Maintenance cost	Minor maintenance	Major over all require

Table 5.2 *Contd...*

S. No	Features	S.I. engines	C.I. engines
13	Super charging	Limited by detonation. Used only in aircraft engines.	Limited by blower power and mechanical and thermal stresses. Widely used.
14	Two stroke operation	Less suitable, fuel loss in scavenging. But small two stroke engines are used in mopeds, scooters and motor cycles due to their simplicity and low cost.	No fuel loss in scavenging. More suitable.
15	High powers	No	Yes
16	Distribution of fuel	A/F ratio is not optimum in multi cylinder engines.	Excellent distribution of fuel in multi cylinder engines.
17	Starting	Easy, low cranking effort.	Difficult, high cranking effort.
18	Exhaust gas temperature.	High, due to low thermal efficiency	Low, due to high thermal efficiency.
19	Weight per unit power	Low (0.5 to 4.5 kg/kW)	High (3.3 to 13.5 kg/kW)
20	Initial capital Cost	Low	High due to heavy weight and sturdy construction; costly construction 1.25-1.5 times
21	Noise and distribance	Less	More idle noise problem
22	Applications	Mopeds, Scooters, motorcycles, Simple engine passenger cars	Buses, trucks, locomotives, tractors, earth moving machinery and stationary generating plants.

5.4.6 Comparison between 2-Stroke and 4-Stroke Engines

The two-stroke engine was developed to obtain a greater output from the same size of the engine. The engine mechanism also eliminates the valve arrangement making it mechanically simpler. Almost all two-stroke engines have no conventional valves but only ports (some have an exhaust valve). This simplicity of the two-stroke engine makes it cheaper to produce and easy to maintain. Theoretically a two-stroke engine develops twice the power of a comparable four stroke engine because of one power stroke every revolution (compared to one power stroke every two revolutions of a four-stroke engine). This makes the two-stroke engine more compact than a comparable four-stroke engine. In actual practice power output is not exactly doubled but increased by only about 30% because of:

(i)   reduced effective expansion stroke and

(ii) increased heating caused by increased number of power strokes that limits the maximum speed.

The other advantages of the two-stroke engine are more uniform torque on crankshaft and comparatively less exhaust gas dilution. However, when applied to the spark-ignition engine the two stroke cycle has certain disadvantages which have restricted its application to only small engines suitable for motor cycles, scooters, lawn mowers, outboard engines etc. In the SI engine, the incoming charge consists of fuel and air. During scavenging, as both inlet and exhaust ports are open simultaneously for some time, there is a possibility that some of the fresh charge containing fuel escapes with the exhaust. This results in high fuel consumption and lower thermal efficiency. The other drawback of two-stroke engine is the lack of flexibility, viz., the capacity to operate with the same efficiency at all speeds. At part throttle operating condition, the amount of fresh mixture entering the cylinder is not enough to clear all the exhaust gases and a part of it remains in the cylinder to contaminate the charge. This results in irregular operation of the engine. The two-stroke diesel engine does not suffer from these defects. There is no loss of fuel with exhaust gases as the intake charge in diesel engine is only air. The two-stroke diesel engine is used quite widely. Many of the high output diesel engines work on this cycle. A disadvantage common to all two-stroke engines, gasoline as well as diesel, is the greater cooling and lubricating oil requirements due to one power stroke in each revolution of the crankshaft. Consumption of lubricating oil is high in two-stroke engines due to higher temperature. A detailed comparison of two-stroke and four-stroke engines is given in the Table 5.3.

Table 5.3 Comparison of two-stroke and four-stroke engines

S. No.	Features	Four Stroke Cycle Engines	Two Stroke Cycle Engines
1	Completion of cycle	The cycle is completed in four strokes of the piston or in two revolutions of the crank shaft. Thus one power stroke is obtained in every two revolutions of the crank shaft.	The cycle is completed in two strokes of the piston or in 1 revolution of the crank shaft. Thus power stroke is obtained in each revolutions of the crank shaft.
2	Flywheel required heavier or lighter	Because of the turning movement is not so uniform and hence heavier flywheel is needed.	More uniform turning movements and hence lighter flywheel is needed.
3	Power produced for same size of engine	Again because of one power stroke for two revolutions, power produced for same size of engine is small or for the same power the engine is heavy and bulky.	Because of one power stroke for one revolution, power produced for same size of engine is more (Theoretically twice, actually about 1.3 times) or for the same power the engine is light and compact.
4	Cooling and lubrication requirements	Because of one power stroke in two revolutions lesser cooling and lubrication requirements. Lesser rate of wear and tear.	Because of one power stroke in one revolutions greater cooling and lubrication requirement. Large rate of wear and tear.
5	Valve and valve mechanism	The four stroke engine contains valve and valve mechanism.	Two stroke engines have no valves but only ports (some two stroke engines are fitted with conventional exhaust valves).
6	Initial cost	Because of heavy weight and complication of valve mechanism, the initial cost is higher.	Because of light weight and simplicity due to absence of valves mechanism, cheaper in initial cost.
7	Volumetric efficiency	Volumetric efficiency more due to more time of induction	Volumetric efficiency less due to lesser time for induction.
8	Thermal and part load efficiency	Thermal efficiency higher, part load efficiency better than two stroke cycle engine.	Thermal efficiency lower, part load efficiency lesser than four stroke cylinder engine.

Table 5.3 *Contd...*

S. No.	Features	Four Stroke Cycle Engines	Two Stroke Cycle Engines
9	Applications	Used where efficiency is important; in cars, buses, trucks, tractors, industrial engines, aeroplane, power generators, etc.	In two strokes petrol engine some fuel is exhausted during scavenging. Used where (a) low cost, and (b) compactness and light weight important. Two stroke (air cooled) petrol engines used in very small sizes only, lawn movers, scooters, motor cycles (lubrication oil mixed with petrol). Two stroke diesel engines used in very large sizes more than 60 cm bore, for ship propulsion because of low weight and compactness.

Solved Numerical Examples

***Example* 1:** The efficiency of otto cycle is 60% and $\gamma = 1.5$. What is the compression ratio?

Solution:

$$Efficiency \quad \eta = 1 - \frac{1}{r^{\gamma-1}}$$

$$0.6 = 1 - \frac{1}{(r)^{0.5}}$$

$$\frac{1}{r^{0.5}} = \frac{1}{0.4} = 2.5$$

$$Compression\ is\ (r) = (2.5)^{\frac{1}{0.5}} = 6.25$$

Example 2: Calculate the ideal air standard cycle efficiency of the petrol engine with a cylinder diameter as 50 mm, a stroke of 75 mm and clearence volume of 1.3 cm^3.

Solution:

$$\text{Stroke volume } (V_s) = \frac{\pi}{4} \times d^2 \times L$$

$$= \frac{\pi}{4} \times (0.05)^2 \times 0.075$$

$$= 1.4726 \times 10^{-4} \, m^{3.}$$

$$\text{Compression ratio } (\eta) = \frac{V_1}{V_2} = \frac{V_c + V_s}{V_c}$$

$$= \frac{21.3 \times 10^{-6} + 1.4726 \times 10^{-4}}{21.3 \times 10^{-6}}$$

$$= 7.91$$

$$\textit{Efficiency } \eta = 1 - \frac{1}{r^{\gamma-1}}$$

$$= 1 - \frac{1}{7.91^{0.4}} = 56.27\%$$

Example 3: In a constant volume "Otto cycle" the pressure at the end of compression is 15 times that at the start. The temperature of air at the beginning of compression is 38 °C and maximum temperature obtained in the cycle is 1950 °C. Determine

       (i) Compression ratio

       (ii) Thermal efficiency

       (iii) Work done

Solution:

(i) $\dfrac{V_1}{V_2} = \left(\dfrac{P_2}{P_1} \right)^{\frac{1}{r}} = r$

$$\text{Compression ratio} = (15)^{\frac{1}{1.4}}$$

$$r = 6.9$$

(ii)   *Efficiency* $\eta = 1 - \dfrac{1}{r^{\gamma-1}}$

$$= 1 - \dfrac{1}{6.9^{0.4}}$$

$$= 53.81\ \%$$

(iii) $\dfrac{T_2}{T_1} = \left(\dfrac{V_1}{V_2}\right)^{\gamma-1}$

$$T_2 = (6.9)^{0.4} \times 311$$

$$= 673.44\ K.$$

Heat supplied $(Q_s) = m\ c_v\ (T_3 - T_2)$

$$= 1 \times 0.719\ (2223 - 673.44)$$

$$= 1111.\ 03\ kJ/kg.$$

$$\eta_{thermal} = \dfrac{W.D}{W_p}$$

$$W.D = \eta_{thermal} \times Q_s$$

$$= 0.5381 \times 1111.03$$

$$= 597.84\ KJ/Kg.$$

Example 4: In an otto cycle operation the compression ratio is changed from 4.5 to 8.5, determine the increase in air standard efficiency of the cycle.

Solution:

$$\eta_1 = 1 - \dfrac{1}{r_1^{\gamma-1}}$$

$$= 1 - \dfrac{1}{4.5^{0.4}}$$

$$= 45.208\ \%$$

$$\eta_2 = 1 - \dfrac{1}{r_2^{\gamma-1}}$$

$$= 1 - \dfrac{1}{8.5^{0.4}}$$

$$= 57.515\ \%$$

Increse in $\eta = \eta_2 - \eta_1$

$$= 57.515 - 45.208$$

$$= 12.30\ \%$$

Example 5: A diesel engine has a compression ratio of 14 and cut-off takes place at 6 % of the stroke. Find the air standard efficiency.

Solution:

$$r = 14$$

$$\frac{V_c + V_s}{V_c} = 14$$

$$1 + \frac{V_s}{V_c} = 14 \qquad \therefore \quad \frac{V_s}{V_c} = 13$$

Fuel cut-off $= 6\ \%$

$$\frac{r_c - 1}{r - 1} = 0.06 \qquad [OR] \qquad r_c = \frac{V_3}{V_2}$$

$$r_c - 1 = 0.06 \times 13 \qquad\qquad = \frac{V_c + 0.06\ V_s}{V_c}$$

$$r_c = 1.78 \qquad\qquad = 1 + 0.06\ \frac{V_s}{V_c}$$

$$= 1.78$$

$$\eta_{e,Diesel} = 1 - \frac{1}{r^{\gamma-1}}\ \frac{r_c^{\gamma} - 1}{\gamma(r_c - 1)}$$

$$= 1 - \frac{1}{1.4(14)^{0.4}}\left[\frac{1.78^{1.4} - 1}{1.78 - 1}\right]$$

$$= 1 - 0.2485 \times 1.5919$$

$$\eta = 60.43\ \%$$

Example 6: The stroke and cylinder diameter of a compression ignition engine are 250 mm and 150 mm respectively. If the clearance volume is 0.0004 m^3 and fuel injection takes place at constant pressure for 5% of the stroke. Determine the efficiency of diesel engine.

Solution:

$$V_s = \frac{\pi}{4} \times d^2 \times L$$

$$= \frac{\pi}{4} \times 0.15^2 \times 0.25$$

$$= 4.4178 \times 10^{-3} \text{ m}^3$$

$$r = \frac{V_1}{V_2} = \frac{V_c + V_s}{V_c}$$

$$= \frac{0.0004 + 4.4178 \times 10^{-3}}{0.0004}$$

$$= 12.04$$

$$\frac{r_c - 1}{r - 1} = 0.05$$

$$r_c - 1 = 0.05 \times 11.04$$

$$= 1.552$$

$$\eta_{e,\text{Diesel}} = 1 - \frac{1}{r^{\gamma-1}} \frac{r_c^{\gamma} - 1}{\gamma(r_c - 1)}$$

$$= 1 - \frac{1}{1.4(12.04)^{0.4}} \left[\frac{1.552^{1.4} - 1}{1.552 - 1} \right]$$

$$= 59.33 \%$$

Example 7: Calculate the air standard efficiency of diesel engine having compression ratio 18 and expansion ratio 10.

Solution:

$$\frac{r}{r_c} = 10 \quad \Rightarrow \quad \frac{18}{r_c} = 10$$

$$r_c = 1.8$$

$$\eta_{e,Diesel} = 1 - \frac{1}{r^{\gamma-1}} \frac{r_c^{\gamma} - 1}{\gamma(r_c - 1)}$$

$$= 1 - \frac{1}{1.4(18)^{0.4}} \left[\frac{1.8^{1.4} - 1}{1.8 - 1} \right]$$

$$= 64.13 \%$$

Example 8: In an ideal diesel cycle the compression ratio is 14: 1 and expansion ratio 8: 1. The pressure and temperature at the beginning of compression are 98 kN/m^2 and 44 °C and the pressure at the end of expansion is 258 KN/m^2. Determine

    (i) Maximum temperature of cycle

    (ii) Thermal η of cycle

Solution:

$$\frac{r}{r_c} = 7.5$$

$$r_c = \frac{r}{7.5} = \frac{15}{7.5} = 2$$

$$\eta_{e,Diesel} = 1 - \frac{1}{r^{\gamma-1}} \frac{r_c^{\gamma} - 1}{\gamma(r_c - 1)}$$

$$= 1 - \frac{1}{1.4(15)^{0.4}} \left[\frac{2^{1.4} - 1}{2 - 1} \right]$$

$$= 69.38$$

$$\frac{T_2}{T_1} = \left(\frac{V_1}{V_2} \right)^{\gamma-1}$$

$$T_2 = (15)^{0.4} \times 317$$

$$= 936.47 \text{ K}$$

$$\frac{T_3}{T_2} = r_c \Rightarrow T_3 = 2 \times 936.47$$

$$T_3 = 1872.94 \text{ K}$$

Example 9: Find the power output of a diesel engine working on a diesel cycle with a compression ratio of 16 and air flow rate of 0.25 kg/sec. The initial condition of air at 1 bar obsolute and 27 °C temperature, heat added per cycle 2500 kJ/kg

Solution:

$$\frac{T_2}{T_1} = (r)^{\gamma-1}$$

$$= (16)^{0.4} \times 300$$

$$= 909.42 \text{ K}$$

Heat supplied

$$Q_s = m \, c_p \, (T_3 - T_2)$$

$$T_3 = \frac{2500}{1 \times 1.005} + 909.42$$

$$T_3 = 3409.42 \text{ K}$$

$$r_c = \frac{T_3}{T_2} = \frac{3409.42}{909.42}$$

$$= 3.749$$

$$\eta_{e,\text{Diesel}} = 1 - \frac{1}{r^{\gamma-1}} \frac{r_c^{\gamma} - 1}{\gamma(r_c - 1)}$$

$$= 1 - \frac{1}{1.4(16)^{0.4}} \left[\frac{3.749^{1.4} - 1}{3.749 - 1} \right]$$

$$= 54.08 \, \%$$

$$\eta = \frac{W.D}{Q_s} \quad W.D$$

$$\eta \times Q_s = 0.5408 \times 2500$$

$$= 1352.03 \text{ KJ/kg.}$$

Review Questions

1.   Draw the Otto cycle on P-V diagram.

2.   What is carnot cycle and its importance?

3.   What is indicator diagram?

4. Why scavenging is important in two stroke engines compared to four stroke engines

5. Write any four advantage of 4 stroke petrol engine over 2 stroke petrol engine.

6. Write the air standard efficiency of an Otto cycle.

7. Compare four stroke and two stroke cycle engines and bring out their relative merits and demerits.

8. Explain with neat sketches the two different types of two stroke engines

9. Discuss the two stroke engine construction and operation with neat sketches

10. Briefly explain the principle of four stroke petrol engine with a simple sketch

11. Draw the ideal and actual indicator diagram for a four stroke petrol engine and explain why they differ each other.

12. What are the advantage and disadvantage of 2 stroke engine?

13. What is thermodynamic cycle?

14. What are the assumptions made for air standard cycle analysis?

15. Mention the various process of diesel cycle.

16. Mention the various process of dual cycle.

17. Sketch Otto cycle on p-V diagram and name all the process.

18. Plot and explain briefly the Diesel cycle on p-V & T-s diagram.

19. Define mean effective pressure as applied to gas power cycle. How it is related to indicate power of an I.C. engine?

20. Write down the comparison between the S.I engine and C.I engine.

21. Define the following terms.

  1. Compression ratio   2. cut off ratio   3. Expansion ratio.

22. Name the factors that affect air standard efficiency of Diesel cycle?

23. For the same compression ratio and heat supplied, state the order of decreasing air standard efficiency of Otto, diesel and dual cycle.

24. What is the range of compression ratio for Otto and Diesel cycle?

25. Write an expression for mean effective pressure for an Otto cycle interns of compression ratio and other parameters.

26.   Differentiate between the Otto cycle and Diesel cycle.

27.   A gas working on the Otto cycle has a cylinder of diameter 200 mm and stroke 250 m. The clearance volume is 1570 CC. Find the air standard efficiency. Assume $c_p = 1.004$ KJ/kg K and $c_v = 0.7171$ KJ/kg K for air.

Important Formulae

Otto Cycle

$$\text{Spoke volume } (V_s) = \frac{\pi}{4} \times d^2 \times L$$

$$\text{Compression ratio } (r) = \frac{V_1}{V_2}$$

$$\eta_{Otto} = 1 - \frac{T_1}{T_2} = 1 - \frac{1}{r^{\gamma-1}}$$

Diesel Cycle

$$\text{Heat supplied } (Q_s) = m\, c_p\, (T_3 - T_1)$$

$$\text{Heat rejected } (Q_R) = m\, c_r\, (T_4 - T_1)$$

$$\text{Thermal } \eta = 1 - \frac{T_4 - T_1}{r(T_3 - T_2)}$$

$$\eta_{e,Diesel} = 1 - \frac{1}{r^{\gamma-1}} \frac{r_c^{\gamma} - 1}{\gamma(r_c - 1)}$$

$$\text{Compression ratio } (r) = \frac{v_1}{v_2}$$

$$\text{Cut-off ratio } (r_c) = \frac{v_3}{v_2}$$